triple lead 1½-6 NC AM. NAT.

Turning
Technology

ENGINE & TURRET LATHES

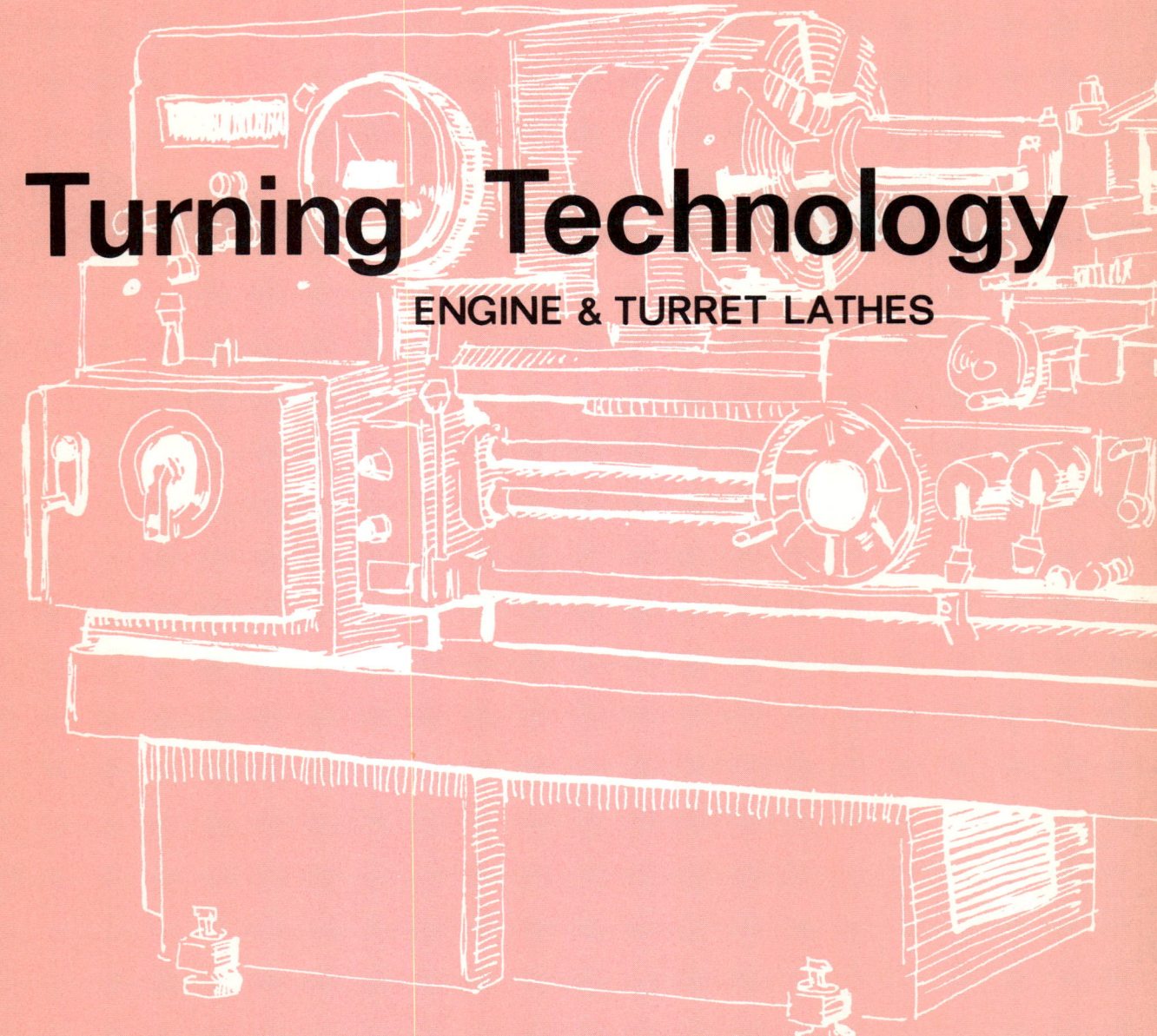

Turning Technology
ENGINE & TURRET LATHES

DELMAR PUBLISHERS
COPYRIGHT © 1971
BY LITTON EDUCATIONAL PUBLISHING, INC.

All rights reserved. No part of this work covered by the copyright hereon may be reproduced or used in any form or by any means — graphic, electronic, or mechanical, including photocopying, recording, taping, or information storage and retrieval systems — without written permission of the publisher.

LIBRARY OF CONGRESS CATALOG CARD NUMBER: 78-153723

PRINTED IN THE UNITED STATES OF AMERICA
PUBLISHED SIMULTANEOUSLY IN CANADA BY
DELMAR PUBLISHERS, A DIVISION OF
    VAN NOSTRAND REINHOLD, LTD.

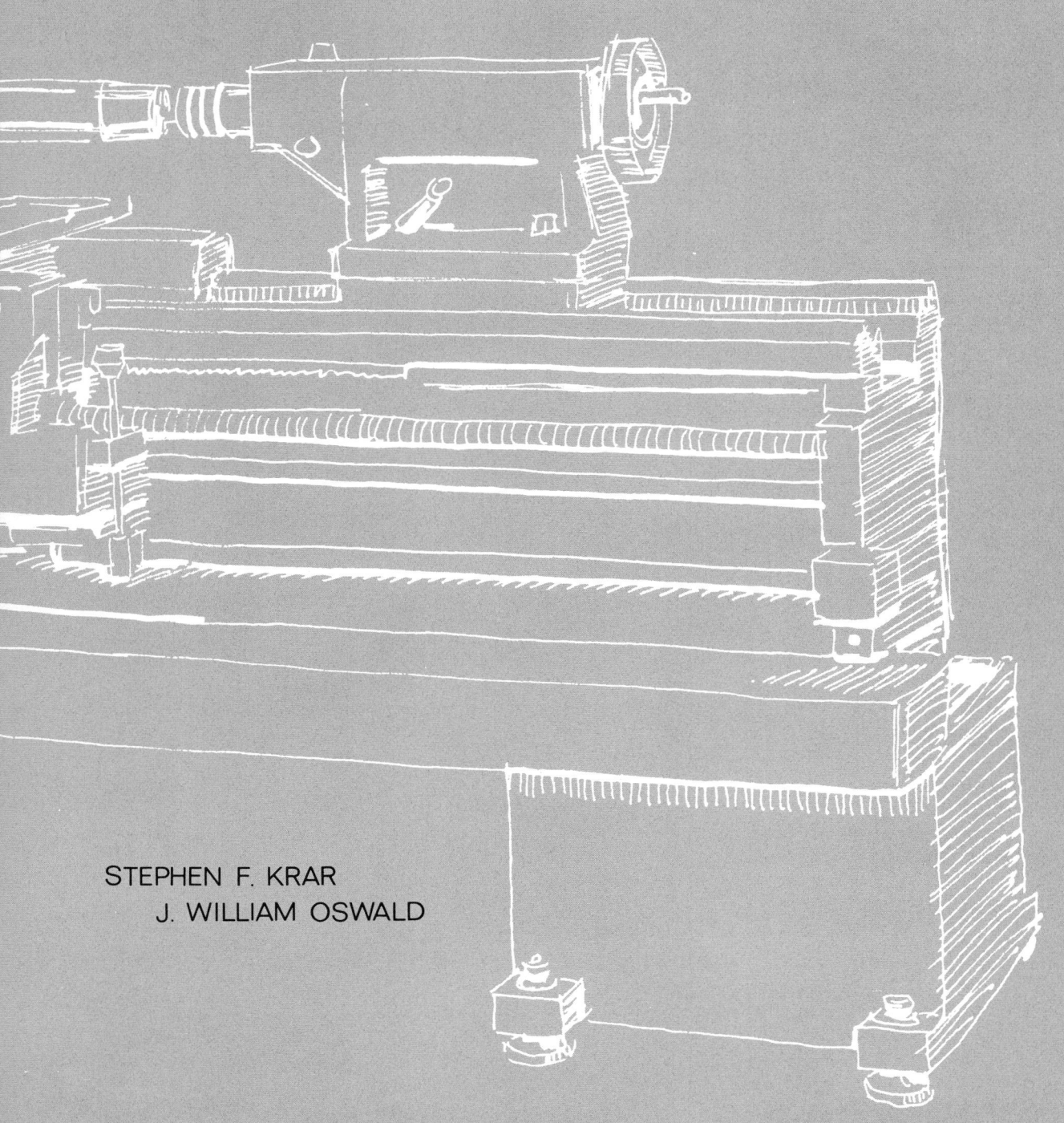

STEPHEN F. KRAR
J. WILLIAM OSWALD

DELMAR PUBLISHERS, ALBANY, NY 12205

Contents

Section 1 LATHE CONSTRUCTION, TOOLS, AND ACCESSORIES

Unit		Page
1	Lathe Types and Construction	1
2	Lathe Accessories	11
3	Lathe Maintenance	22
4	Lathe Safety	29
5	Cutting Tools and Toolholders	31
6	Grinding Lathe Cutting Tools	42
7	Cutting Speeds and Feeds	48

Section 2 MACHINING BETWEEN CENTERS

8	Centering Work	53
9	Alignment of Lathe Centers	60
10	Mounting Work Between Centers	64
11	Facing	71
12	Setting the Depth of Cut	74
13	Parallel Turning	78
14	Shoulder Turning	82
15	Filing and Polishing	87
16	Knurling	93
17	Grooving	97
18	Taper Turning	100
19	Form Turning	115
20	Threads	119
21	Thread Cutting	126
22	Thread Measurement	143
23	Steady and Follower Rest	150
24	Mandrels	155
25	Eccentrics	161

Section 3 MACHINING IN A CHUCK

26	Mounting and Removing Chucks	165
27	Mounting Work in a Chuck	171
28	Center Hole Drilling and Truing	178
29	Facing in a Chuck	180
30	Turning in a Chuck	183
31	Cutting-Off or Parting	185
32	Drilling	188
33	Boring	191
34	Counterboring	194
35	Reaming	196
36	Tapping	202
37	Taper Turning	206
38	Internal Threading	210
39	Faceplate Work	213

Section 4 SPECIAL OPERATIONS

40	Special Form Turning	218
41	Turning a Commutator	222
42	Grinding	225
43	Spring Winding	231
44	Tracer Attachments	234
45	Digital Readout System	237
46	Surface Finishes	242
47	Numerical Control Lathes	244
48	Fits and Allowances	249

Section 5 THE TURRET LATHE

49	Fundamentals of the Turret Lathe	250
50	Basic Cuts	256

TABLES 273

INDEX 276

ACKNOWLEDGMENTS

The authors wish to express their sincere thanks to Alice H. Krar for the countless hours she spent typing, proofreading, and checking the manuscript for this text. Without her patience, encouragement, and untiring efforts this text would not have been possible.

We are grateful to the many teachers and industrial personnel who were kind enough to review portions of the manuscript and offer suggestions for improvement. Wherever possible, the text was revised to incorporate these suggestions. We sincerely thank them for their assistance in making this text as accurate and up-to-date as possible.

Special thanks are extended to the Warner and Swasey Company for providing the excellent section on Turret Lathes. We would also like to express our appreciation to the McGraw-Hill Company for providing various illustrations from the texts *Machine Shop Training,* 2nd Edition, and *Technology of Machine Tools.*

The following firms, who were kind enough to supply technical information, illustrations, and helpful suggestions, have assisted in making this text as up-to-date as possible.

Armstrong Bros. Tool Co.	**Lodge and Shipley Co.**
Bendix Corporation	**Logan Engineering Co.**
Automation and Measurement Div.	**McGraw-Hill**
Brown and Sharpe Mfg. Co.	**Mimik Ltd.**
Butterfield Division	
Litton Industries	**Monarch Machine Tool Co.**
Canadian General Electric Co. Ltd.	**Neill, James and Co. (Sheffield) Ltd.**
Carborundum Co.	**Nicholson File Co. of Canada Ltd.**
Cincinnati Milacron Co.	**Norton Co.**
Clausing Corporation	**Pratt and Whitney, Inc., Machine Tool Div.**
Cleveland Twist Drill Co.	**Sheldon Machine Co.**
Colchester Lathe Co.	**Shell Canada Ltd.**
Cushman Industries, Inc.	**South Bend Lathe**
Dumore Co.	**Standard-Modern Tool Co.**
Enco Mfg. Co.	**Starrett, L.S. Co.**
Greenfield Tap and Die Co.	**Taft-Peirce Mfg. Co.**
Hardinge Brothers, Inc.	**Taper Micrometer Corporation**
Jacobs Mfg. Co.	**Warner and Swasey Co.**
Jones and Lamson Co.	**Wickman, A.C. Ltd.**
LeBlond, R.K. Machine Tool Co.	**Williams, J.H. and Co.**

SECTION 1

Lathe Construction, Tools, and Accessories

Unit 1 LATHE TYPES AND CONSTRUCTION

The engine lathe, one of the first machine tools developed, is still one of the most versatile and widely used in industry. Many machine tools, such as turret lathes, automatic screw machines, vertical boring mills, and numerically-controlled lathes, are a direct development of the engine lathe. The lathe and its various modifications have played a large part in the industrial and economic development of this continent.

LATHE FUNCTION

The principal function of an engine lathe is to produce circular shapes by revolving the workpiece against a cutting tool, which may be controlled to produce the desired form, figure 1-1. However, other operations such as facing, tapering, form turning, threading, drilling, boring, reaming, and grinding may also be performed on a lathe.

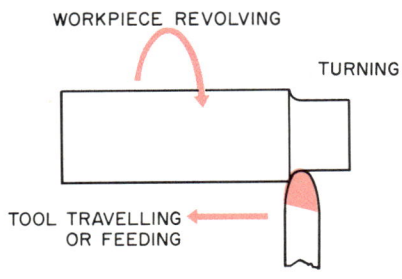

FIG. 1-1 LATHE CUTTING ACTION

TYPES OF LATHES

The most common lathes are generally classified under three categories: engine lathes, manufacturing lathes, and production lathes.

Engine Lathes

Engine lathes include bench, speed, precision, toolroom, and gap-bed lathes in various sizes.

A bench lathe, figure 1-2, is a small engine lathe which can be mounted on a bench or a metal cabinet. Bench lathes are generally small in size and used for light machining on relatively small workpieces.

FIG. 1-2 A BENCH LATHE MOUNTED ON A METAL CABINET. (Courtesy Logan Engineering Co.)

Section 1 Lathe Construction, Tools, and Accessories

FIG. 1-3 A SPEED LATHE USED FOR LIGHT MACHINING AND FINISHING OPERATIONS. (Courtesy Hardinge Brothers, Inc.)

A speed lathe, figure 1-3, is usually mounted on a bench or cabinet and is noted for the fast setup and changeover of work, ease of operation, and low maintenance. Speed lathes are used for light machining operations, turning, polishing, and finishing on small precision work.

The toolroom lathe, figure 1-4, similar in appearance to an engine lathe, is equipped with special attachments and accessories to allow a variety of precision operations to be performed. It is generally used to produce tools and gages which are used in tool and die work. The precision lathe is very similar to a toolroom lathe.

The gap-bed lathe, figure 1-5, has a section of the bed, below the faceplate, which can be removed easily to increase the swing or maximum work diameter that can be revolved.

FIG. 1-4 A TOOLROOM LATHE. (Courtesy Monarch Machine Tool Co.)

Unit 1   Lathe Types and Construction

FIG. 1-5   A GAP-BED LATHE. (Courtesy Colchester Lathe and Tool Co.)

Manufacturing Lathes

Manufacturing lathes, figures 1-6A and B, are basically engine lathes which have been modified by the addition of a tracer attachment or a digital readout system. Tracer lathes are used to reproduce accurately parts which may be too difficult or costly to produce on other types of lathes. Figure 1-6A illustrates a hydraulic tracer attachment on a lathe and the part being produced. Lathes equipped

FIG. 1-6A   A HYDRAULIC TRACER ATTACHMENT MOUNTED ON A LATHE. (Courtesy Cincinnati Milacron Co.)

FIG. 1-6B A LATHE EQUIPPED WITH A DIGITAL READ-OUT SYSTEM. (Courtesy Bendix Corp.)

Section 1   *Lathe Construction, Tools, and Accessories*

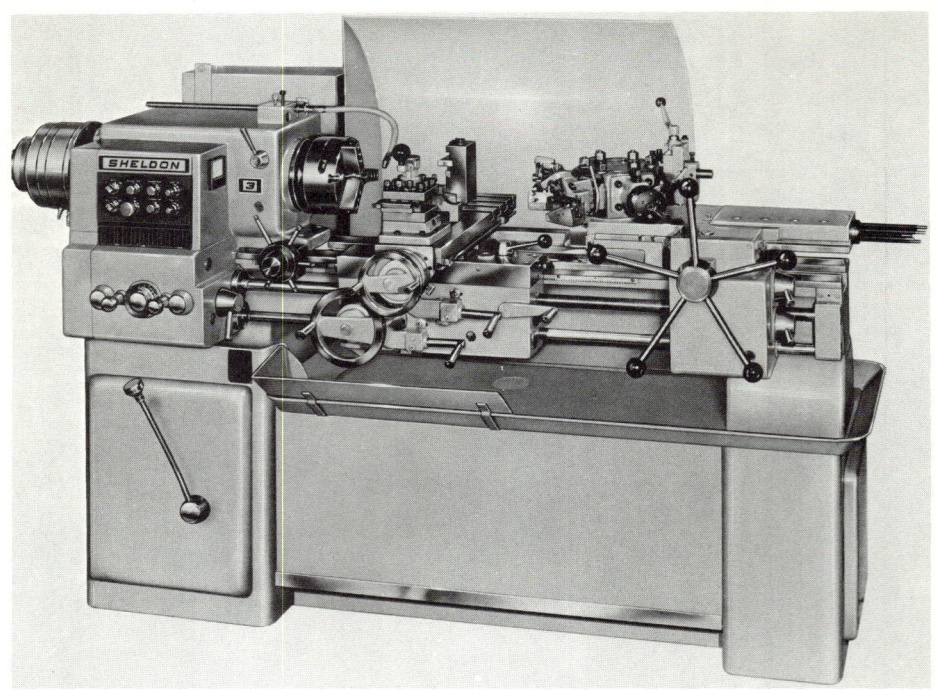

FIG. 1-7   A TURRET LATHE IS USED TO MASS-PRODUCE PARTS. (Courtesy Sheldon Machine Co.)

with digital readout systems are used to speed the production of parts normally produced on an engine lathe, figure 1-6B.

Production Lathes

Production lathes are generally used when a large number of duplicate parts must be produced. Turret lathes, single-spindle automatic lathes, and the numerically-controlled lathes are the common production machines in this group.

The turret lathe, figure 1-7, is used to produce a large number of duplicate parts which may require several operations, such as turning, drilling, boring, reaming, facing, and threading. On some models of turret lathes, as many as twenty different tools may be mounted on a ram- or saddle-type turret, and each tool may be rotated into position quickly and accurately. Once the tools have been set, each part can be quickly and accurately produced.

A single-spindle automatic lathe, figure 1-8, is designed to automatically mass-produce parts which require primarily turning and fac-

FIG. 1-8   A SINGLE-SPINDLE AUTOMATIC LATHE. (Courtesy Jones and Lamson Co.)

Unit 1   Lathe Types and Construction

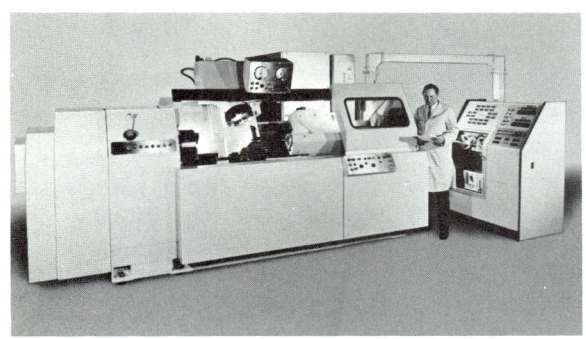

FIG. 1-9   A NUMERICALLY-CONTROLLED LATHE.
(Courtesy Cincinnati Milacron Co.)

ing operations. Automatic lathes generally have a front and a rear toolslide mounted on the carriage. The front-slide tooling is used for turning and boring operations. The rear-slide tooling is used for facing, undercutting, chamfering, and necking operations.

The numerically-controlled lathe, figure 1-9, a recent development, is one of the latest modifications of the basic engine lathe. This lathe, controlled by numerical tape, is used primarily for turning operations and economically and automatically produces shafts of almost any shape. On the model illustrated, a circular or crown turret which can be indexed holds eight carbide tools rigidly for performing various turning operations. The numerically-controlled lathe can outperform most types of lathes, and provides savings in tooling, setup, and cycle time.

LATHE SIZE

The size of a lathe generally is determined by the largest work diameter which can be revolved over the ways (A), and the longest piece of work which can be turned between centers (B), figure 1-10. Some manufacturers designate lathe size by the largest work diameter which can be revolved (A), and the length of the lathe bed (C).

Lathes are manufactured in a variety of sizes. The most common swing sizes range from 9" to 30", with the distance between centers ranging from 18" to 120". A 12" x 24" lathe is capable of handling work 12" in diameter and 24" long.

LATHE PARTS

The primary function of a lathe is to provide a means of removing metal by rotating a workpiece against a cutting tool. All lathes are basically the same, and the various parts and accessories serve three functions. They provide:

1. A support for the lathe components or the workpiece.

2. A means of holding and rotating the workpiece.

3. A method of holding and moving the cutting tool.

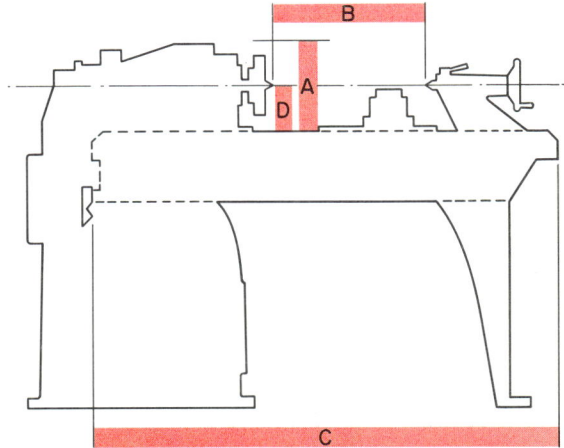

A — SWING
B — DISTANCE BETWEEN CENTERS
C — LENGTH OF BED
D — RADIUS (ONE-HALF OF SWING)

FIG. 1-10   THE SIZE OF A LATHE IS INDICATED BY THE MAXIMUM SWING AND THE MAXIMUM DISTANCE BETWEEN CENTERS.

Section 1 Lathe Construction, Tools, and Accessories

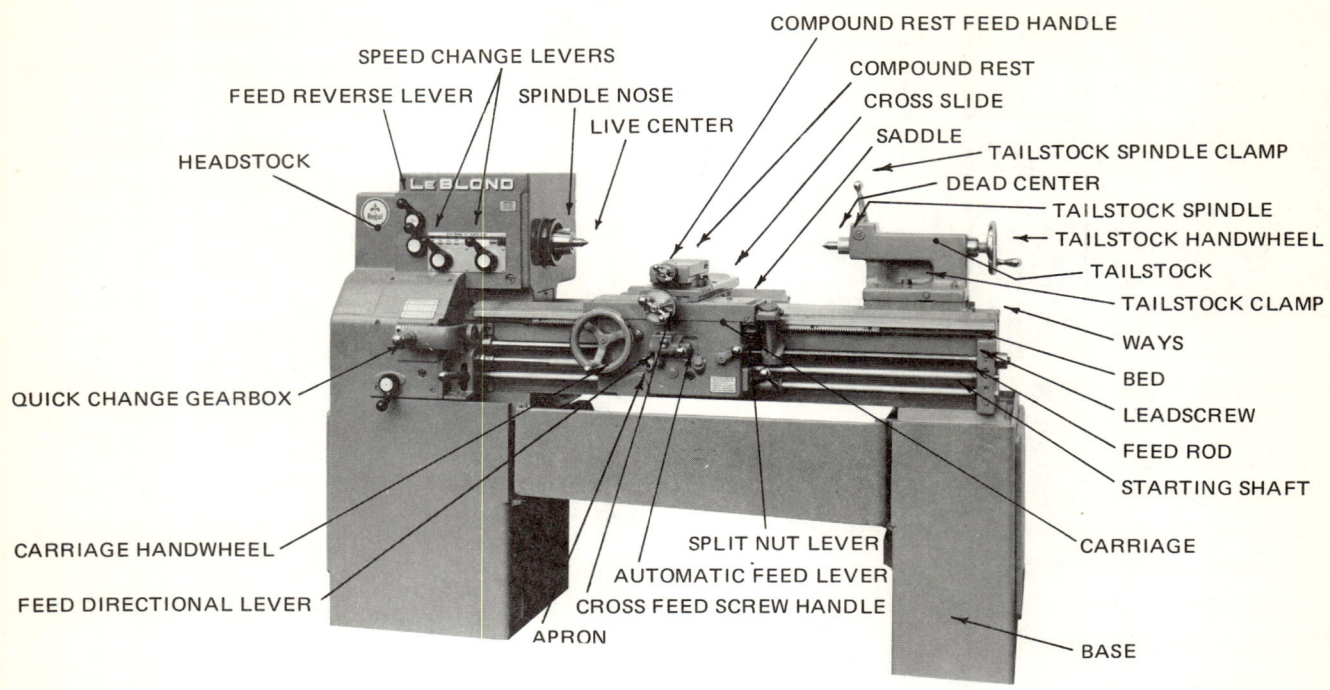

FIG. 1-11 THE MAIN PARTS OF AN ENGINE LATHE (Courtesy R.K. LeBlond Machine Tool Co.)

Bed

The *lathe bed,* figure 1-12, is a heavy, rugged casting which provides the foundation or base to which other parts are fitted. On its top section, *V-ways* are machined to guide and align the major parts of the lathe, such as the headstock, tailstock, and carriage. Most lathes have either two or three raised ways which are precision finished and in many cases hardened to prevent wear.

Headstock

The *headstock,* figure 1-13, is the housing fastened to the left-hand end of the ways and bed. It contains the gearing mechanism which

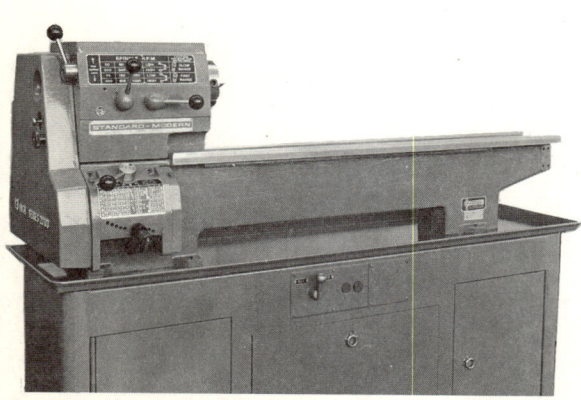

FIG. 1-12 THE LATHE BED IS THE FOUNDATION OF A LATHE. (Courtesy Standard-Modern Tool Co.)

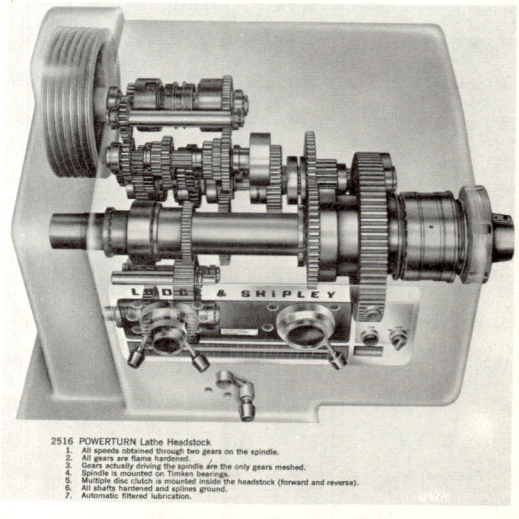

FIG. 1-13 A GEAR DRIVE HEADSTOCK. (Courtesy Lodge and Shipley Co.)

transmits a drive from the motor to the spindle in the form of controlled speeds. The *headstock spindle* is a hollow, cylindrical shaft which is supported by bearings. Work-holding devices, such as chucks, centers, fixtures, and face- or driveplates are mounted on the right-hand end of the spindle to support and drive the workpiece.

Headstock spindles may be driven by a cone pulley and belt, a variable-speed drive, or by the transmission gears in the headstock. The speed of a belt-driven lathe may be altered by changing the belt to different-sized pulleys. The speed of a variable-speed drive may be varied mechanically or electrically while the lathe is in operation. The speed of a gear-driven lathe is set by changing the position of the speed change levers while the lathe is stopped.

The *feed reverse lever,* mounted on the left-hand end of the headstock, is used to reverse rotation of the feed rod and leadscrew.

Tailstock

The *tailstock,* figure 1-14, used mainly to support the right-hand end of the workpiece, may be locked in any position along the lathe bed. It is made in two sections; the base is machined to fit the ways of the lathe, and the top section may be set for parallel or taper turning by adjusting the two setscrews in the base.

The *tailstock spindle* has an internal Morse taper to receive the *dead center* which supports the right-hand end of the work. Drills, reamers, tools, and various types of other holders may be fitted into the tapered spindle so that different machining operations may be performed.

The *tailstock spindle clamp* is used to lock the spindle in a fixed position. By turning the *tailstock handwheel,* the spindle may be moved in or out of the tailstock to provide a hand feed for various machining operations such as drilling and reaming.

Quick-Change Gearbox

The *quick-change gearbox,* figure 1-15, located on the front of the bed directly below the headstock, contains a number of different-sized gears to provide a drive for the feed rod and leadscrew. The *feed rod* provides automatic feed to the carriage for various turning operations. The *leadscrew* is used to advance the carriage for thread-cutting operations.

The *feed and thread chart,* mounted on the quick-change gearbox, indicates the lever

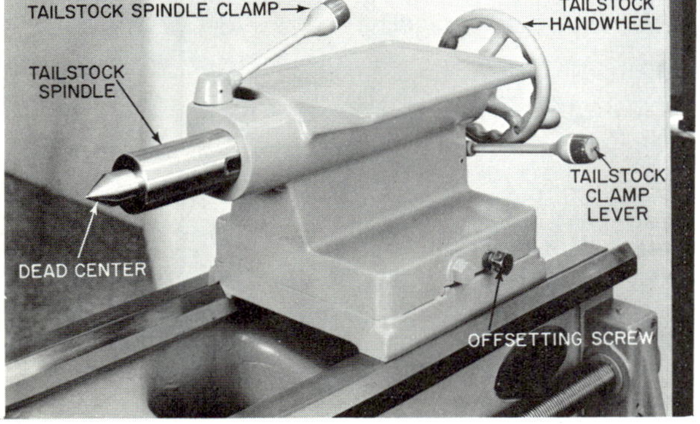

FIG. 1-14   THE TAILSTOCK OF AN ENGINE LATHE. (Courtesy Cincinnati Milacron Co.)

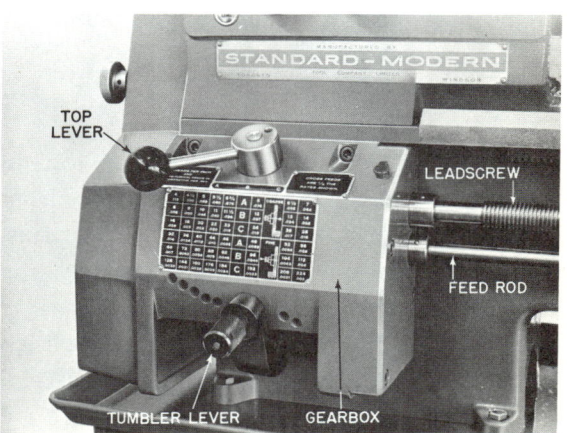

FIG. 1-15   THE QUICK-CHANGE GEARBOX ENABLES FAST SETTINGS OF FEEDS AND THREADS. (Courtesy Standard-Modern Tool Co.)

Section 1   Lathe Construction, Tools, and Accessories

FIG. 1-16   THE MAIN PARTS OF A LATHE CARRIAGE ARE THE SADDLE, APRON AND CROSS-SLIDE. (Courtesy Standard-Modern Tool Co.)

settings for various feeds. The top figure in each block refers to the number of threads per inch that are cut if the leadscrew is used to engage the carriage. The bottom number (in thousandths) refers to the distance the carriage advances for each spindle revolution if the feed rod is engaged.

Carriage

The *carriage,* figure 1-16, is used to control the movement of the cutting tool either automatically or manually. The carriage assembly consists of three main parts — the saddle, cross-slide, and apron.

The *saddle* is an H-shaped casting which slides along the ways between the headstock and tailstock. The apron and cross-slide are attached to the saddle.

The *apron,* which is bolted to the saddle, is on the front of the lathe and contains the carriage and cross-slide controls. The *apron handwheel,* connected to a gear which meshes with a rack on the lathe bed, is used to move the carriage manually along the lathe bed. The *automatic feed lever* engages a clutch to provide automatic feed to the carriage and cross-slide, depending upon the position of the *feed directional plunger.* The *split-nut lever* can be engaged for thread cutting when the feed directional plunger is set in the neutral position. The threads of the split-nut lever engage the threads of the leadscrew and cause the carriage to move at a predetermined rate.

The *cross-slide* is a casting, mounted on the top of the saddle, which provides a cross movement for the cutting tool. The *crossfeed screw handle* is used to move the cross-slide manually. The *compound rest* is mounted on top of the cross-slide and is used to provide a support for the cutting tool. The base of the compound rest may be swivelled to any angle in a horizontal plane for cutting short tapers. The compound rest may be fed manually by turning the *compound rest screw handle.* Both the crossfeed and compound rest screw handles are equipped with *graduated micrometer collars* to allow accurate settings of the cutting tool in thousandths of an inch.

REVIEW QUESTIONS

1. What are three machine tools which are direct developments of the engine lathe?
2. What is the function of an engine lathe?
3. What are six operations which may be performed on a lathe?

Types of Lathes

4. Describe three types of engine lathes. What is the purpose of each type?
5. What are two duplicating lathes and what is the purpose of each?
6. A turret lathe is more suitable than an engine lathe for producing a large number of duplicate parts. Why is this true?
7. How does a single-spindle automatic lathe compare with a numerically-controlled lathe?

Lathe Size

8. How is the size of a lathe determined?
9. What is meant by a 10" x 20" lathe?

Lathe Parts

10. What three functions do the various lathe parts and accessories serve?
11. What is the purpose of the V-ways on the lathe bed?
12. What are the names of three types of headstock drives, and how does each type operate?
13. What is the purpose of the feed reverse lever?
14. Describe the tailstock. What purpose does it serve?
15. Describe each of the following items. What is the purpose of each item?
    a. The quick-change gearbox
    b. The feed and thread chart
16. What are the three main parts of the carriage assembly?
17. What are three ways in which a carriage can be moved along the lathe bed?

Section 1   Lathe Construction, Tools, and Accessories

18. What are the four parts of a lathe which are used for a thread-cutting operation?

19. Locate the following items. What is the purpose of each item?

    a.   The cross-slide

    b.   The compound rest

20. What is the purpose of the graduated micrometer collars on the compound rest and crossfeed screws?

Unit 2 LATHE ACCESSORIES

The versatility of a lathe is greatly increased by many accessories which are available. These accessories may be attached to different parts of the lathe and are described under the categories of accessories for the headstock, tailstock, carriage, and bed.

HEADSTOCK ACCESSORIES

The most common headstock accessories are those which hold or drive the work such as chucks, collets, centers, faceplates, and drive plates.

Chucks

Probably the most common work-holding devices are chucks, which grip work of various shapes and sizes by means of adjustable jaws. There are various types of chucks such as the two-jaw universal, three-jaw universal, four-jaw independent, combination, collet, and magnetic chuck.

The two-jaw universal chuck, figure 2-1, has two adjustable jaws, always equidistant from the center, whose movement is controlled by one adjusting screw. The adjusting screw, with a right-hand thread on one end and a left-hand thread at the other end, moves both jaws simultaneously. One feature of this chuck is that work of various shapes and sizes may be held for machining by means of special-shaped jaws machined to suit the workpiece.

The three-jaw universal geared scroll chuck, figure 2-2, is normally used for holding round or hexagonal stock. The large body houses a scroll plate, the back of which meshes with a gear, figure 2-2. The three jaws, whose teeth mesh with the scroll, move in or out simultaneously as the chuck wrench turns the gear. Three-jaw universal chucks are usually provided with two sets of jaws, one for outside chucking and one for internal chucking. When changing jaws on chucks of this type, the jaws must be assembled in the proper order so that the work is held concentrically.

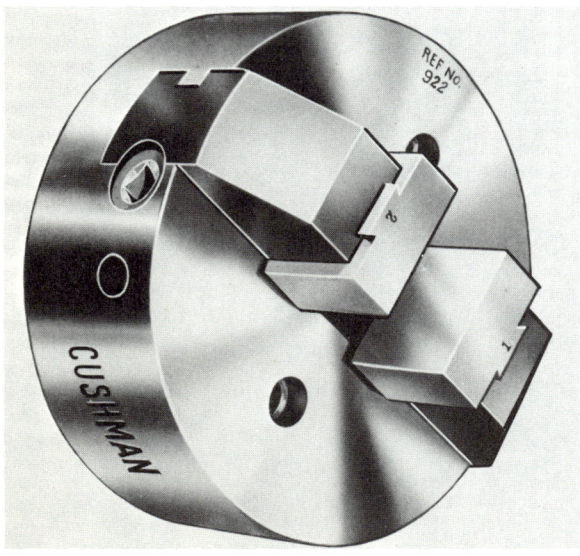

FIG. 2-1 A TWO-JAW UNIVERSAL CHUCK. (Courtesy Cushman Industries, Inc.)

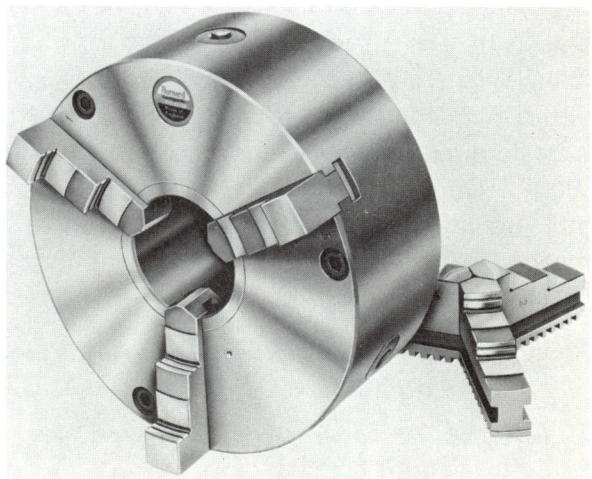

FIG. 2-2 A THREE-JAW UNIVERSAL CHUCK. (Courtesy Clausing Corp.)

Section 1   Lathe Construction, Tools, and Accessories

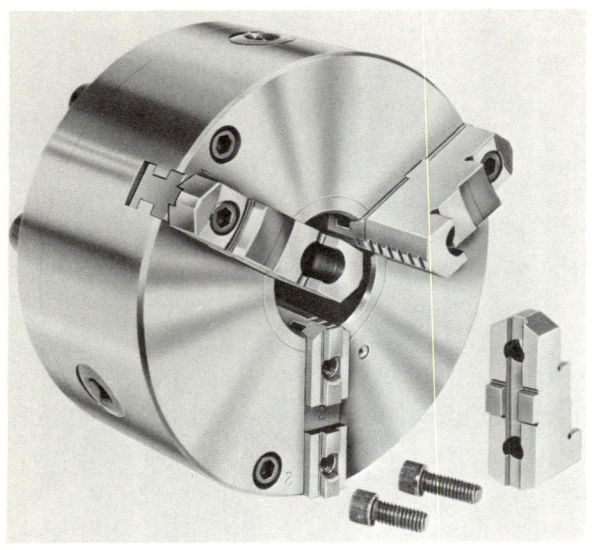

FIG. 2-3A   A THREE-JAW UNIVERSAL CHUCK WITH REVERSIBLE TOPS.   (Courtesy Clausing Corp.)

FIG. 2-3B(i)   A THREE-JAW GRIPTRU CHUCK (Courtesy Clausing Corp.)

Some types of three-jaw universal chucks have jaws which are fastened to the top of three slides by cap screws, figure 2-3A. Work can be gripped internally by reversing the jaws on the slides. Three-jaw chucks hold work to within .002 – .003 of concentricity when the chuck is new and kept in good condition. As the jaws or the scroll threads become worn, the chuck loses this accuracy, and the jaws may have to be reground.

A relatively new development in lathe chucks is the three- and six-jaw "Griptru" chuck, figure 2-3B(i). Each chuck is equipped with a microadjustment mechanism which permits holding workpieces concentric to within .0002. Once a "Griptru" chuck is set for a job, no further adjustment of the micro-mechanism is required to chuck duplicate parts to within ± .00025.

Procedure: Chucking a Workpiece in a "Griptru" Chuck

1. Advance the chuck jaws to the workpiece by turning the chuck pinion.
2. Back off the microadjusting screws, figure 2-3B(ii).

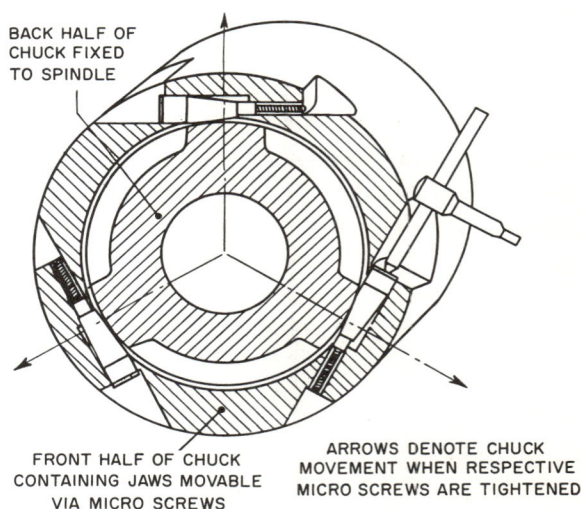

FIG. 2-3B(ii)   THE PRINCIPLE OF A GRIPTRU CHUCK (Courtesy Clausing Corp.)

3. Mount a dial indicator on the lathe and bring it into contact with the work diameter until the needle registers approximately one-half a turn.
4. Rotate the lathe spindle until the indicator registers the lowest reading.
5. Tighten the microadjusting screw nearest the low reading until the indicator needle moves one-half the total eccentric reading.

Unit 2 Lathe Accessories

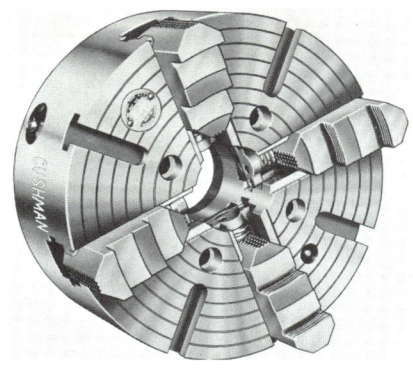

FIG. 2-4 A FOUR-JAW INDEPENDENT CHUCK. (Courtesy Cushman Industries, Inc.)

FIG. 2-5 A COMBINATION CHUCK. (Courtesy Cushman Industries, Inc.)

6. Continue this procedure until the indicator needle does not move when the spindle is rotated.

   Note: After the first piece has been aligned, succeeding pieces having the same diameter may be chucked and run true to within ± .00025 without further adjustment of the chuck.

The four-jaw independent chuck, figure 2-4, has four jaws, each of which may be adjusted independently to suit the workpiece. Round, square, hexagonal, octagonal, and odd-shaped workpieces may be held in a four-jaw chuck. The jaws may be reversed to grip work internally. When work must run absolutely true, it should be mounted in a four-jaw chuck and set up with a dial indicator.

Combination chucks. When the accuracy of a four-jaw chuck and the convenience of a three-jaw chuck are desired, a *combination chuck,* figure 2-5, may be used. The scroll plate of this chuck causes the three jaws to move in or out simultaneously. Each jaw is further equipped with an adjusting screw so that work may be centered accurately by adjusting one or all of the jaws independently as required.

Collet chucks, figures 2-6A and B, are used when high precision and production are required. Their design permits rapid and accurate mounting of workpieces, which should be scale-free and close to the shape and size of the collet. Collets are available for holding small round, square, hexagonal, and special-shaped

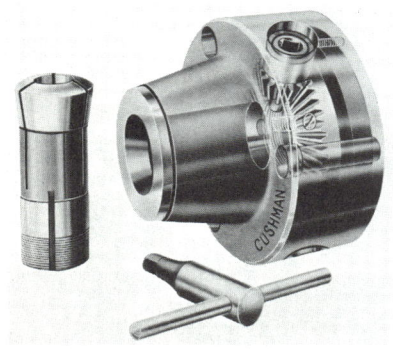

FIG. 2-6A A SPRING COLLET CHUCK. (Courtesy Cushman Industries, Inc.)

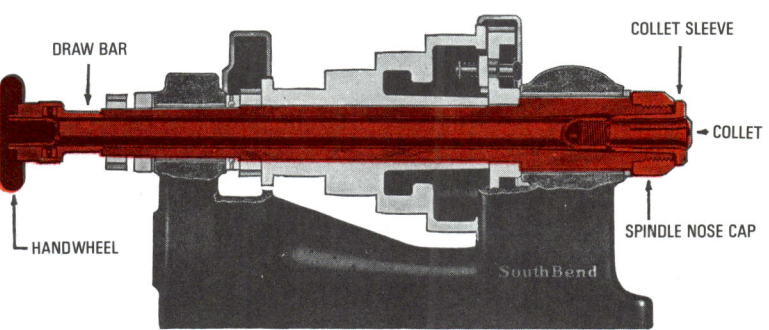

FIG. 2-6B A HEADSTOCK VIEW SHOWING THE CONSTRUCTION OF A DRAWN-IN COLLET CHUCK. (Courtesy South Bend Lathe, Inc.)

Section 1 Lathe Construction, Tools, and Accessories

FIG. 2-6C A RUBBER-FLEX COLLET CHUCK HAS A WIDER RANGE THAN OTHER COLLETS. (Courtesy Jacobs Manufacturing Co.)

work. Collet chucks may be of two types, the spring collet and the rubber-flex collet.

Spring collet chucks, figure 2-6A, have a range of only a few thousandths over or under the size stamped on the nose of the collet. When the chuck wrench is turned, the collet is drawn into a tapered hole of the chuck, causing the jaws to close evenly around the workpiece. Spring collet chucks hold the work securely and exactly true for machining purposes.

Another form of spring collet is the *draw-in collet attachment* which fits through the spindle of the lathe, figure 2-6B. Due to the necessity of the draw bar fitting into the spindle of the lathe, this type is more limited in the size of the work that can be held.

The *rubber-flex* or the *Jacobs spindle-nose chuck,* figure 2-6C, is a more convenient type of collet chuck since it can handle a wide range of sizes. The collet adaptor is mounted on the spindle and is equipped with a set of eleven collets ranging from 1/8" to 1 3/8" diameter. Each rubber-flex collet has a range of almost 1/8"; thus a wider variety of work can be held with fewer collets than with the draw-in spring type. When the handwheel is turned clockwise, the rubber-flex collet is forced into the taper in the adaptor, causing the metal inserts to tighten on the workpiece. The workpiece is released when the handwheel is rotated counterclockwise.

Magnetic lathe chucks, figure 2-7, may be used for holding irregularly-shaped work or thin parts which may be distorted when held by any other means. These chucks are used for holding work for light machining or grinding operations. The work is first set up visually with the concentric rings, and the chuck is partially energized by turning the wrench about 1/4 turn. This causes the chuck to hold the workpiece lightly so that it can be centered. After the work has been properly centered, the chuck is fully energized by turning the chuck wrench to the full-on position. Care should be taken that the chuck is fully on for machining operations. Light cuts should be taken when machining or grinding work held by this type of chuck. Thin-walled workpieces such as piston rings may be held in position on the chuck face by a spigot or steel plate machined to the inside diameter of the ring. The large bearing surface of the spigot takes the thrust and holds the ring in place for machining purposes. This arrangement provides a convenient means for rapidly centering rings for machining.

Drive and Faceplates

The *driveplate,* figure 2-8A, is a round steel plate, having one or more slots on its face, which may be attached to the headstock of the lathe. It is used to drive the lathe dog when machining work supported between the *60° live center* in the headstock and the dead center in the tailstock.

Work that cannot be held in a chuck or between centers is often fastened to a *faceplate,* figure 2-8B. The slots in the faceplate permit the use of bolts to secure the work to the face-

Unit 2   Lathe Accessories

FIG. 2-7   WORK MAY BE HELD ON A MAGNETIC CHUCK FOR LIGHT MACHINING OPERATIONS.   (Courtesy James Neill and Co.)

plate. An angle plate, bolted to the faceplate, may also be used to hold certain odd-shaped workpieces. Counterbalances must be used when large workpieces are mounted off-center to prevent excess vibration during machining.

TAILSTOCK ACCESSORIES

The main accessory used in the tailstock is the *dead center* which supports the right-hand end of the workpiece. The other end may be held in a chuck or supported on a

FIG. 2-8A   A DRIVEPLATE IS USED WHEN MACHINING WORK BETWEEN CENTERS.

FIG. 2-8B   WORK MAY BE FASTENED TO FACEPLATE FOR VARIOUS MACHINING OPERATIONS.

Section 1 Lathe Construction, Tools, and Accessories

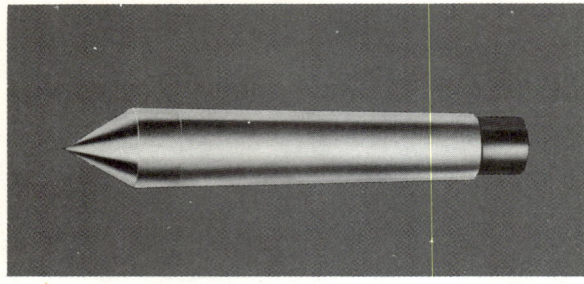

FIG. 2-9A A 60° LATHE CENTER. (Courtesy Hardinge Brothers, Inc.)

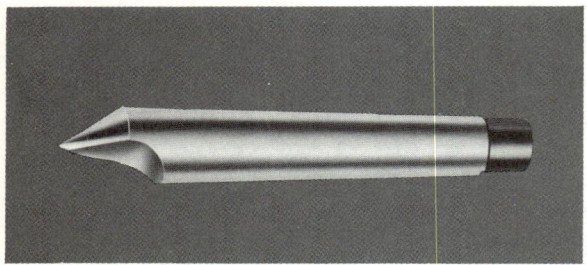

FIG. 2-9B A HALF-CENTER USED FOR FACING AND GRINDING OPERATIONS. (Courtesy Hardinge Brothers, Inc.)

Dead Centers

The 60° dead center, figure 2-9A, may be made of carbon steel, high-speed steel, or with a carbide-tipped insert. Carbon steel centers, although easily made, require much care and attention to prevent them from burning and damaging both the work and the center.

High-speed steel centers are more suitable for most lathe operations where work must be supported by a dead center. The high-speed steel permits higher work speeds since it can withstand higher heat. If properly lubricated and adjusted, these centers stand up well and there is less danger of damaging the workpiece should it become overheated, as compared with carbon steel centers.

The half-center, figure 2-9B, is used in facing operations, particularly on small-diameter workpieces where it is necessary to bring the facing tool close to the center hole of the workpiece.

The revolving dead center, figure 2-10A, is mounted in the tailstock spindle and revolves with the workpiece, thus eliminating friction between the work and the lathe center. The

live center. There are several types of dead centers, the most common of which are the standard 60° center, the half-center, the revolving dead center, and the Micro-Set adjustable center.

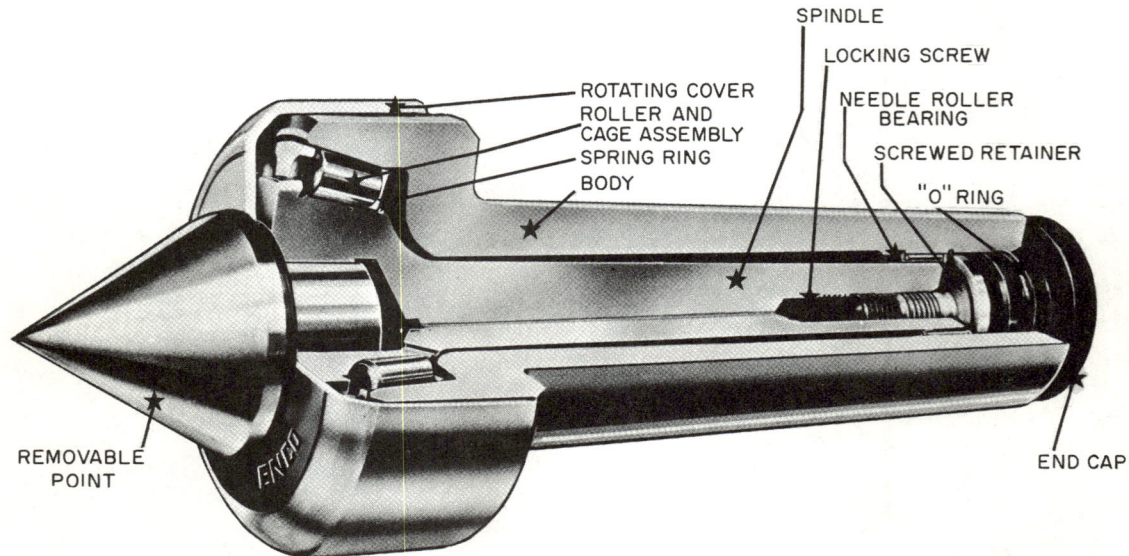

FIG. 2-10A A REVOLVING DEAD CENTER ELIMINATES FRICTION BETWEEN THE WORK AND LATHE CENTER. (Courtesy Enco Manufacturing Co.)

Unit 2   Lathe Accessories

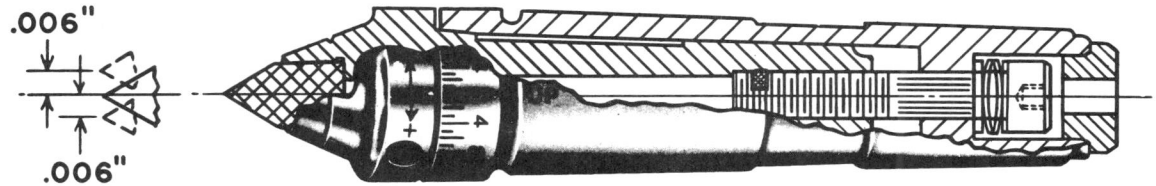

FIG. 2-10B   THE MICRO-SET CENTER ALLOWS THE LATHE CENTERS TO BE ALIGNED QUICKLY. (Courtesy Enco Manufacturing Co.)

use of this center permits the use of higher spindle speeds. This center is useful when taking finish cuts on long material.

The Micro-Set adjustable center, figure 2-10B, provides a quick and easy means of aligning lathe centers or producing a slight taper on a workpiece. A very limited offset can be achieved using a small eccentric, figure 2-10B, or a dovetail slide.

CARRIAGE ACCESSORIES

The main *carriage accessories* are the follower rest and the toolpost. Special attachments, which can be mounted to the carriage, are available for contouring, milling, and grinding operations.

Follower Rest

The *follower rest,* figure 2-11, can be mounted on the saddle of the lathe to support long, slender work during machining. Most follower rests have two adjustable soft-faced jaws, one bearing on top of the workpiece and the other behind the workpiece opposite the cutting tool. This arrangement prevents the work from springing up or away from the cutting tool during the machining operation.

Toolposts

Toolposts are used for securing toolholders or lathe tools for various machining operations. Several types of toolposts are available.

The round toolpost, figure 2-12A, generally is supplied with the lathe as standard equip-

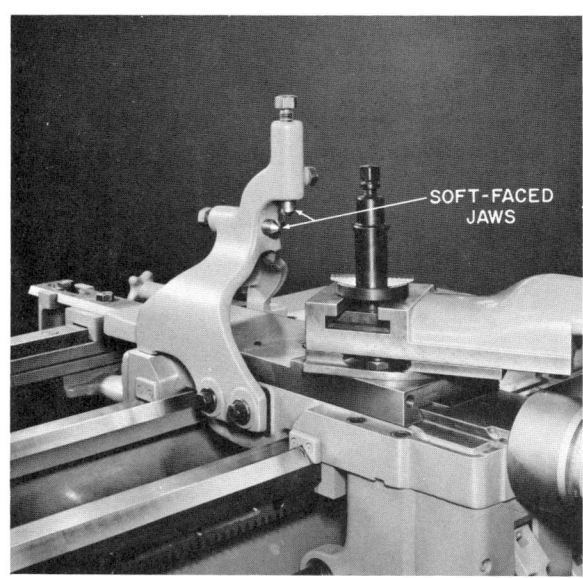

FIG. 2-11 A FOLLOWER REST PREVENTS WORK FROM SPRINGING AS THE CUTTING TOOL MOVES ALONG.
(Courtesy Cincinnati Milacron Co.)

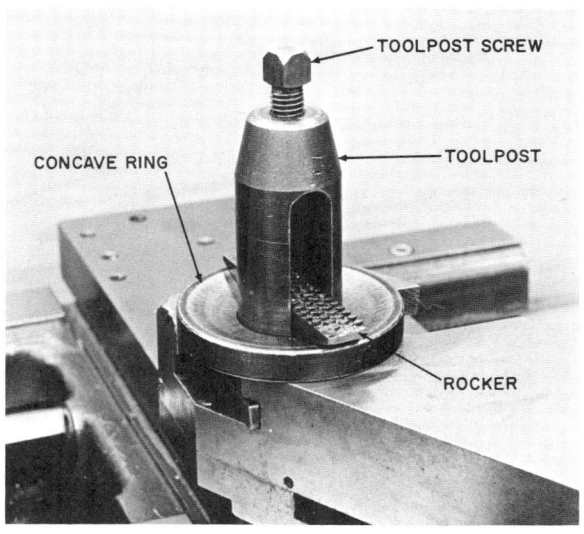

FIG. 2-12A   A STANDARD TOOLPOST FOUND ON AN ENGINE LATHE.

Section 1   Construction, Tools, and Accessories

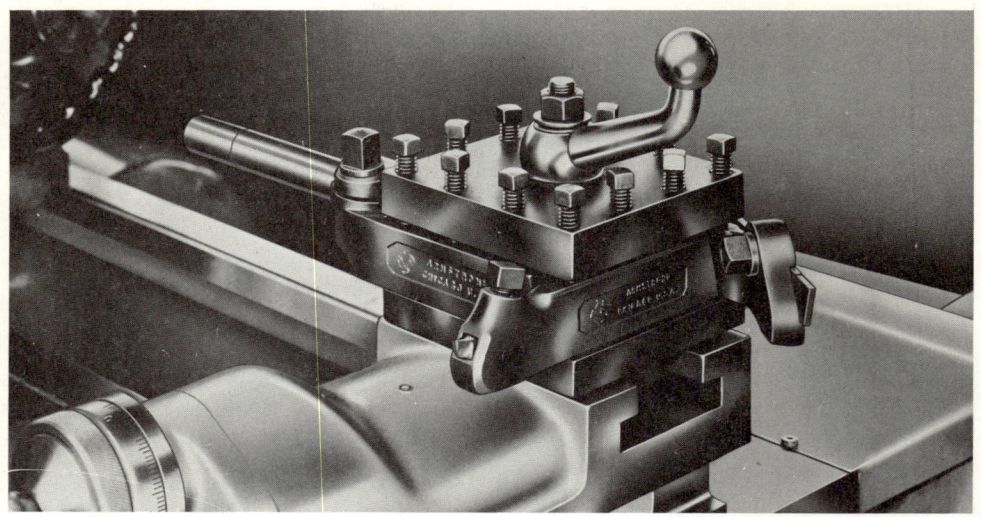

FIG. 2-12B   A TURRET TOOLPOST HOLDS FOUR CUTTING TOOLS AT ONE TIME. (Courtesy Cincinnati Milacron Co.)

ment and is mounted in the T-slot of the compound rest. The toolholder is adjusted to center height by means of a rocker or wedge which fits into a concave ring. The toolholder is held in position by a single heat-treated screw. Care must be taken when using this type of toolpost that the cutting tool does not swing into the work as the cut progresses, and cause damage to the workpiece.

The four-way turret toolpost, figure 2-12B, provides a convenient and efficient way of holding up to four cutting tools which may be used when several operations such as turning, grooving, threading, and parting must be performed on a workpiece. The cutting tools or toolholders are held in position by screws on the top edge of the turret. Shims may be required under the toolbit to bring it up to center height. The four-way turret can be indexed to twelve positions (at 30° intervals) by loosening the locking handle and rotating the turret to the desired position.

The heavy-duty toolpost, figure 2-13A, holds a single tool or toolholder by means of two screws on the top. Shims or packing strips may be required under the tool to set it to center height. The toolpost may be swung to any position and locked in place by a nut.

The quick-change toolpost, figure 2-13B, provides a rigid mount for a special dovetailed toolholder into which a cutting tool has been mounted. This type of toolholder is often used on a lathe equipped with a digital-readout system. The tools are ground and accurately set in the dovetailed toolholder in the toolroom and used as required on the lathe. Toolholders containing other cutting tools set for a particular operation can be quickly interchanged by loosening the locking handle.

LATHE BED ACCESSORIES

Lathe accessories, fastened to the bed of the machine, are the taper attachment, the micrometer carriage stop, and the steady rest.

Taper Attachment

The *taper attachment* permits the machining of internal and external tapers on a workpiece without disturbing the alignment of the lathe centers. There are two types of taper

Unit 2   Lathe Accessories

FIG. 2-13A   A HEAVY-DUTY TOOLPOST USED WHEN TAKING HEAVY CUTS. (Courtesy Cincinnati Milacron Co.)

FIG. 2-13B   THE QUICK-CHANGE TOOLPOST PERMITS RAPID CHANGING OF VARIOUS CUTTING TOOLS. (Courtesy Cincinnati Milacron Co.)

attachments available: the plain taper and the telescopic taper.

When using the *plain taper attachment,* figure 2-14A, the depth of cut is set with the compound rest screw handle, since the binding screw is removed from the crossfeed nut and replaced in the sliding block on the taper attachment.

The *telescopic taper attachment,* figure 2-14B, is a more convenient attachment since the crossfeed screw is not disengaged. This permits the depth of cut to be set

FIG. 2-14A   A PLAIN TAPER ATTACHMENT. (Courtesy Standard-Modern Tool Co.)

FIG. 2-14B   A TELESCOPIC TAPER ATTACHMENT. (Courtesy Monarch Machine Tool Co.)

19

Section 1 Lathe Construction, Tools, and Accessories

FIG. 2-15 A MICROMETER CARRIAGE STOP IS USED FOR CUTTING TO ACCURATE LENGTHS. (Courtesy Standard-Modern Tool Co.)

FIG. 2-16 THE STEADY REST SUPPORTS LONG, SLENDER WORK. (Courtesy South Bend Lathe Co.)

with the crossfeed handle. This type of taper attachment must be clamped to the bed by means of an anchor bracket when taper turning.

Both types of taper attachments are usually graduated in inches per foot at one end of the swivel bar, and in degrees at the other end.

Micrometer Carriage Stop

The *micrometer carriage stop,* figure 2-15, is a device which may be clamped to the bed of the lathe to control the travel of the carriage accurately in thousandths of an inch. It is used to advantage when several duplicate parts must be made.

When the micrometer stop is mounted on the bed to the right of the carriage, it controls the starting position of the cutting tool. When the micrometer stop is mounted on the bed to the left of the carriage, it controls the travel of the carriage and therefore the length of cut. When the carriage stop is used in this position, care must be taken to shut off the feed before the carriage comes up to the stop. The last small section of the cut should be made by using hand feed until the carriage comes into contact with the stop.

Steady Rest

The *steady rest,* figure 2-16, is used to support long work which may be held between centers or in a chuck. Three equally spaced jaws, usually bronze or plastic tipped, can be adjusted to support work of any diameter within the capacity of the steady rest. Long work, which is held in a chuck and cannot be supported by the tailstock center, can be supported by the steady rest for machining operations. Work can also be supported anywhere along its length by the steady rest to prevent springing, provided that the rest does not interfere with carriage travel.

REVIEW QUESTIONS

Headstock

1. What is the purpose and advantage of each of the following chucks?

    a. Two-jaw universal

    b. Three-jaw universal

    c. Four-jaw independent

    d. Combination

    e. Collet

2. How is a three-jaw universal chuck constructed? How does it operate?

3. What are two types of collet chucks? Describe each type.

4. What is the disadvantage of a draw-in type collet?

5. What advantages does the rubber-flex collet have over the spring collet type?

6. How should work be centered on a magnetic chuck?

7. How may thin-walled work be held on a magnetic chuck for machining purposes?

8. What precaution must be observed when mounting large workpieces off-center on a faceplate?

Tailstock

9. What is the purpose and what are the advantages of each of the following?
   a. The half center
   b. The revolving dead center
   c. The Micro-Set adjustable center

Carriage

10. What is the purpose of a follower rest?

11. How is the follower rest adjusted to the workpiece?

12. What are four types of toolposts and what are the advantages of each?

Lathe Bed

13. What advantage does the telescopic taper attachment have over the plain taper attachment?

14. What is the purpose of the micrometer carriage stop?

15. What precaution must be observed when using a micrometer carriage stop to the left of the carriage?

16. What are three applications of the steady rest?

Unit 3 LATHE MAINTENANCE

Proper care and maintenance are very important to the life and accuracy of a lathe and are a reflection upon the operator. A good operator or toolmaker takes pride in keeping the lathe clean and lubricated regularly. It is good practice to follow a regular maintenance program which should include such things as: lubrication, adjustment of gibs and clutches, checking spindle bearings, adjusting the level of the machine, and thorough cleaning. A well-maintained lathe retains its accuracy indefinitely with a minimum amount of "down-time," resulting in increased production.

LATHE CLEANLINESS

The time spent in keeping a lathe clean is more than repaid by the added life of the machine. Chips, scale and grit produced by the cutting tool must be brushed away frequently to prevent them from mixing with the oil on the slides and ways of the lathe to form a sludge. This abrasive sludge, which forms between sliding parts, increases friction which makes the lathe hard to operate and causes wear between the parts.

If chips or grit are allowed to get under the carriage or tailstock, or into the headstock and tailstock spindle tapers, the accuracy of the lathe may be impaired. Small chips from tool steel are very hard and sharp, and can easily cut the lathe ways. Sand and scale from cast iron, bronze, or other cast metals, because of their abrasive qualities, can damage the slides and lathe ways.

Procedure: Cleaning a Lathe

1. Always shut off the electrical switch before attempting to clean a lathe.

2. *Compressed air is not recommended* for cleaning a lathe because chips and dirt may be blown into oil holes and between bearing surfaces. There is also danger that flying chips may cause injury to others nearby.

3. *Never handle steel chips by hand* – they are razor sharp. Pliers or a metal hook should be used to remove long, continuous chips from the lathe.

FIG. 3-1 A PAINTBRUSH IS HANDY FOR REMOVING CHIPS AND DIRT FROM A LATHE.

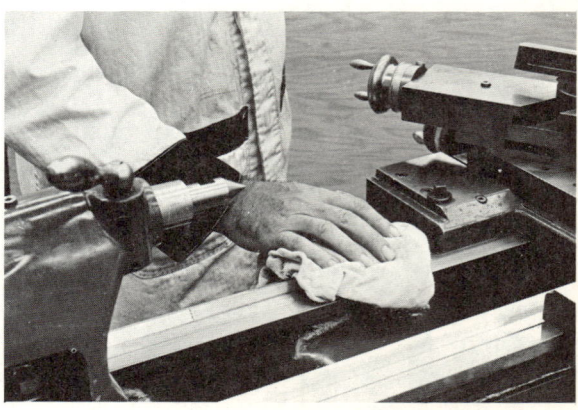

FIG. 3-2 USE A CLEAN CLOTH TO REMOVE DIRT AND SLUDGE.

4. A relatively inexpensive 3" paintbrush is recommended for removing dirt and chips from a lathe, figure 3-1. This brush is small enough to get into most places, and yet large enough to remove chips quickly.

5. Use a clean cloth, after brushing, to remove any traces of dirt and sludge. A little oil on the cloth protects the finished surfaces of the machine from rust.

6. Once every week, the lathe should be thoroughly cleaned with kerosene or varsol to remove gum and oil stains. Caustic or inflammable cleaners should be avoided.

LUBRICATION

The importance of an adequate lubrication program cannot be overstressed. It pays dividends by prolonging the life and accuracy of a lathe. Given proper care, a lathe retains its accuracy for years; if lubrication is neglected, bearing surfaces are soon damaged. As a result, the accuracy is destroyed and the life of the lathe is shortened.

A good lubricant must perform several functions in a machine tool.

- First, and probably the most important function, is that of becoming an actual part of the machine tool. The lubricant fills in the clearance spaces, that have necessarily been allowed between moving parts, with a hydrostatically controlled *fluid shim* so that the machine maintains its built-in precision.

- The lubricant must *prevent undue wear* of the various bearing surfaces so that the precision built into the machine tool by the manufacturer can be preserved to extend the useful life of a machine, figure 3-3.

- The lubricant must largely *overcome friction* within the machine, thus preserving its precision by preventing distortion due to the overheating of parts.

- The lubricant must also serve as a *cooling agent* by taking up and removing heat from the areas in which it is generated. This helps to prevent the expansion of metal parts which could cause distortion, figure 3-3.

- The lubricant must preserve the precision-built bearing surfaces of the machine against *corrosion or rusting* brought about by atmospheric conditions or the presence of other corrosive agents.

Frequency of Lubrication

Any machine tool should be lubricated and have the level of the oil reservoirs checked daily. This should be the task of the first operator. *Never attempt to lubricate any machine unless it is stopped.*

Each new lathe is supplied with an operator's manual in which the manufacturer lists the type and grade of lubricant recommended for each part of the machine. It is good practice to follow these recommendations. Generally, a good grade of SAE 20 or 30 machine oil is recommended for the general lubrication of the lathe.

The oil reservoirs in the headstock, quick-change gearbox, and apron should be drained, flushed, and filled with fresh lubricant once every six months. *The use of an incorrect grade of oil in the headstock is liable to cause overheating and possible damage.* All other lubrication points, such as oil holes and oil cups, should be filled or inspected daily.

Section 1   Lathe Construction, Tools, and Accessories

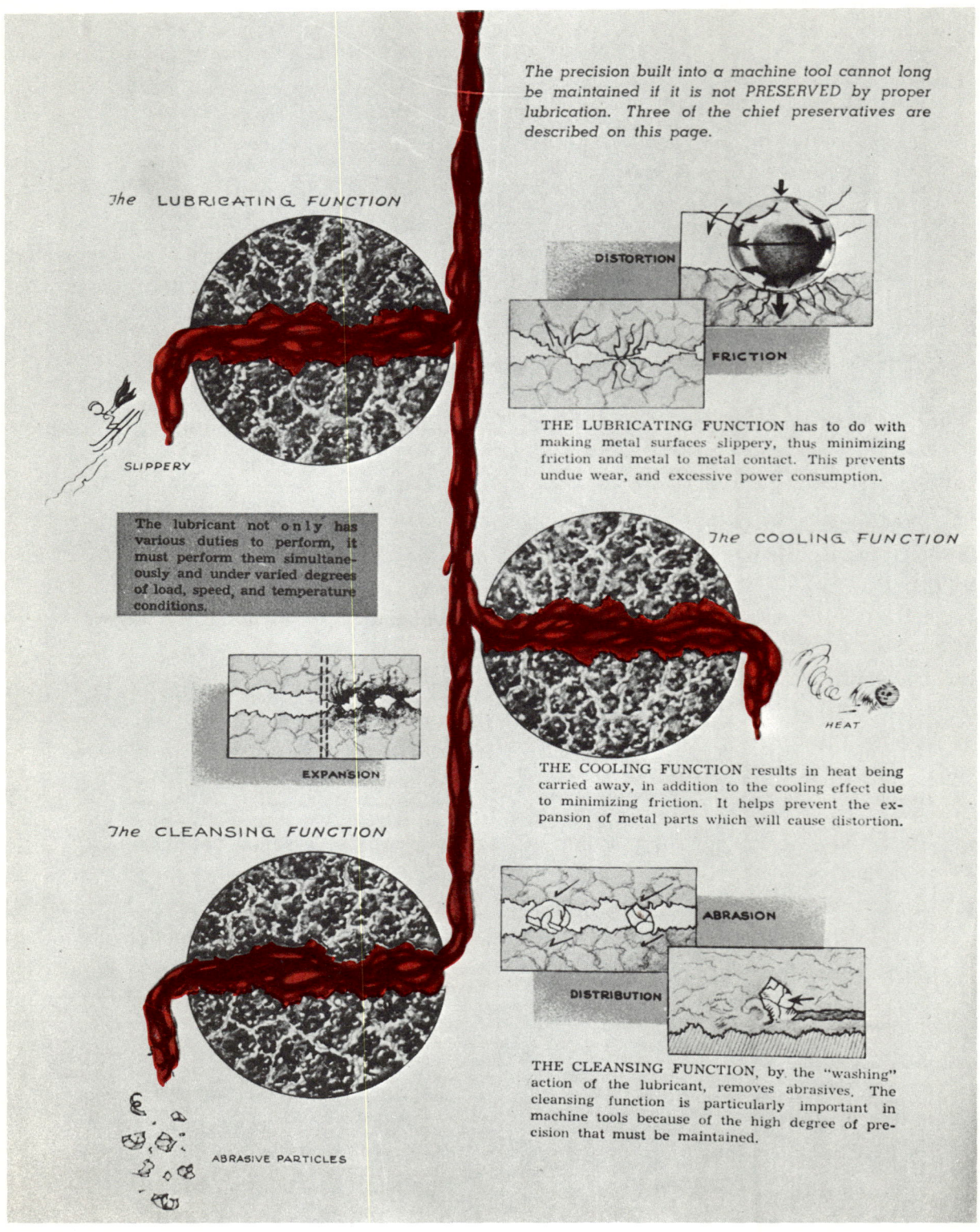

FIG. 3-3 A GOOD LUBRICANT MUST HELP TO PRESERVE THE PRECISION BUILT INTO A MACHINE TOOL.
(Courtesy Shell Oil Ltd.)

Unit 3 Lathe Maintenance

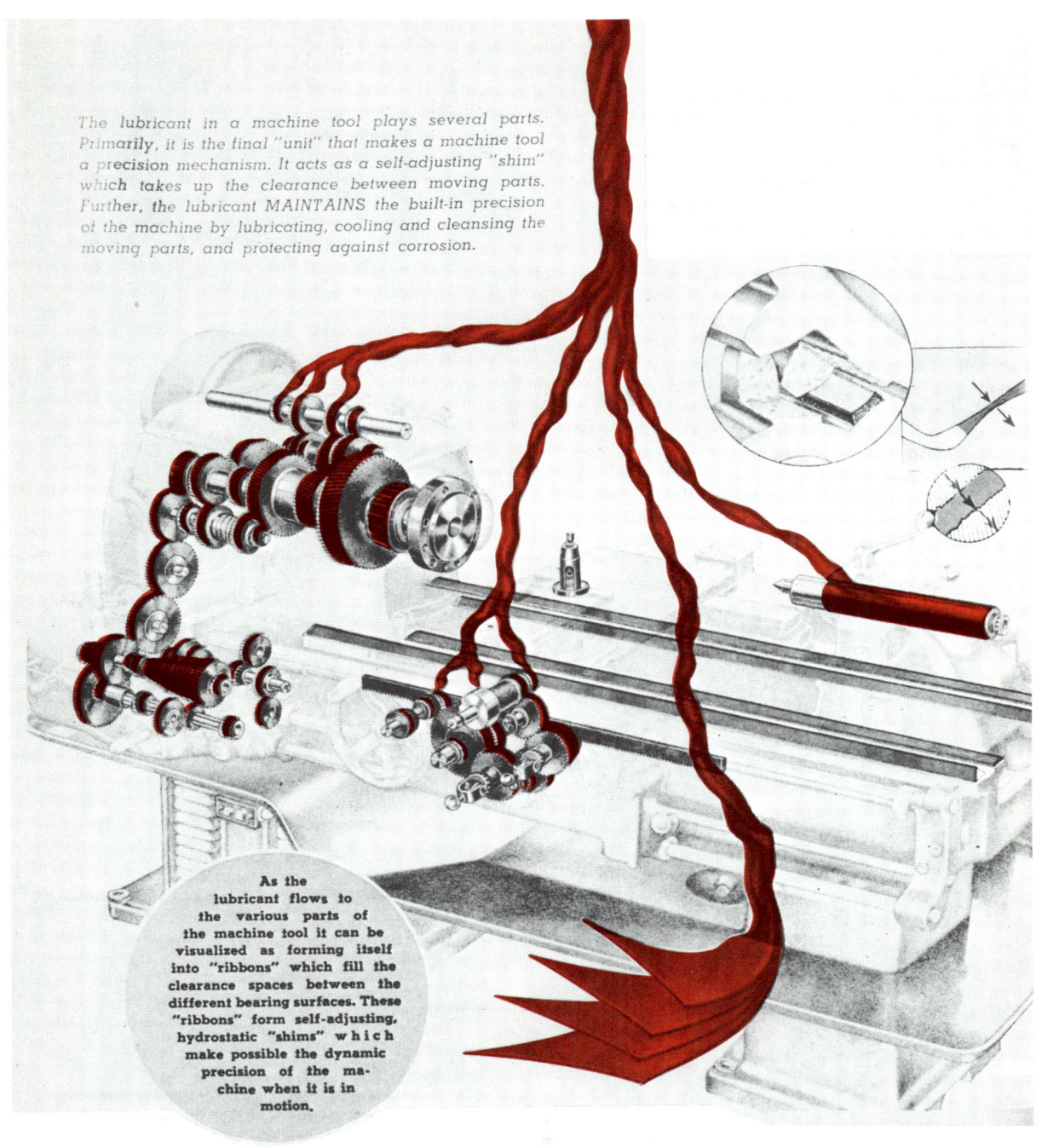

The lubricant in a machine tool plays several parts. Primarily, it is the final "unit" that makes a machine tool a precision mechanism. It acts as a self-adjusting "shim" which takes up the clearance between moving parts. Further, the lubricant MAINTAINS the built-in precision of the machine by lubricating, cooling and cleansing the moving parts, and protecting against corrosion.

As the lubricant flows to the various parts of the machine tool it can be visualized as forming itself into "ribbons" which fill the clearance spaces between the different bearing surfaces. These "ribbons" form self-adjusting, hydrostatic "shims" which make possible the dynamic precision of the machine when it is in motion.

FIG. 3-4 THE LUBRICANT FILLS IN CLEARANCE SPACES BETWEEN MOVING PARTS, FORMING SELF-ADJUSTING HYDROSTATIC SHIMS TO MAKE POSSIBLE THE DYNAMIC PRECISION OF THE LATHE. (Courtesy Shell Oil Ltd.)

Section 1 Lathe Construction, Tools, and Accessories

LATHE MAINTENANCE

A lathe is a precision-built machine tool, and with reasonable care should require very few adjustments to keep it in good condition. However, after a period of time, parts become worn and, to maintain the accuracy of the lathe, slight adjustments may be necessary. Always consult the operator's manual before making any adjustments to a lathe.

Gibs

Wear may occur between the bearing surfaces on the slides of the cross-slide, compound rest, taper attachment, or carriage. All of these parts are equipped with *flat or tapered gibs* which can be adjusted to compensate for the wear between mating parts.

Tapered gibs generally have one adjusting screw at each end so that the gib can be adjusted to eliminate looseness between mating parts. Tapered gibs are usually found on the dovetail slides of the cross-slide and saddle, figure 3-5A. *Straight gibs* are adjusted by the screws on the side, figure 3-5B.

Gib Adjustment

Before gibs are adjusted, the sliding surfaces must be thoroughly cleaned and lubricated. Gibs should be adjusted so that no play exists between the two sliding parts, and so that a slight drag is felt when the feed screws or controls are operated.

SHEAR PINS AND CLUTCHES

Safety devices are built into many lathes to prevent damage to gears or mechanisms in case of an accident or overload. The most common safety devices are shear pins and slip clutches.

Shear Pins

Shear pins are safety devices provided by some manufacturers in the feed screw, lead screw, or gear train of a lathe to prevent damage to costly machine parts, figure 3-6. The shear pins, generally 1/8" or 3/16" diameter brass, are designed to withstand a certain

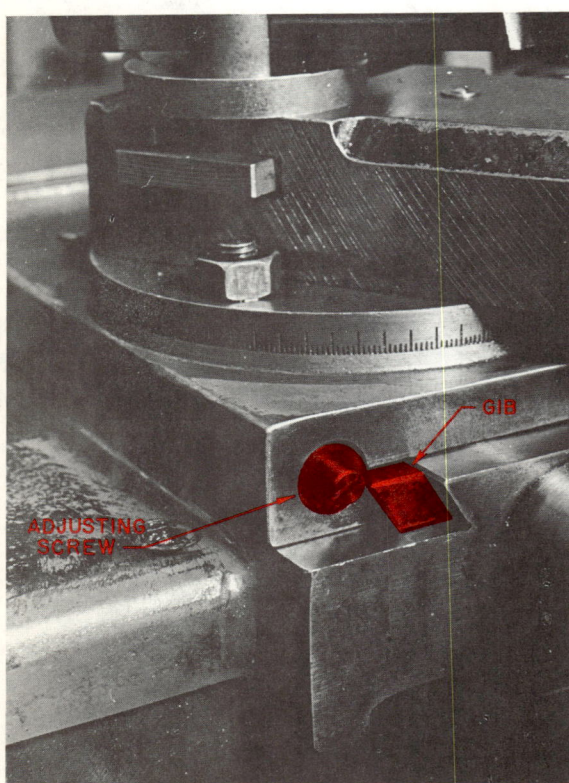

FIG. 3-5A THE ADJUSTING SCREW MOVES A TAPERED GIB TO COMPENSATE FOR WEAR BETWEEN THE CROSS-SLIDE AND SADDLE.

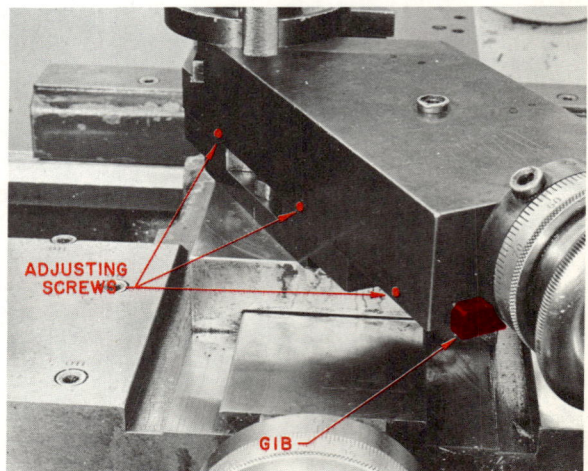

FIG. 3-5B THE ADJUSTING SCREWS ARE USED TO MOVE A STRAIGHT GIB TO COMPENSATE FOR WEAR ON THE COMPOUND REST.

amount of strain. When this strain is exceeded, the pin breaks (shears) and stops motion in that part.

Since shear pins have either straight or tapered diameters, caution should be exercised when removing or replacing them. *In addition, never replace a brass shear pin with a steel pin.* If a steel pin is used, damage to the lathe may result.

Slip Clutches

Some manufacturers equip the feed rod with a slip clutch to protect the lathe from being damaged. When the strain on the feed rod exceeds a certain point, the spring-ball mechanism, figure 3-7, releases the feed rod and automatically re-engages it when the strain is removed. Clutches of this type are preset by the manufacturer. The instructions in the operator's manual should be followed if these clutches require adjustment.

FIG. 3-6 A SHEAR PIN IN THE END GEAR TRAIN PREVENTS DAMAGED GEARS IN CASE OF AN OVERLOAD. (Courtesy Colchester Lathe Co., Inc.)

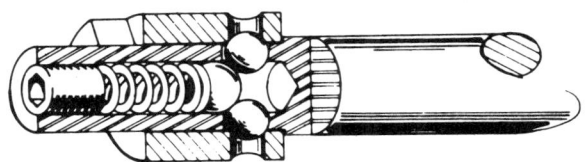

FIG. 3-7 A SPRING-BALL CLUTCH WILL SLIP WHEN EXCESSIVE STRAIN IS APPLIED TO THE FEED ROD.

REVIEW QUESTIONS

1. What points should be included in a regular maintenance program?

2. What are three advantages of a regular maintenance program?

Lathe Cleanliness

3. How does each of the following affect a lathe?

   a. Chips　　　　　　b. Scale　　　　　　c. Grit

4. What are three important points which should be observed when cleaning a lathe?

5. What procedure should be followed when cleaning a lathe?

Lubrication

6. Why is a regular lubrication program important?

7. What are the five functions of a good lubricant? Describe each function.

8. Why is it important to use the correct grade of oil in the headstock reservoir?

Section 1   Lathe Construction, Tools, and Accessories

9. What parts of the lathe should be lubricated:

    a. Daily?　　　　　　　　　b. Every six months?

Lathe Maintenance

10. What purpose do gibs serve on a lathe?

11. Where are the following gibs used?

    a. Tapered gibs　　　　　　b. Straight gibs

12. What is the procedure for adjusting the gib.

Shear Pins and Clutches

13. Describe a shear pin. What is the purpose of a shear pin?

14. What precaution should be observed when removing and replacing a shear pin?

15. What is the operating principle of a slip clutch found on the feed rod of a lathe?

Unit 4 LATHE SAFETY

Most machine tools can be very hazardous if not handled properly. Even though modern machines are equipped with certain safety features, it is still the operator who must prevent most accidents. A good lathe operator takes pride in keeping his machine and its immediate area neat and tidy. He realizes that a clean, orderly area is a safe place in which to work.

Accidents don't just happen; they are generally caused by someone's carelessness and can therefore be avoided. The most common machine shop accidents occur because of:

- Horseplay or carelessness
- Loose clothing being caught in a machine, figure A
- Handling metal chips by hand
- Being struck by flying chips or objects, figure B
- Slipping on oil spots on the floor, figure C
- Tripping over material left on the floor
- Wearing rings and watches which become caught, figure D

SAFETY PRECAUTIONS

To minimize the hazards of a lathe and reduce accidents, the following precautions should be observed:

1. Always wear safety glasses when operating any machine tool.
2. Remove all rings and watches.
3. Roll sleeves up to the elbow, remove ties, and tuck in any loose clothing.

(Courtesy — The Clausing Corporation)

Section 1   Lathe Construction, Tools, and Accessories

4. Avoid all horseplay when operating machine tools.
5. Stop the lathe to do any measuring, cleaning, oiling, or adjusting.
6. Never attempt to stop a lathe spindle with your hands; let it stop of its own accord.
7. Do not use a rag on a lathe in operation. It may be drawn into the machine.
8. Do not attempt to start or operate a lathe unless its mechanism is fully understood.
9. Never operate a lathe unless all safety guards are in place.
10. Tighten the lock ring securely when a chuck or driveplate is being mounted on lathes equipped with taper-nose spindles.
11. Always remove the chuck wrench from the chuck.
12. Keep the floor free from oil, grease, tools, workpieces, and metal cuttings.
13. Before starting a lathe, revolve the spindle by hand one complete turn to see that all parts will clear without jamming.

REVIEW QUESTIONS

1. What are the benefits of a good safety program?
2. What are five of the most important causes of lathe accidents?
3. Why are the following precautions important?
    a. Wearing safety glasses
    b. Removing loose clothing and rolling sleeves to the elbow
    c. Stopping the lathe before measuring or cleaning
    d. Keeping the work area free from oil and metal cuttings
4. How can metal chips be removed from a lathe safely?

Unit 5   CUTTING TOOLS AND TOOLHOLDERS

One of the chief factors which determines the efficiency of a machine tool is the shape and type of cutting tool employed. During the first part of this century, cutting tools were made mainly of carbon steel. In order to increase production, industry required new cutting tool materials which would maintain a sharp cutting edge and resist wear longer than those made of carbon steel.

CUTTING TOOL MATERIALS

Research has led to the development of more efficient cutting tool materials such as: high-speed steel, cast alloys, cemented carbides, ceramics, and diamonds.

High-speed steel may contain tungsten, molybdenum, vanadium, chromium, and cobalt, in addition to carbon. This type of toolbit is capable of taking heavy cuts, withstanding shock, and maintaining a cutting edge even under red heat. High-speed steel toolbits can machine metals at twice the speed of carbon steel tools.

Cast nonferrous alloy toolbits, known under the trade names of Stellite®, Tantung®, Rexalloy®, and J-Metal®, permit cutting speeds from 30% to 100% above those of high-speed steel tools. The composition of cast nonferrous alloys varies with each manufacturer; however, a typical toolbit may contain 25% to 35% chromium, 4% to 25% tungsten, and 1% to 3% carbon; the remainder is cobalt, which is used as a binder. Toolbits of this type have high hardness, high wear resistance, and excellent red-hardness qualities which permit them to retain the cutting edge when operating at a red heat.

These cast toolbits are quite weak and brittle, and are not nearly as tough as high-speed steel. Stellite toolbits can be used to advantage for heavy cuts on hard-to-machine materials such as chilled cast iron and cast steels.

Cemented carbide tools are produced by the powder metallurgy process in which powdered carbides of tungsten, titanium, or tantalum are combined with a suitable binder. The mixture is pressed into shape and sintered. In the sintering process, the carbides are cemented together to form a very hard, brittle cutting tool which operates extremely well at high cutting speeds.

Carbide toolbits may be of two types: throwaway inserts which are indexible, figure 5-1, and tips which are brazed to a steel shank, figure 5-2.

FIG. 5-1 COMMON THROWAWAY CARBIDE INSERTS
(Courtesy J.C. Williams and Company)

FIG. 5-2 BRAZED CARBIDE TOOLS ARE AVAILABLE IN A VARIETY OF SHAPES AND SIZES. (Courtesy of A.C. Wickman Ltd.)

Section 1 Lathe Construction, Tools, and Accessories

Titanium and tantalum carbides are used for machining various types of steel, while tungsten carbides are used on cast iron and nonferrous metals. Cemented carbide toolbits are usually operated at speeds three to four times greater than high-speed steel toolbits. Due to the extreme brittleness, care must be taken in the application and use of these types of toolbits. Feeds are generally lighter than those of high-speed steel or cast alloy toolbits.

Ceramic cutting tools are produced by sintering aluminum oxide with small quantities of titanium or magnesium oxide. After the inserts have been formed and sintered, they are ground to shape with a diamond-impregnated wheel. Ceramic cutting tools are available in a variety of shapes, in throwaway and cemented types, figure 5-3.

FIG. 5-3 TYPES OF CERAMIC CUTTING TOOLS. (Courtesy Carborundum Company)

Ceramic cutting tools may be used to machine hard ferrous materials and cast iron. Although these tools are very brittle, they can be operated at twice the speed of carbides for certain applications. Ceramic cutting tools have been used successfully on highly abrasive castings, hard steels, and on operations where other types of tools wear rapidly.

Single-point diamond-tipped cutting tools are being used to a limited degree on lathe work for machining nonferrous metals and non-metallic materials, where extremely high finishes and close tolerances are required. They resist abrasion, can be operated at very high speeds, (2 to 10 times faster than other cutting tools), and are capable of taking very fine cuts (as light as .0005). Diamond cutting tools are available in a variety of shapes and sizes to suit certain turning operations. They are expensive, extremely brittle, and must be handled with care.

LATHE TOOLHOLDERS AND TOOLPOSTS

Lathe cutting tools are generally held by two methods:

- In *toolholders* which provide a means of rigidly holding the cutting tool. The most common types of toolholders are those used for turning, cutting-off, threading, and boring operations.

- In *toolposts* which provide a means of holding either a toolholder or a cutting tool. The most common types used are the standard (round), turret, heavy-duty (open side), and quick-change toolposts.

TOOLHOLDERS

Toolholders for High-Speed Steel Toolbits

Toolholders for high-speed toolbits are manufactured in three styles:

The left-hand offset toolholder, figure 5-4A, is designed for machining work close to the headstock, facing operations, and cutting from right to left. It is identified by the letter **L**.

The straight toolholder, figure 5-4B, is designed for taking cuts in either direction. It is a general-purpose toolholder, and is identified by the letter **S**.

The right-hand offset toolholder, figure 5-4C, is designed for machining work close to the tailstock, facing operations, and cutting from left to right. It is identified by the letter **R**.

All toolholders of this type accommodate square toolbits, in sizes ranging from 3/16" to 3/4" square, which are held in place by a setscrew on the top of the toolholder. The square

Unit 5 Cutting Tools and Toolholders

A. LEFT-HAND OFFSET

B. STRAIGHT

C. RIGHT-HAND OFFSET

FIG. 5-4 TOOLHOLDERS FOR HIGH-SPEED STEEL TOOLBITS. (Courtesy Armstrong Bros. Tool Co.)

hole which accommodates the toolbit is at an angle of 15° to 20° to the base of the toolholder. This provides the proper back rake for high-speed steel toolbits when machining on a lathe.

Toolholders for Brazed Carbide-Tipped Toolbits

Toolholders of this type are similar in appearance to those used for high-speed steel toolbits. They are available in five sizes to accept toolbit shanks from 1/4'' to 5/8'' square. Since carbide cutting tools require little or no back rake, the hole in the toolholder is parallel to the base, figure 5-5.

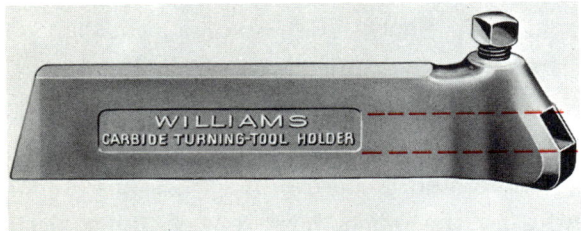

FIG. 5-5 CARBIDE TOOLHOLDERS POSITION THE TOOLBIT PARALLEL TO THE BASE OF THE HOLDER. (Courtesy J.H. Williams Co.)

Carbide turning toolholders are available in straight, right-hand, and left-hand offset types.

Toolholders for Throwaway Carbide Inserts

These toolholders are manufactured in a variety of types and sizes to accommodate various shapes of inserts such as square, triangular, diamond-shaped, and round. The insert may be locked on by means of a cam,

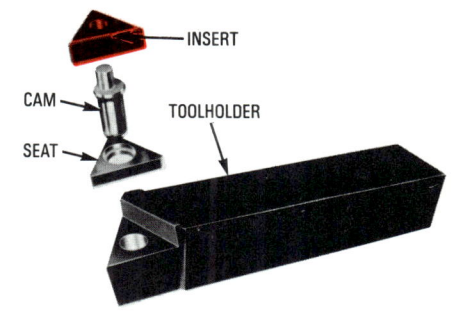

FIG. 5-6 INSERTS ARE HELD SECURELY IN THE HOLDER BY MEANS OF A CAM ACTION. (Courtesy Canadian General Electric Co. Ltd.)

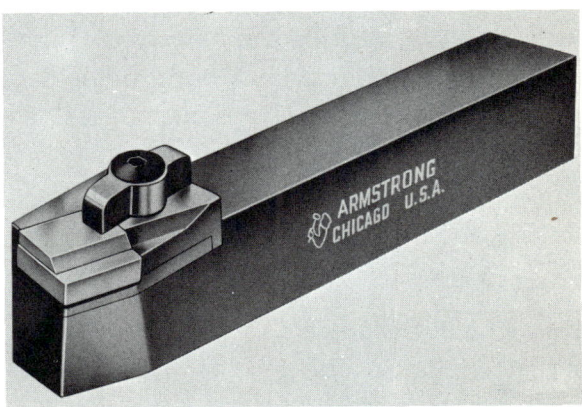

FIG. 5-7 TYPICAL TOOLHOLDER WHICH EMPLOYS A CLAMP TO SECURE THE INSERT. (Courtesy Armstrong Bros. Tool Co.)

figure 5-6, or held securely to the toolholder face by means of a clamp, figure 5-7.

Throwaway carbide insert toolholders are available for use with conventional toolposts, and for turret-type and heavy-duty toolposts.

Toolholders for ceramic inserts are similar to those used for carbide inserts. When a ceramic insert is clamped in a holder, care

Section 1 Lathe Construction, Tools, and Accessories

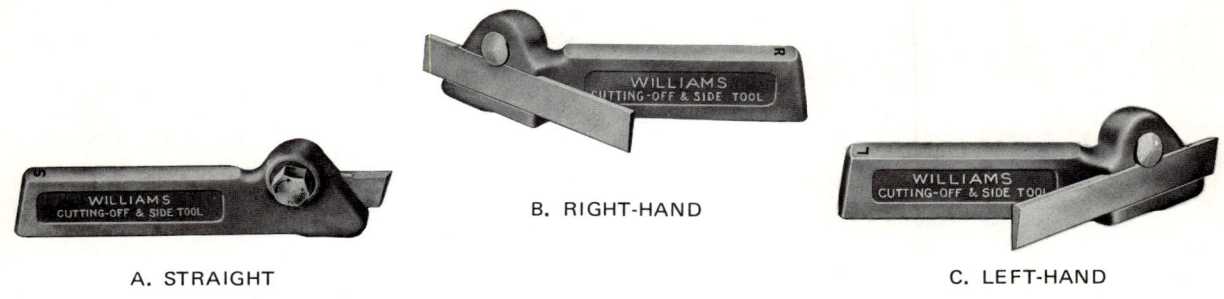

FIG. 5-8 TYPES OF SOLID CUTTING-OFF TOOLHOLDERS. (Courtesy J. H. Williams and Company)

must be taken to see that the face of the holder is clean. This insures that the insert is not broken.

Cutting-Off Toolholders

A cutting-off or parting tool is generally used when work is to be parted off, grooved, or undercut. Parting tools are long, thin blades provided with suitable side clearance to prevent them from binding when in use. They are locked tightly into place by means of a cam or wedging action provided in the toolholder.

Cutting-off toolholders may be of the solid type or the spring type.

Solid-type toolholders, figure 5-8, are available in straight, right-hand offset, and left-hand offset shanks.

Spring-type toolholders, figure 5-9, are designed to relieve excess pressure on the cutting blade. The "gooseneck" design of the holder reduces cutting tool chatter, and minimizes tool breakage caused when work "climbs" onto the cutting tool. Spring-type toolholders are made with straight and right-hand offset shanks. This type of toolholder is suited for use with automatic infeed.

Threading Toolholders

Threading in a lathe is generally performed by holding a toolbit, ground to the desired thread form, in a toolholder. A special toolholder, figure 5-10, with a cam-shaped cutter ground to 60°, is available for thread cutting. The cutter is form relieved and should only be ground on the top face when sharpening is required. The tool height is adjustable, and after the height has been set, the tool is locked into position by means of hardened locking screws.

Boring Toolholders

Boring toolholders are made in several styles to accommodate the wide range of in-

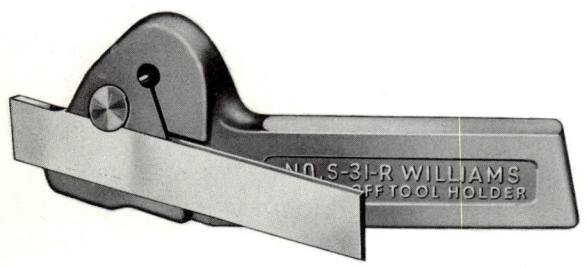

FIG. 5-9 SPRING-TYPE TOOLHOLDERS REDUCE THE CHANCES OF BROKEN TOOLS AND DAMAGED WORK ON PARTING-OFF OPERATIONS. (Courtesy J. H. Williams and Company)

FIG. 5-10 A PREFORMED THREAD-CUTTING TOOLHOLDER. (Courtesy J. H. Williams and Company)

Unit 5 Cutting Tools and Toolholders

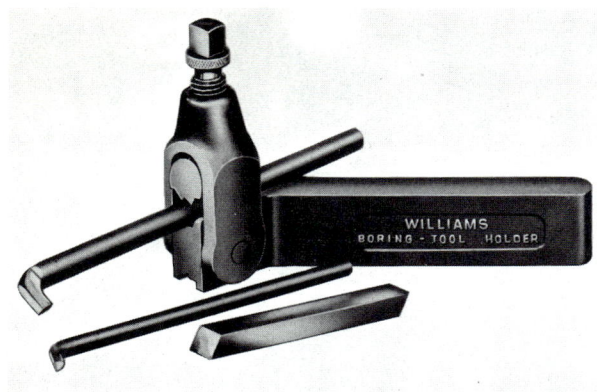

FIG. 5-11A A LIGHT BORING TOOLHOLDER. (Courtesy J. H. Williams and Company)

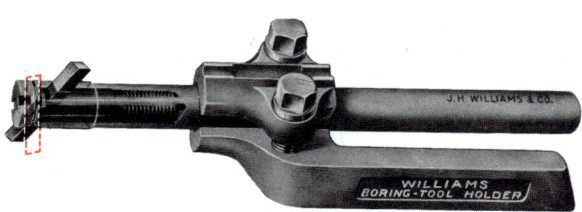

FIG. 5-11B A BORING BAR TOOLHOLDER. (Courtesy J.H. Williams and Company)

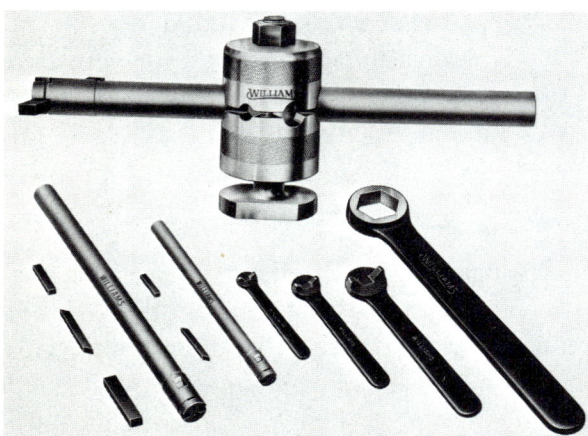

FIG. 5-11C A HEAVY-DUTY BORING TOOL. (Courtesy J. H. Williams and Company)

ternal machining which can be performed on a lathe.

Light boring toolholders, figure 5-11A, mounted in the toolpost, are used for small holes and light cuts. The boring tools are generally made of round stock with forged ends. They may be ground for boring or internal threading as required.

A boring tool suitable for heavier work, figure 5-11B, is held in the toolpost. This arrangement consists of a boring bar mounted in a toolholder, and a toolbit fitted into the end of the boring bar. Boring bars generally allow toolbits to be held at 45° or 90° to the axis of the bar.

The heavy-duty boring bar holder, figure 5-11C, is mounted on the compound rest. This type of toolholder can accommodate three bars of different diameters, which permits the use of the largest bar possible for the job. A toolbit is held in the end of the bar, either at 45° or 90° to the axis of the bar.

TOOLPOSTS

Standard (Round) Toolpost

The toolpost usually supplied with an engine lathe is the standard or round type, figure 5-12. This toolpost, which fits into the T-slot of the compound rest, provides a means of holding and adjusting a toolholder or a cutting tool. A concave ring and wedge (rocker) provide a means of adjusting the cutting tool height.

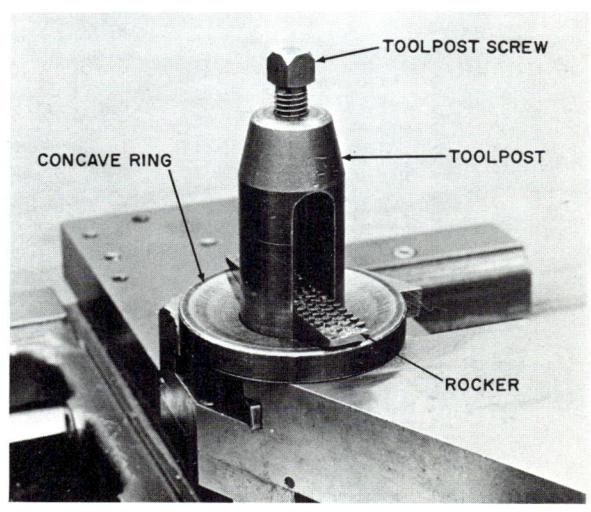

FIG. 5-12 A STANDARD TOOLPOST

Section 1 Lathe Construction, Tools, and Accessories

FIG. 5-13 TURRET-TYPE TOOLPOST PROVIDES AN EFFICIENT MEANS OF PERFORMING SEVERAL OPERATIONS ON A WORKPIECE. (Courtesy Monarch Machine Co.)

Turret-Type Toolpost

Turret-type toolposts, figure 5-13, are designed to hold four cutting tools, any of which can be easily indexed for use as required. Several operations, such as turning, threading, grooving, and parting may be performed on a workpiece by merely loosening the locking handle and rotating the holder

FIG. 5-15 THE DOVETAILED TOOLPOST PERMITS DIFFERENT TYPES OF CUTTING TOOLS TO BE RAPIDLY CHANGED AND ACCURATELY SET. (Courtesy Cincinnati Milacron Co.)

FIG. 5-14 THE HEAVY-DUTY TOOLPOST PERMITS RIGID SETTING OF THE CUTTING TOOL FOR HEAVY CUTS. (Courtesy Cincinnati Milacron Co.)

until the desired toolbit is in the cutting position. This reduces the setup time for various toolbits, thereby increasing production.

The turret-type toolpost, which fits into the slot on the compound rest, is designed for use with solid-type tools and toolholders for throwaway carbide inserts.

Heavy-Duty Toolpost

The heavy-duty toolpost, figure 5-14, also fits into the slot of the compound rest. It is designed to hold a single cutting tool or toolholder and is used when taking very heavy cuts.

Quick-Change Toolpost and Holders

Quick-change toolholders are made in several different styles to accommodate different types of cutting tools. Each holder is dovetailed and fits on a dovetailed toolpost, figure 5-15, which is mounted on the compound rest.

The tool is held in position by setscrews and is generally sharpened in the holder, after

Unit 5 Cutting Tools and Toolholders

FIG. 5-16A A QUICK-CHANGE TURNING AND FACING TOOLHOLDER. (Courtesy Cincinnati Milacron Co.)

FIG. 5-16B A QUICK-CHANGE, HEAVY-DUTY BORING BAR HOLDER. (Courtesy Cincinnati Milacron Co.)

which it is preset. After a tool becomes dull, the unit (the holder and tool) may be replaced with another preset unit. This is useful where many parts of one size are being machined since the cutting point on the toolbit, having been preset in the toolroom, is in exactly the same position as the tool it replaces. Each toolholder, figure 5-16, fits onto the dovetail on the toolpost and is locked in position by means of a clamp. A knurled nut on each holder provides for vertical adjustment of the unit.

SINGLE-POINT CUTTING TOOL NOMENCLATURE

Most cutting tools used on a lathe are of the single-point type. Although the shape of the tool may vary according to its use, the same terms or nomenclature apply to all lathe cutting tools, figure 5-17.

The *base* is the bottom surface of the toolbit blank.

The *cutting edge* is the leading edge of the toolbit which does the cutting.

The *face* is the upper surface of the toolbit along which the chip slides as it is separated from the workpiece.

The *flank* is the side of the tool which is adjacent to and below the cutting edge.

Note: The junction of the face and the flank forms the cutting edge.

The *nose* is that section of the toolbit formed by the junction of the end and side cutting edges.

The *nose radius* is the radius which is ground on the nose of the toolbit. The amount of

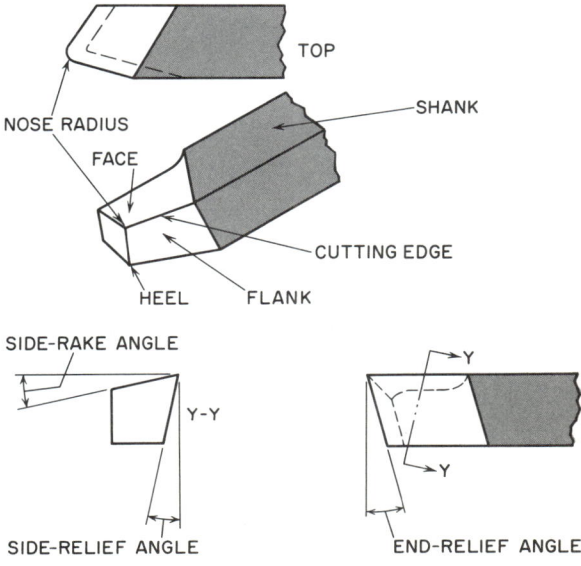

FIG. 5-17 CUTTING TOOL NOMENCLATURE

Section 1   Lathe Construction, Tools, and Accessories

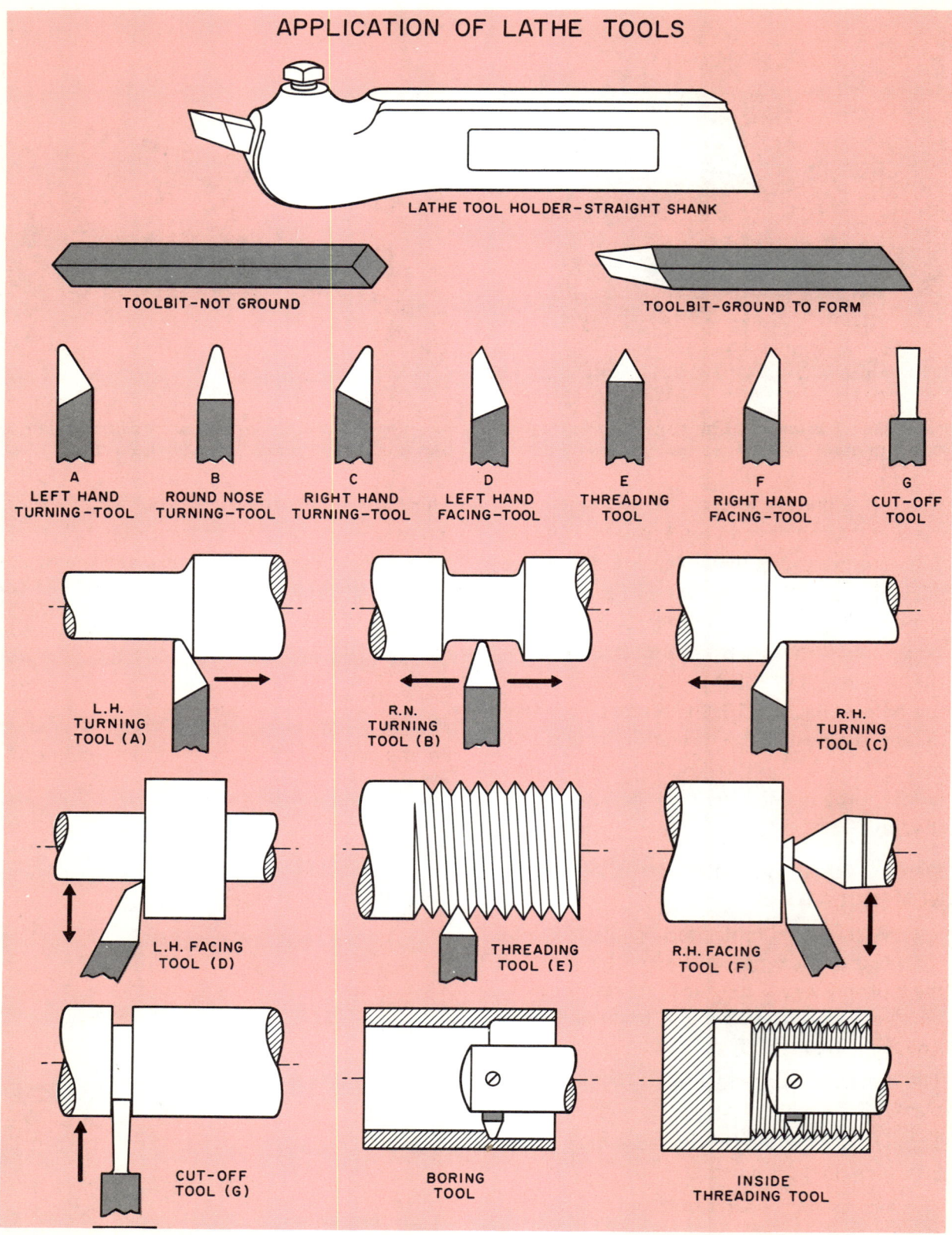

FIG. 5-18   SINGLE-POINT CUTTING TOOLBITS AND THEIR APPLICATIONS. (Courtesy South Bend Lathe Co.)

radius ground on the toolbit affects the surface finish of the workpiece. A small radius (about 1/64") is ground on a toolbit for roughing cuts, while a larger radius (up to 1/8") is used for finish cuts.

The *point* is the end portion of the toolbit which has been ground.

The *shank* is the body of the toolbit which is held in the toolholder.

TYPES AND APPLICATIONS OF SINGLE-POINT CUTTING TOOLS

The selection of the proper cutting tool for the job is an important task of the lathe operator. The use of improper or "makeshift" tools is not only poor practice, it is also uneconomical.

The most commonly used toolbit shapes and their applications are shown in figure 5-18. The lathe operator should start with these basic tool shapes and modify them as required to suit the operation being performed. Once a particular toolbit shape has proved satisfactory, the same shape and clearance angles should be duplicated on successive tools having the same application.

LATHE TOOLBIT ANGLES AND CLEARANCES

A lathe toolbit can be expected to perform properly only if the rake and clearance angles have been correctly ground. Although these angles vary for different materials, the nomenclature is the same for all types of toolbits.

Side-cutting edge angle: the angle which the cutting edge forms with the centerline of the toolbit shank, figure 5-19. This angle varies from 10° to 25°, depending on the application. If the side-cutting edge angle becomes too great, the toolbit tends to chatter.

End-cutting edge angle: the angle formed by the end-cutting edge and a line at right angles to the centerline of the toolbit shank, figure 5-19. This angle varies with the particular application. For roughing cuts, an angle of 5° to 15° is used. The smaller angle provides greater support to the end of the toolbit. For normal turning operations, the angle varies from 15° to 30°. The larger angle is particularly suited for shoulder turning and taking cuts close to the chuck or lathe dog, since this toolbit can be swung more to the left than one with a smaller end-cutting edge angle.

Side-relief (clearance) angle: the angle (generally 6° to 10°) ground on the flank of the cutting tool below the cutting edge, figure 5-19. This clearance permits the cutting tool to be fed lengthwise into the work so that the flank does not rub against the workpiece.

End-relief (clearance) angle: the angle ground on the end of the toolbit below the end-cutting edge, figure 5-19. This clearance is generally 10° to 15° and is measured when the toolbit is held in the toolholder. The end-relief angle, which permits the toolbit to be fed into the work, varies with the hardness and type of material being cut. The smaller angle, which provides more support for the end of the toolbit, is used when machining hard materials.

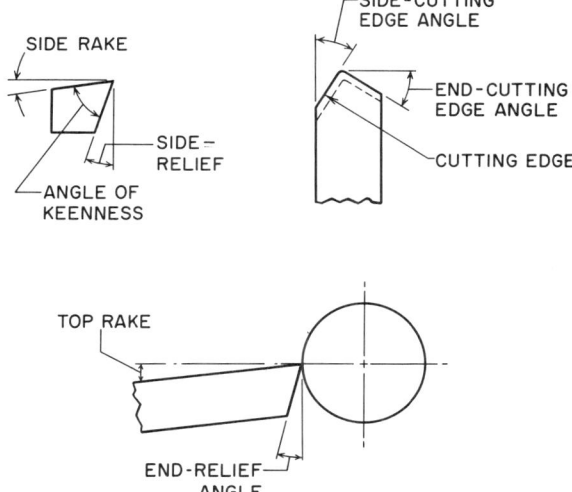

FIG. 5-19 RELIEF AND RAKE ANGLES ON A GENERAL-PURPOSE CUTTING TOOL.

Section 1 Lathe Construction, Tools, and Accessories

Side rake angle: the angle at which the face (top) of the toolbit slopes away from the cutting edge, figure 5-19. This angle, generally 12° to 15°, varies according to the hardness and the type of material being machined. Side rake creates a keener cutting edge and permits the chips to flow away quickly. The included angle between the flank and the face of the toolbit is known as the *angle of keenness*. This angle should be smaller for soft materials and larger for harder materials, since more support is required for the cutting edge when cutting hard materials.

Back (top) rake: the slope of the tool face from the nose to the back of the toolbit, figure 5-19. Back rake permits the chip to flow freely from the point of the toolbit. The back rake angle is generally 16° to 20° for high-speed steel turning toolbits. It is controlled by the angle of the hole in the toolholder.

REVIEW QUESTIONS

Cutting Tool Materials

1. What are five common materials which are used in the manufacture of cutting tools?

2. Compare high-speed steel toolbits with cast nonferrous alloy toolbits.

3. How are cemented carbide tools manufactured?

4. What are two grades of cemented carbides called, and what is the purpose for each?

5. How are ceramic cutting tools manufactured and where are they used most successfully?

6. What are the advantages of single-point diamond-tipped cutting tools and where are they used?

Lathe Toolholders

7. What are two methods of holding lathe cutting tools?

8. Describe each of the following items. What is the purpose of each item?

    a. the left-hand toolholder   b. the right-hand toolholder

9. Why is the hole in a standard toolholder at an angle to the base?

10. How does the hole in a carbide toolholder compare with the hole in a standard toolholder?

11. How are throwaway carbide inserts fastened to the toolholder face?

12. What are the advantages of a spring-type versus the solid-type cutting-off toolholder?

13. What are the three types of boring toolholders and what is the purpose of each?

Toolposts

14. Describe a standard toolpost. How is the cutting tool height adjusted on such a toolpost?

15. What are the advantages of turret-type toolposts?

16. What is the purpose of the heavy-duty toolpost?

17. Describe the construction of quick-change toolholders. How are these toolholders held on the toolpost?

18. What are the advantages of quick-change toolholders?

Single-Point Tool Nomenclature

19. Where is each of the following parts located on the cutting tool?
    a. Cutting edge
    b. Face
    c. Flank
    d. Nose radius

20. Make a neat diagram of a single-point cutting tool, and label the parts in question 19.

Types and Applications of Single-Point Cutting Tools

21. What are five single-point cutting tools? What is the purpose of each tool?

Lathe Toolbit Angles and Clearances

22. What is the definition of,
    a. Side-cutting edge angle?
    b. End-cutting edge angle?

23. What end-cutting edge angle is used for normal turning operations?

24. Why is a side-relief angle necessary on a toolbit?

25. What is the purpose of the end-relief angle and why is it necessary to the cutting action?

26. What are the recommended side- and end-relief angles for a general-purpose cutting tool?

27. How do the side rake angle and back rake assist in the cutting action of a toolbit?

Unit 6 GRINDING LATHE CUTTING TOOLS

The proper grinding of lathe toolbits is an operation with which the lathe operator should become familiar. The angles at which a toolbit performs best should be noted, and these should be duplicated on cutting tools of the same type. The importance of relief (clearance) and rake angles should not be overlooked. A lathe tool cuts if it is provided with end- and side-relief angles. However, it cuts better if it is provided with the *proper* amount of end- and side-relief. Too much relief angle (side and end), while providing a keener cutting edge, reduces the amount of support at the cutting edge and weakens a toolbit. Too little side relief prohibits the use of roughing feeds, while too little end relief limits the rate of crossfeed.

Rake angles are ground on the toolbit to enable the chip to clear more readily, and to increase the keenness of the cutting edge. Too much side rake creates a keener edge but weakens the edge so that it breaks down rapidly. Too little side rake produces a poorly formed chip and requires more turning power (torque) to take a cut. A good lathe operator should be able to determine the correct toolbit angles after observing how the toolbit performs.

Table 6-1 lists the recommended rake and relief angles for single-point, high-speed steel and cemented-carbide lathe toolbits. It may be necessary to deviate slightly from these recommendations to compensate for factors such as:

- The size and condition of the lathe
- The shape and hardness of the workpiece
- The type of cutting operation being performed

PREPARING THE GRINDER

Lathe toolbits are generally ground or sharpened freehand on a bench or pedestal grinder, figure 6-1 The most common grinders

MATERIAL	RAKE AND RELIEF ANGLES IN DEGREES FOR SINGLE-POINT LATHE TOOLBITS							
	Side Relief		End Relief		Side Rake		Back Rake	
	H.S.S.	Carbide	H.S.S.	Carbide	H.S.S.	Carbide	H.S.S.	Carbide
Aluminum	12	6 to 10	8	6 to 10	15	10 to 20	35	0 to 10
Brass	10	6 to 8	8	6 to 8	5 to -4	8 to -5	0	0 to -5
Bronze	10	6 to 8	8	6 to 8	5 to -4	8 to -5	0	0 to -5
Cast Iron	10	5 to 8	8	5 to 8	12	6 to -7	5	0 to -7
Machine Steel	10 to 12	5 to 10	8	5 to 10	12 to 18	6 to -7	8 to 15	0 to -7
Tool Steel	10	5 to 8	8	5 to 8	12	6 to -7	8	0 to -7
Stainless Steel	10	5 to 8	8	5 to 8	15 to 20	6 to -7	8	0 to -7

Table 6-1

Unit 6 Grinding Lathe Cutting Tools

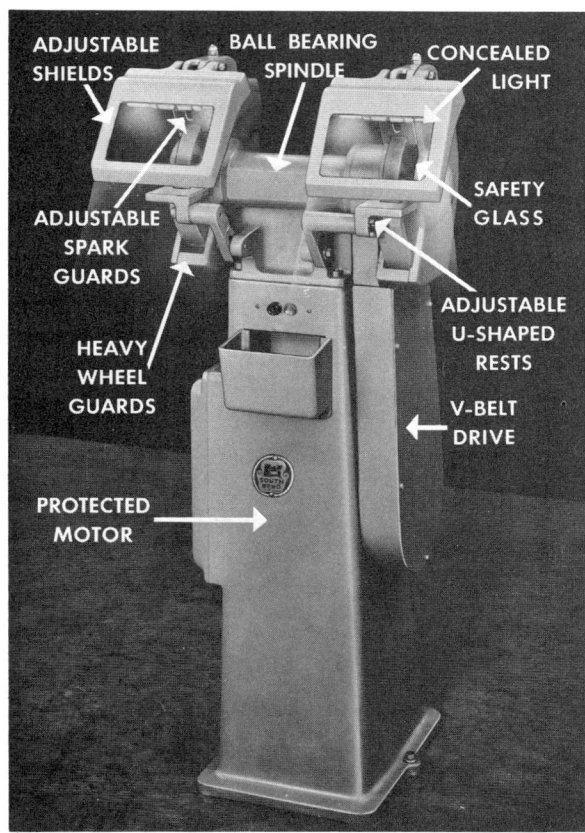

FIG. 6-1 A PEDESTAL GRINDER. (Courtesy South Bend Lathe, Inc.)

found in a machine shop have a coarse-grained roughing wheel mounted on one end of the spindle, and a fine-grained finishing wheel on the other end. Aluminum oxide grinding wheels are usually mounted on these grinders, since the majority of the toolbits to be ground are made of high-speed steel.

Note: Special silicon carbide (green grit) wheels should be used when sharpening cemented-carbide lathe tools.

Safety Precautions

Before operating any grinder, it is wise to observe the following safety precautions:

1. Keep the work rests adjusted to within 1/16" of the wheel to prevent work from being jammed between the work rest and the grinding wheel.

2. Stand to one side of the grinding wheel when starting a grinder. Most grinding wheel breakage occurs as the wheel approaches maximum speed.

3. Always wear safety glasses while grinding, even though the machine may be equipped with eyeshields.

Dressing a Grinding Wheel

Several things may occur to the grinding wheel whenever metal is ground.

- Small metal particles from the workpiece tend to imbed themselves in the wheel, causing it to become clogged or loaded.

- Some abrasive grains are worn smooth, which causes the wheel to lose its cutting action.

- Grooves may be worn in the face of the grinding wheel.

Whenever any of these conditions exist, the grinding wheel should be dressed. *Dressing* is the process of removing the imbedded metal particles, squaring the face of the wheel, and restoring the sharpness of the grinding surface.

A mechanical wheel dresser, figure 6-2, commonly called a star dresser, is generally used to dress a bench or pedestal grinding

FIG. 6-2 DRESSING A GRINDING WHEEL TO MAKE IT CUT BETTER. (Courtesy McGraw-Hill)

Section 1 Lathe Construction, Tools, and Accessories

wheel. These dressers consist of a number of hardened, pointed discs mounted in a holder in which they can revolve.

Procedure: Dressing a Grinding Wheel

1. Hold the handle of the star dresser in one hand, while the other hand firmly holds the dresser on the work rest, figure 6-2.

2. Press the dresser firmly against the revolving grinding wheel.

3. Slowly move the dresser from side to side across the face of the wheel, being careful not to run off the edge of the wheel. Always keep the dresser firmly against the wheel during the dressing operation.

Procedure: Grinding a General-Purpose Turning Toolbit

Since all lathe cutting tools, regardless of their shape, must have rake and relief angles, only the grinding of a general-purpose lathe toolbit will be explained.

1. Grasp the toolbit firmly and support the hands on the grinder toolrest.

2. Adjust the position of the toolbit blank and grind the cutting edge angle, figure 6-3.

3. At the same time, tilt the bottom of the toolbit in and grind the 10° side-relief angle.

4. Continue grinding until the side cutting edge is about 1/2" long and the point is over about 1/4 the width of the toolbit, figure 6-3.

Note: When grinding, the toolbit should be moved back and forth across the face of the wheel. This speeds up the grinding and prevents grooves from being formed on the face of the wheel.

5. Cool the toolbit frequently in water so that it does not overheat during the grinding operation. Overheating may damage the toolbit.

Note: Stellite and cemented-carbide toolbits should never be quenched when being ground.

6. Grind the end cutting edge so that it forms an angle of less than 90° with the side cutting edge, figure 6-4. The toolbit should be held so that the back end is lower than the point. This forms the end-relief angle of 15° at the same time, figure 6-4.

Note: Some machinists prefer to hold the toolbit blank in a toolholder when *rough grinding* the side cutting edge and the end cutting edge. By supporting the toolholder on the grinder toolrest, the approximate relief angles are ground on the toolbit because the toolbit is tilted to approximately the same angle as it will be when mounted in the lathe. Care should be taken so that the toolbit blank extends sufficiently from the toolholder to prevent grinding the toolholder.

7. Hold the toolbit so that it is approximately 45° to the axis of the wheel and tilt the bottom of the toolbit in so that the side rake of about 14° is ground on the top of the toolbit, figure 6-5. Care should be taken that the top of the cutting edge is not ground below the top of the toolbit.

8. Round the point slightly, maintaining the same end- and side-relief angle.

9. Hone the cutting edge and the point of the toolbit with a fine oilstone. This produces a keener edge and a better surface finish and increases the life of the toolbit.

Note: End- and side-relief angles should be measured when the toolbit is in the toolholder, figure 6-6. A toolbit gage, which indicates end and side relief as well as the angle of keenness for cast iron and steel, may be used for checking these angles.

Unit 6   Grinding Lathe Cutting Tools

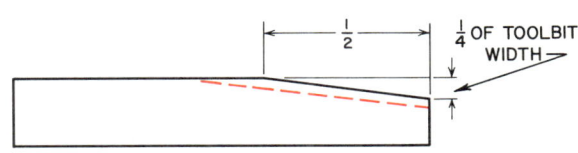

FIG. 6-3   GRINDING THE SIDE CUTTING EDGE AND SIDE-RELIEF ANGLES ON A TOOLBIT

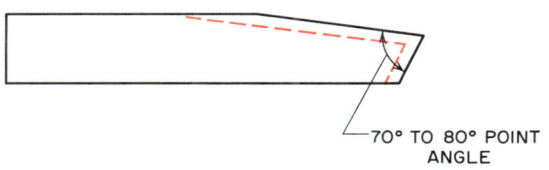

70° TO 80° POINT ANGLE

FIG. 6-4   GRINDING THE END-RELIEF ANGLE ON A LATHE TOOLBIT

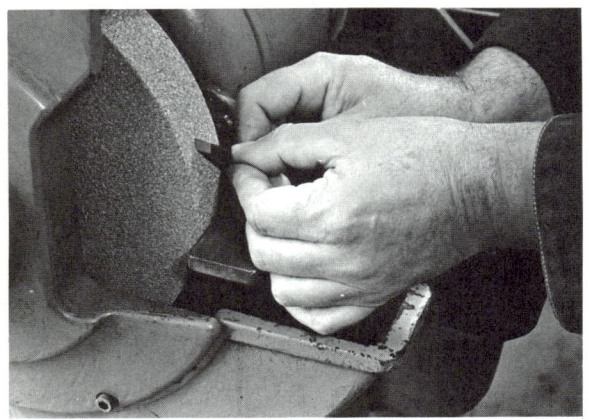

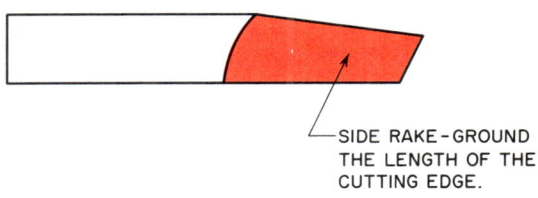

SIDE RAKE – GROUND THE LENGTH OF THE CUTTING EDGE.

FIG. 6-5   GRINDING THE SIDE RAKE ON A LATHE TOOLBIT

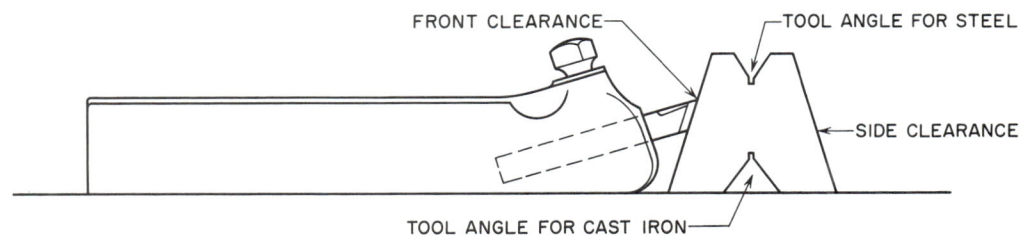

FIG. 6-6   CHECKING THE END-RELIEF ANGLE OF A TOOLBIT (Courtesy South Bend Lathe Co., Inc.)

Section 1 Lathe Construction, Tools, and Accessories

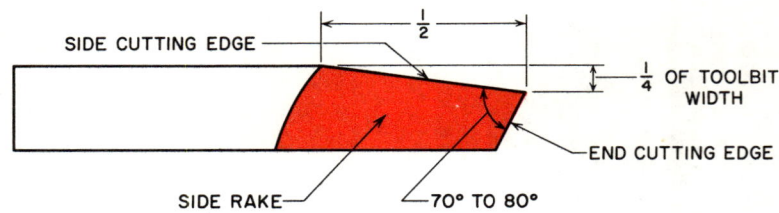

FIG. 6-7 SHAPE AND DIMENSIONS OF A GENERAL – PURPOSE TOOLBIT (TOP VIEW)

Sharpening a General-Purpose Toolbit

A general-purpose toolbit, ground to the dimensions shown in figure 6-7, can be quickly and accurately resharpened by grinding *only the end cutting edge*. After the worn portion is removed, the point radius should be ground and honed.

The advantages of grinding a general-purpose toolbit to this shape and dimensions are:

- There is less chance of changing the shape or angles of the toolbit since only one surface must be ground.
- The toolbit can be reground quickly and accurately.
- The beginning lathe operator finds it relatively easy to resharpen a toolbit.
- The toolbit can be resharpened many times before the side cutting edge becomes too short and must be reground.
- Since the top of the tool is not ground, the toolbit is not weakened.
- Less toolbit material is removed for each resharpening.

Once the side cutting edge becomes too short as a result of repeated sharpenings, it is necessary to regrind the side cutting edge and side rake to the original dimensions.

REVIEW QUESTIONS

1. What is the purpose of end and side relief on a toolbit?
2. What are the disadvantages of each of the following,
   a. Too little side relief?
   b. Too much end relief?
3. What is the purpose of rake angles on a lathe toolbit?
4. What are the disadvantages of too little side rake?
5. What factors may affect the rake and relief angles which are ground on a toolbit?
6. What are the rake and relief angles for a high-speed steel toolbit and for a carbide toolbit if both are to be used for cutting machine steel?

Preparing the Grinder

7. What type of grinding wheels should be used to grind:
   a. High-speed steel tools?
   b. Cemented carbide tools?
8. How close to the grinding wheel should the work rests be set? Explain why.

9. What are three things which may happen to the face of a grinding wheel when metal is ground?

10. What is the purpose of dressing a grinding wheel?

11. Describe a mechanical wheel dresser.

12. What is the procedure for dressing a wheel?

Grinding a General-Purpose Toolbit

13. How are the cutting-edge angle and the side-relief angle ground on a toolbit?

14. Why should a toolbit be quenched frequently while it is being ground?

15. How much side rake should be ground on a general-purpose toolbit? How is it ground?

16. What precaution should be taken when grinding side rake?

17. What is the purpose of honing the cutting edge of a toolbit?

18. How are the end- and side-relief toolbit angles measured?

Sharpening a General-Purpose Toolbit

19. How is a general-purpose toolbit resharpened?

20. What are the advantages of grinding a general-purpose toolbit to the form shown in figure 6-7?

Unit 7 CUTTING SPEEDS AND FEEDS

Modern industrial practices require the most efficient use of both time and material. Since time is one of the most costly items of a manufactured product, it is imperative that metals be cut as quickly as possible. Each material is cut most efficiently if the work passes the cutting tool at the proper rate of speed. This rate is referred to as the cutting speed of the material.

CUTTING SPEED

Cutting speed may be defined as the rate, in feet per minute (f.p.m.), at which a point on the circumference of the work travels past the cutting tool. For example, if steel has a cutting speed of 100 feet per minute, the spindle speed (r.p.m.) should be set so that 100 feet of the work circumference passes the cutting tool in one minute. If the work is revolved too slowly, valuable time is lost. If the work is revolved too fast, the cutting tool breaks down quickly, and much time is wasted in regrinding the cutting tool.

Cutting speed varies for each metal and for each type of cutting tool material. The recommended cutting speeds for various metals, tables 7-1 and 7-2, have been developed after considerable research on the part of the metal and cutting tool manufacturers.

R.P.M. CALCULATIONS

The cutting speed of the metal should not be confused with the number of turns the workpiece makes in one minute (r.p.m.)

To determine the r.p.m. for any material, it is necessary to know the diameter of the work and the cutting speed of the material. The proper spindle speed (r.p.m.) can then be calculated by dividing the cutting speed of the material, in inches, by the circumference of the work, in inches.

$$\text{r.p.m.} = \frac{\text{cutting speed} \times 12}{3.1416 \times \text{diameter of work (inches)}}$$

$$= \frac{CS \times 12}{3.1416 \times D}$$

Since it is impossible to set most lathes to the exact calculated speed, a simpler formula, may be applied. In this modified formula the 3.1416 in the denominator is changed to 3. The new formula then becomes:

$$\text{r.p.m.} = \frac{CS \times 12}{3 \times D}$$

$$= \frac{CS \times 4}{D}$$

where CS = cutting speed of the material

D = diameter of the workpiece

The formula can be applied as in the following example: Calculate the r.p.m. required to rough-turn a piece of 3" diameter machine steel using a high-speed steel toolbit. (By referring to table 7-1, the recommended rough-cutting speed for machine steel is 100 f.p.m.).

$$\text{r.p.m.} = \frac{CS \times 4}{D}$$

$$= \frac{100 \times 4}{3}$$

$$= \frac{400}{3}$$

$$= 133.3$$

The spindle speed change levers on the lathe can then be set to produce the closest speed to 133 r.p.m.

Unit 7 Cutting Speeds and Feeds

CUTTING SPEEDS IN F.P.M. USING HIGH-SPEED STEEL TOOLS						
MATERIAL	TURNING, BORING, FACING		THREADING	REAMING	PARTING	KNURLING
	Rough Cut	Finish Cut				
Aluminum	200	300	50	50	200	40
Brass	180	220	45	20	150	40
Bronze	180	220	45	20	150	40
Cast Iron	70	80	20	35	80	30
Machine Steel	100	110	40	40	80	30
Tool Steel	70	80	25	35	70	30
Stainless Steel	75-90	100-130	35	35	70	20

TABLE 7-1

CUTTING SPEEDS IN F.P.M. FOR CARBIDE AND CERAMIC TOOLS				
MATERIAL	ROUGH CUT		FINISH CUT	
	Carbide	Ceramics	Carbide	Ceramics
Aluminum	300-450	400-2000	700-1000	600-3000
Brass	500-600	400-800	700-800	600-1200
Bronze	500-600	150-800	700-800	200-1000
Cast Iron	200-250	200-800	350-450	200-2000
Machine Steel	400-500	250-1200	700-1000	400-1800
Tool Steel	300-400	300-1500	500-750	600-2000
Stainless Steel	250-300	300-1000	375-500	400-1200

TABLE 7-2

When setting the spindle speed, it is advisable to set the machine to the closest speed *under* the calculated spindle speed. If the machine performs satisfactorily, the speed may be increased. If the cutting tool and the work vibrate, due to loose bearings or gibs, it is necessary to reduce the speed and possibly increase the feed.

Although the figures in table 7-1 are recommended for the various materials shown, it may not be possible to use the calculated r.p.m., since the speed at which the work revolves depends on several other factors such as:

- The condition of the material, i.e., hard spots or scale
- The shape of the work material
- The design of the cutting tool
- The rate of feed
- The condition of the machine

Section 1 Lathe Construction, Tools, and Accessories

FEED

Feed is defined as the distance the toolbit advances along the work for each revolution of the spindle. If a feed of .010 is used, the cutting tool travels .010 along the work during each revolution. Therefore, it requires 100 revolutions of the spindle for the tool to travel 1" (100 x .010) along the work.

When machining any workpiece, it is advisable to use as few cuts as possible; one roughing and one finishing cut should be sufficient in most cases.

Since the purpose of the roughing cut is to remove metal as quickly as possible, and surface finish is not important, a coarse feed should be used. A fine feed should be used for finish cuts since a better surface finish and higher accuracy are produced.

Table 7-3 lists suggested feeds for various materials when using high-speed steel or carbide cutting tools.

FEEDS FOR HIGH-SPEED STEEL AND CARBIDE TOOLBITS

Material	Rough Cut	Finish Cut
Aluminum	.015–.030	.005–.010
Brass	.015–.025	.003–.010
Bronze	.015–.025	.003–.010
Cast Iron (med.)	.015–.025	.005–.012
Machine Steel	.010–.020	.003–.010
Tool Steel	.010–.020	.003–.010
Stainless Steel	.015–.030	.005–.010

TABLE 7-3

DEPTH OF CUT

The third factor that should be considered for efficient machining is the depth of cut, since the number of cuts determines the time required to produce a finished workpiece.

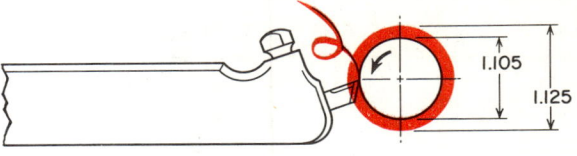

TOOL FEED ⟶ .010
MATERIAL REMOVED .020

FIG. 7-1 THE DIAMETER OF THE WORKPIECE IS REDUCED BY TWICE THE DEPTH OF CUT.

The amount of material removed in each cut is always twice the depth of the cutting tool setting, figure 7-1.

Since the lathe speed (r.p.m.), feed, and depth of cut affect efficiency of metal removal, the following points should be considered.

1. When the calculated r.p.m. falls between two speeds available on a lathe, set the machine for the lower speed. Increase the speed if the condition of the machine, the cutting tool, and the workpiece allow.

2. The feeds given in table 7-3 are approximate and should be varied to suit the conditions of the machine, the type of toolbit, and the rigidity of the workpiece.

3. Roughing cuts should be as deep and coarse as possible. Leave approximately .030 on the diameter for a finish cut.

4. When selecting the rough stock for any job, the diameter should be approximately 1/8" larger than the largest finished diameter of the workpiece.

CALCULATING MACHINING TIME

To calculate the time required to machine any workpiece, factors such as speed, feed, and depth of cut must be considered. By applying the following formula, the time required to take a cut can be readily calculated.

$$\text{Time required} = \frac{\text{length of cut}}{\text{feed} \times \text{r.p.m.}}$$

Example: Calculate the time required to turn a 12" long machine steel shaft to 2.000. The rough stock used is 2 1/8" in diameter.

Unit 7 Cutting Speeds and Feeds

Solution: Rough Cut

$$\text{r.p.m.} = \frac{CS \times 4}{D}$$
$$= \frac{100 \times 4}{2\ 1/8}$$
$$= 188$$

Roughing feed = .020

Rough cut time $= \dfrac{\text{length of cut}}{\text{feed} \times \text{r.p.m.}}$

$$= \frac{12}{.020 \times 188}$$
$$= 3.2 \text{ minutes}$$

Finish Cut

$$\text{r.p.m.} = \frac{CS \times 4}{D}$$
$$= \frac{110 \times 4}{2}$$
$$= 220$$

Finishing feed = .003

Finish cut time $= \dfrac{12}{.003 \times 220}$

$$= 18.2 \text{ minutes}$$

Total machining time = 3.2 + 18.2 = 21.4 min.

REVIEW QUESTIONS

Cutting Speed

1. Why is the selection of the proper spindle speed important when machining any workpiece?

2. Define cutting speed.

3. Define r.p.m.

4. What r.p.m. is required to rough turn the following workpieces:
    a. 3" diameter machine steel?   c. 3 1/2" diameter aluminum?
    b. 2" diameter cast iron?

5. At what r.p.m. should a 1" piece of machine steel be revolved for:
    a. Threading?   b. Knurling?

6. What factors will affect the speed at which a piece of work is revolved?

7. How may vibration be removed when turning a piece of work in a lathe?

Feed

8. Define feed.

9. If a feed of .015 is used, how many revolutions must the work revolve so that the tool will travel 3"?

10. What is the purpose of:
    a. The roughing cut?   b. The finish cut?

Section 1   Lathe Construction, Tools, and Accessories

Depth of Cut

11. A 2" diameter workpiece is to be turned to 1 3/4". What should be the depth of:

    a. The rough cut?   b. The finish cut?

12. A steel bar 3 1/8" in diameter is to be machined to 3.000. What are the depth of cuts and the feed that should be used to machine this workpiece?

13. What is the time required to machine a 16" long aluminum bar, from 2 1/4" to 2" diameter?

SECTION 2

Machining Between Centers

Unit 8 CENTERING WORK

Work to be machined in a lathe is generally divided into two categories: work held in a chuck or collet, and work mounted between lathe centers. Schools tend to machine much of the lathe work between centers because:

- The workpiece can be quickly set up in a lathe.
- It can be removed any time.
- The work can be reset quickly any number of times and still maintain the same degree of accuracy.

LOCATING CENTER HOLES

Work that is to be turned between centers must have a center hole drilled in each end to provide bearing surfaces for the lathe centers. Before center holes can be drilled, the center of the work must be located. The layout method used is determined by the workpiece and the accuracy required. Some of the more common methods of laying out the centers of round stock are by the use of a center head, hermaphrodite caliper, surface gage, or bell center punch.

Procedure: Center Head Method

The center head method of laying out the center of round stock, figure 8-1, is most commonly used since it provides a fast and fairly accurate method of locating centers.

1. Place the work in a vise and remove the burrs from each end with a file.

2. Apply layout dye or chalk to both ends of the workpiece.

3. Hold the center head firmly against the body of the work while keeping the rule flat on the end, figure 8-1.

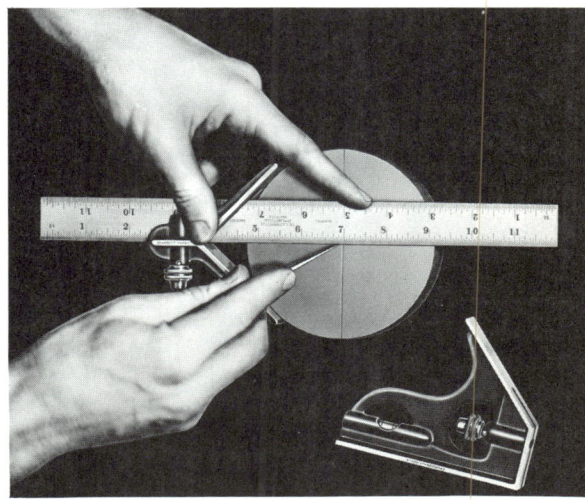

FIG. 8-1 USING THE CENTER HEAD TO LOCATE THE CENTER OF ROUND WORK. (Courtesy L.S. Starrett Co.)

Section 2 Machining Between Centers

4. With a *sharp* scriber, draw a line along the edge of the rule.

5. Move the center head one-quarter of a turn (90°) and scribe another line.

6. Repeat the procedure on the other end of the work.

7. With a *sharp* center punch, lightly punch where the two lines intersect at each end.

8. Check and correct the accuracy of the center punch mark (see the section on Checking the Accuracy of Center Layout).

Procedure: Hermaphrodite Caliper Method

The hermaphrodite caliper method is generally used to locate the centers of work which has been cast or is not quite round.

1. Prepare the workpiece as in steps 1 and 2 of the center head method.

2. Set the hermaphrodite caliper to approximately one-half the work diameter.

3. With the thumb of one hand, hold the bent leg just below the edge of the workpiece and scribe an arc.

4. Scribe four arcs, figure 8-2, by moving the caliper leg one-quarter of a turn for each.

5. Lightly center punch the center of the arcs and check the accuracy of the layout.

Procedure: Surface Gage Method

1. Remove the burrs and apply layout dye to both ends of the workpiece.

2. Place the work in a V-block on a surface plate.

3. Set the surface gage scriber point to approximately the center of the workpiece.

4. Scribe a line across each end of the stock.

5. Scribe four lines on each end of the workpiece, figure 8-3, by rotating the work about 90° (one-quarter turn) for each line.

6. Lightly center punch the center of the square and check accuracy of the layout.

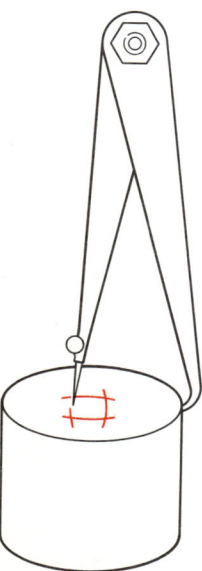

FIG. 8-2 A HERMAPHRODITE CALIPER BEING USED TO LOCATE THE CENTER OF ROUGH WORK. (Courtesy South Bend Lathe, Inc.)

FIG. 8-3 A SURFACE GAGE BEING USED TO LOCATE THE CENTER OF A ROUND WORKPIECE.

Unit 8   Centering Work

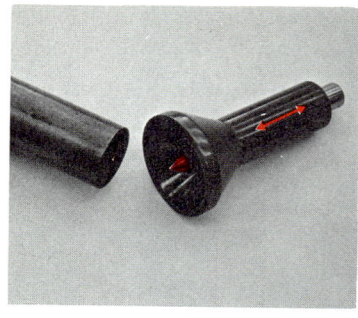

FIG. 8-4   A BELL CENTER PUNCH CAN LOCATE THE CENTER OF ROUND WORK (SQUARE ENDS), QUICKLY AND ACCURATELY.

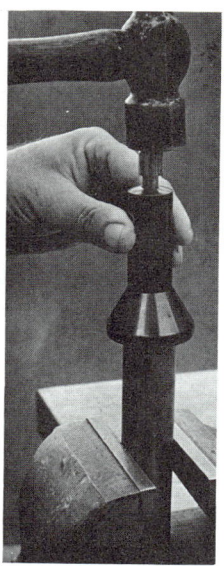

Procedure: Bell Center Punch Method

A quick and fairly accurate method of locating the centers of a piece of round work is by using a bell center punch. The bell center punch consists of a bell-shaped housing with a round, pointed rod (center punch) in its center, figure 8-4. The center punch is free to slide in the housing.

1. Face both ends of the workpiece and remove the burrs to insure an accurate layout.
2. Place the work in a vise.
3. Place the bell center punch over the end of the work, being sure to hold it vertically.
4. Sharply strike the center punch with a hammer and repeat the procedure on the other end.

CHECKING THE ACCURACY OF CENTER LAYOUT

It is considered good practice to check any layout for accuracy before metal is cut. This is especially true with center hole layout since an error in this layout causes the center hole to be drilled off center.

Procedure: Divider Method

1. Place one leg of the divider in the light center punch mark.
2. Adjust the divider so that the other leg is on a line and exactly on the edge of the workpiece, figure 8-5.
3. Revolve the divider one-half turn (180°) and check to see that the leg of the divider is in the same relation to the edge of the workpiece.
4. If the divider leg is not in the same relation to the edge of the workpiece at each end of the line, lightly move the center punch mark to correct the error.
5. Repeat this procedure on the other line until the punch mark is exactly in the center of the work.

Procedure: Lathe Center Method

1. Enlarge the center punch marks slightly and mount the work between centers.

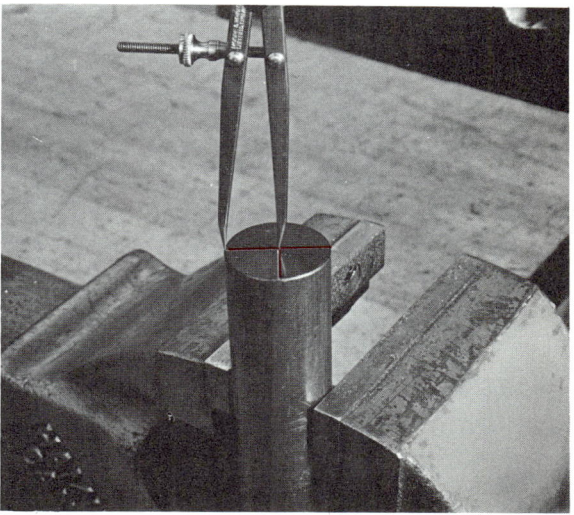

FIG. 8-5 CHECKING THE ACCURACY OF CENTER LAYOUT WITH A PAIR OF DIVIDERS

Section 2 Machining Between Centers

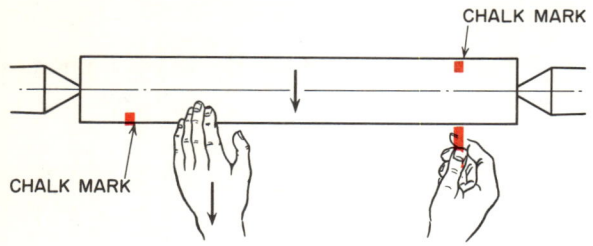

FIG. 8-6A TESTING THE ACCURACY OF THE CENTER LAYOUT. (Courtesy South Bend Lathe, Inc.)

to the correct size to assure an adequate bearing surface on the lathe centers. The holes should also be as smooth as possible to reduce friction when the work revolves on the lathe center.

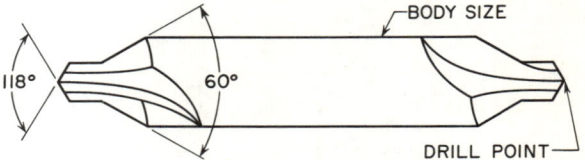

FIG. 8-7A A PLAIN-TYPE CENTER DRILL

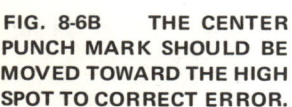

FIG. 8-6B THE CENTER PUNCH MARK SHOULD BE MOVED TOWARD THE HIGH SPOT TO CORRECT ERROR.

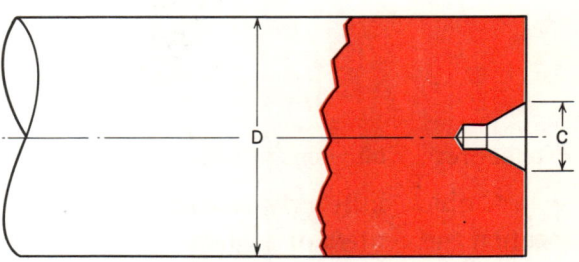

FIG. 8-7B A CROSS SECTION OF A CENTER HOLE

2. Spin the workpiece and lightly mark the high spots at each end with chalk, figure 8-6A.

3. Place the work in a vise, tip the center punch, and with a hammer move the punch mark toward the chalk mark, figure 8-6B.

4. Repeat operations 2 and 3 until the work runs true and the chalk marks are uniform around the work circumference.

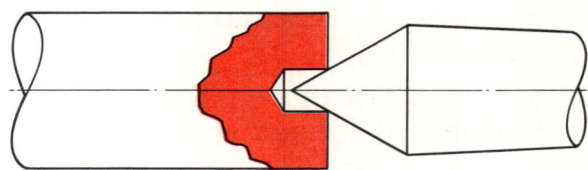

FIG. 8-8A SHALLOW CENTER HOLES PROVIDE VERY LITTLE SUPPORT FOR WORK BETWEEN CENTERS.

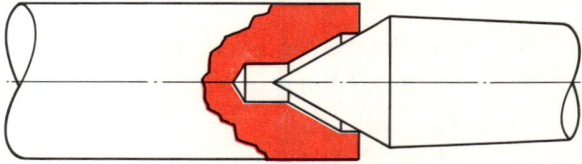

FIG. 8-8B THE TAPERS OF THE LATHE CENTER AND CENTER HOLE CANNOT CONTACT EACH OTHER WHEN HOLES ARE DRILLED TOO DEEP.

CENTER DRILLS

Work that is to be machined between centers must have a hole drilled in each end to fit onto the lathe centers. A *center drill*, figure 8-7A, formerly known as a combination drill and countersink, is used to provide center holes in work. Center holes should be drilled

A center hole which has not been drilled deep enough is shown in figure 8-8A. This does not provide an adequate bearing surface for the work and may result in damage to the work and lathe center.

CENTER DRILL SIZES				
Size	Work Diameter D	Diameter of Countersink C	Drill Point Diameter	Body Size
1	3/16 to 5/16	3/32	3/64	1/8
2	3/8 to 1/2	9/64	5/64	3/16
3	5/8 to 3/4	3/16	7/64	1/4
4	1 to 1 1/2	15/64	1/8	5/16
5	2 to 3	21/64	3/16	7/16
6	3 to 4	3/8	7/32	1/2
7	4 to 5	15/32	1/4	5/8
8	6 and over	9/16	5/16	3/4

TABLE 8-1

A center hole which has been drilled too deep is illustrated in figure 8-8B. This prevents the taper of the lathe center from contacting the center hole taper, and results in very poor support for the workpiece.

Center drills are available in various sizes to suit various diameters of work. Consult table 8-1 for the proper center drill to be used with each work diameter.

Procedure: Drilling Center Holes on a Drill Press

Center holes are generally drilled in a workpiece on a drill press or lathe. If work is to be center drilled in a lathe, the operation of laying out and locating centers is not required.

1. Select the proper center drill to suit the diameter of work to be drilled, table 8-1.
2. Fasten the center drill in the drill chuck using a chuck key. *Note:* To prevent center drill breakage, do not have the center drill extend more than 1/2" beyond the chuck.
3. Mount the work to be center drilled in a drill vise as illustrated in figure 8-9. *Note:* The work should be pressed firmly against the bottom of the vise.
4. Fasten a clamp or table stop on the left-hand side of the drill press table. This prevents the vise from swinging during the drilling operation.

FIG. 8-9 WORK BEING HELD IN A VISE FOR DRILLING CENTER HOLES (Courtesy McGraw-Hill)

Section 2 Machining Between Centers

5. Set the drill press at a moderate speed (1200 to 1500 r.p.m. for most center hole drilling).

6. Start the drill press and slowly lower the spindle while locating the center punch mark of the work with the drill point.

7. Carefully feed the drill into the work.

    CAUTION: Too fast a feed may break off the center drill point.

8. Raise the drill from the work frequently and apply a few drops of cutting lubricant.

9. Continue drilling until the top of the countersunk hole (C) is the correct size, table 8-1. This is generally when approximately *three-quarters of the tapered portion* of the center drill has entered the work.

10. Examine the condition of the center-drilled hole; it should be fairly smooth and free from scratches or rings. If it is not, apply a drop of cutting fluid and lightly bring the center drill into the hole.

Procedure: Drilling Center Holes in a Lathe

Round work can be quickly and accurately center drilled in a lathe without the necessity of the layout operation.

1. Remove the headstock lathe center.

2. Mount a three-jaw universal chuck on the lathe.

3. Fasten the work to be drilled in the three-jaw chuck. *Note:* Work held in a three-jaw chuck should be held short. Generally, no more than a distance equal to the work diameter should extend beyond the chuck jaws.

FIG. 8-10 DRILLING CENTER HOLES IN A LATHE WHILE THE WORK IS HELD IN A THREE-JAW CHUCK

4. Remove the tailstock center, and replace it with a drill chuck, figure 8-10.

5. Mount a suitable center drill in the drill chuck.

6. Set the lathe to a moderate speed (approximately 1200 to 1500 r.p.m. for most center drilling).

7. Slide the tailstock until the center drill is close to the workpiece, and then lock it in position.

8. Start the lathe, *lightly* tighten the tailstock spindle clamp, and slowly feed the center drill into the work by turning the tailstock handwheel. *Note:* Sometimes it is advisable to end-face the workpiece before drilling center holes.

9. Remove the drill frequently from the work and apply cutting fluid.

10. Continue drilling until the countersunk portion (C) is the correct size for the work diameter.

REVIEW QUESTIONS

1. Why is a great deal of the lathe work in a school shop machined between centers?

Locating Center Holes

2. Why are center holes necessary?

3. What are four methods of laying out the centers of round stock?

4. What is the procedure for locating the center of a workpiece by the center head method?

5. Describe a bell center punch. How is this punch used to locate centers?

Checking the Accuracy of Center Layout

6. Why is it important that the layout be checked before center holes are drilled?

7. How is a pair of dividers used to check center hole layout?

8. In which direction should the punch mark be moved to correct the center layout?

Center Drills

9. Describe a center drill and state the purpose for which it is used.

10. Why is it important that center holes be smooth and free from scratches?

11. What occurs when center holes are drilled:
    a. Too shallow? b. Too deep?

12. What size center drill should be used for the following work diameters: 7/16, 1, 2 1/4?

Drilling Center Holes

13. What are two methods of drilling center holes?

14. What is the purpose of a clamp or table stop on the drill press table?

15. How deep should the center drill enter a workpiece?

16. How should work to be center drilled be held in a three-jaw chuck?

17. What is the procedure for drilling a center hole in a lathe?

18. What diameter should the countersunk portion (C) be for a 3/4" diameter workpiece?

Unit 9   ALIGNMENT OF LATHE CENTERS

In order to machine a parallel diameter on work mounted between centers, it is important that the headstock and tailstock centers be in a straight line and true to the centerline of the lathe. If the lathe centers are not aligned, the diameter of the workpiece being turned will be tapered (one end will be larger than the other, figure 9-1.) Facing while the centers are not aligned produces surfaces which are not flat and square, figure 9-2.

The lathe tailstock is constructed in two halves, the baseplate and the tailstock body. The *baseplate* is machined to fit on the ways of the lathe. The *body* is fitted on the baseplate and may be adjusted either towards or away from the cutting tool by means of two adjusting screws. This allows the tailstock center to be adjusted for alignment with the headstock center.

Before attempting to adjust the alignment of lathe centers, it is important that both the headstock and tailstock centers be checked as follows:

1. Remove both centers and remove any nicks or burrs from the tapered shanks.
2. Examine the condition of the points of both centers.
3. With a cloth, clean the tapered holes in the headstock and tailstock spindles. Check for and remove any burrs found in the spindles.
4. Replace the centers.
5. Check the live center to see if it is running true. (See unit 10.)

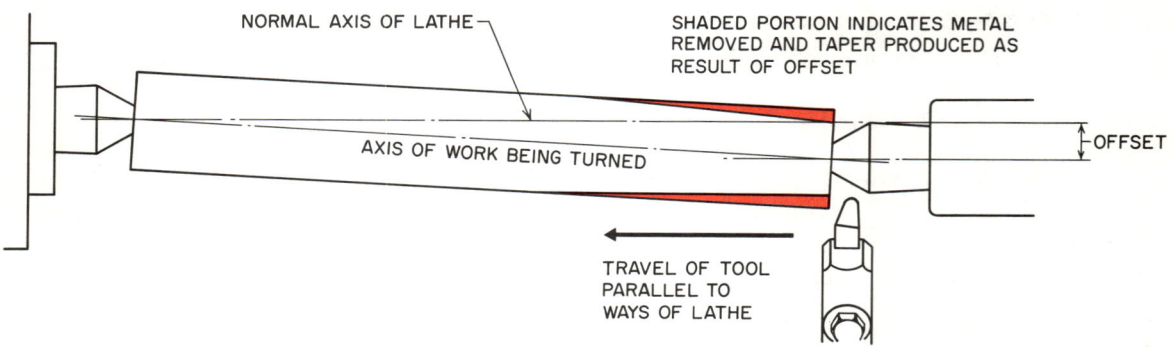

FIG. 9-1   IF CENTERS ARE NOT IN LINE, TAPERED WORK WILL RESULT.

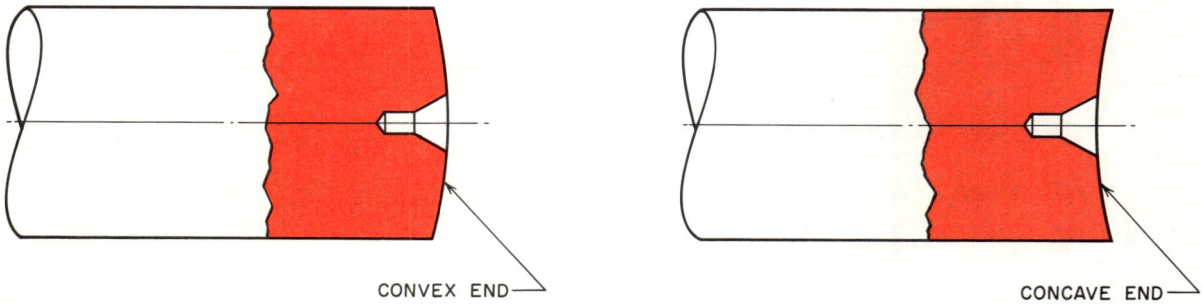

FIG. 9-2   WORK WILL NOT BE FACED SQUARE IF CENTERS ARE NOT ALIGNED.

After the centers are properly mounted and checked, they are aligned by any of the following methods, depending on the accuracy required.

- By visually *aligning the centerlines* on the end of the tailstock, figure 9-3.
- By visually *aligning the points of the live and dead centers,* figure 9-4.

  Note: The *above* methods should not be used where accuracy is required.

- By the *trial cut method,* figure 9-5, in which a cut is taken at each end of the work, and the finished diameters are checked with a micrometer.
- By using a *parallel test bar and dial indicator,* figure 9-6. This is a fast and accurate method of aligning lathe centers.
- By the use of a *micro-set adjustable center,* figure 9-7, which is constructed to permit the center to be offset slightly.

Procedure: Aligning Centers by Means of the Tailstock Graduations

1. *Loosen the tailstock clamp lever or nut.*
2. Loosen one of the adjusting screws and tighten the other, depending on the direction of movement required. Continue adjusting the screws until the line on the tailstock body is in line with the line on the baseplate, figure 9-3.

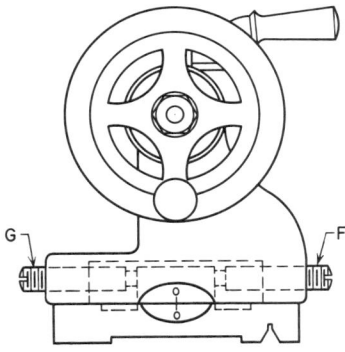

FIG. 9-3 DEAD CENTER IS ALIGNED APPROXIMATELY BY THE TAILSTOCK LINES. (Courtesy South Bend Lathe Company)

3. Tighten the loose adjusting screw to hold the upper part of the tailstock in this position.

Procedure: Aligning the Centers by the Visual Method

1. *Loosen the tailstock clamp lever or nut.*
2. Turn the tailstock handwheel until the tailstock spindle extends about 1".
3. Slide the tailstock to the left until the dead center is approximately 1/2" from the live center.
4. Turn the tailstock handwheel until the points of the centers almost touch.
5. Place a sheet of paper on the lathe bed under the centers.
6. Look down on top of the centers and check the alignment of the lathe center points, figure 9-4.

FIG. 9-4 LIVE AND DEAD CENTERS MAY BE ALIGNED APPROXIMATELY BY THE VISUAL METHOD.

7. Adjust the tailstock by means of the two adjusting screws until the points of both centers are in line.
8. Tighten the loosened adjusting screw to keep the top section in position.

Procedure: Aligning Centers by the Trial Cut Method

1. Mount the workpiece between the centers of the lathe.

Section 2 *Machining Between Centers*

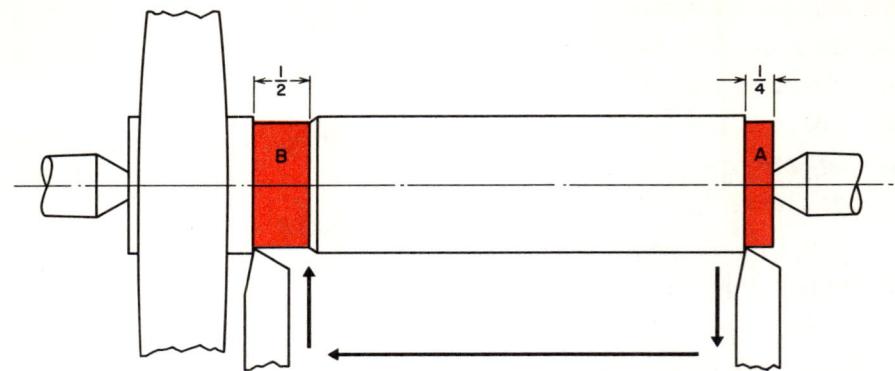

FIG. 9-5 ALIGNING THE CENTERS BY THE TRIAL CUT METHOD IS QUITE ACCURATE.

2. Take a light cut at the tailstock end of the work (section A), deep enough to produce a true diameter, figure 9-5. The cut should be about 1/4" long.

3. Stop the automatic feed and note the reading on the *graduated crossfeed collar.*

4. Back the cutting tool away from the work using the crossfeed handle.

5. Move the carriage until the cutting tool is about 1" from the lathe dog. Be sure that the lathe dog will not strike the compound rest.

6. Turn the crossfeed handle until the graduated collar is at the same setting as it was for section A.

7. Machine section B for about 1/2" long.

8. Stop the machine and measure both turned diameters with a micrometer.

9. The lathe centers are in alignment if the diameters at sections A and B are the same size. If these diameters are different, the tailstock must be adjusted one-half the difference in these two diameters:

   a. *Towards the cutting tool* if the diameter at the tailstock end is *larger*

   b. *Away from the cutting tool* if the diameter at the tailstock end is *smaller*

10. Take another light cut from both sections A and B, using the same graduated collar setting for each cut and measure the diameters.

11. Continue readjusting the tailstock and taking *light* trial cuts until both diameters are the same size.

Procedure: Aligning the Centers Using a Test Bar and Dial Indicator

1. Clean the centers of the lathe and those in the test bar.

2. Mount the test bar snugly between the centers and tighten the tailstock spindle clamp.

3. Mount a dial indicator in the toolpost or on the lathe carriage. *Note:* The contact point should be on center and the indicator plunger should be in a horizontal position.

4. Adjust the cross-slide so that the indicator needle registers about one-half a revolution on diameter A at the tailstock end, figure 9-6.

5. Move the carriage to the left by hand until the indicator registers on diameter B at the headstock end. Note the indicator reading, figure 9-6.

6. If the readings are not the same, move the carriage until the indicator again registers on diameter A.

7. Loosen the tailstock clamp nut.

Unit 9 Alignment of Lathe Centers

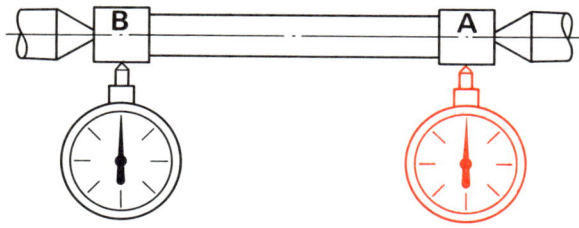

FIG. 9-6 CENTERS MAY BE ALIGNED QUICKLY AND ACCURATELY BY MEANS OF A TEST BAR AND DIAL INDICATOR.

8. By means of the tailstock adjusting screws, move the tailstock in the proper direction the difference between the indicator readings at sections A and B.

9. Tighten the loose adjusting screw to lock the upper part of the tailstock in place.

10. Tighten the tailstock clamp nut and recheck to make sure that the test bar still fits snugly between centers.

11. Repeat steps 4 to 10 until the indicator readings at sections A and B are the same.

Procedure: Aligning the Centers Using a Micro-set Adjustable Center

A micro-set adjustable center, figure 9-7, which fits into the tailstock spindle, provides a fast and accurate method of aligning lathe centers. Some of these centers contain an eccentric, while others contain a dovetail slide which permits a slight adjustment of the center itself to correct alignment.

1. Use either the trial cut or test bar and dial indicator method to determine the accuracy of the center alignment.

2. Adjust the micro-set adjustable center the amount the centers are out of line.

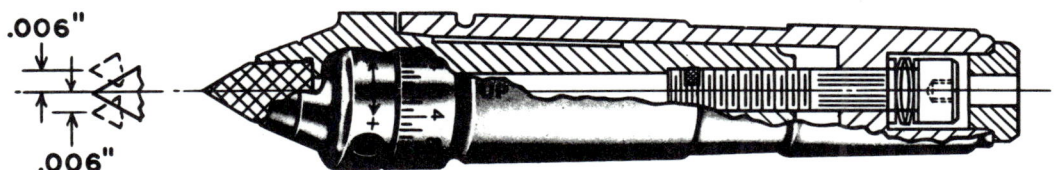

FIG. 9-7 A MICRO-SET ADJUSTABLE CENTER ALLOWS LATHE CENTERS TO BE ALIGNED QUICKLY AND ACCURATELY. (Courtesy Enco Manufacturing Co.)

REVIEW QUESTIONS

1. If the centers are not in line, what is the effect on
   a. The diameter?
   b. The ends of the work?

2. How does the tailstock construction assist in aligning lathe centers?

3. How should the lathe centers be checked prior to alignment?

4. What are two methods of approximately aligning lathe centers?

5. What are two methods of accurately aligning the lathe centers?

6. If the centers are not in line, how far should the tailstock be moved for
   a. The trial cut method?
   b. The test bar and indicator method?

7. What is the procedure for aligning lathe centers using the test bar and indicator method?

8. Describe a micro-set adjustable center.

Unit 10 MOUNTING WORK BETWEEN CENTERS

In various types of training programs, a large percentage of work machined on a lathe is held between centers. This allows the workpiece to be machined, removed from the lathe, and reset for additional machining while still maintaining the trueness of the machined diameters in relation to the center holes. When mounting work between centers, the cutting tool and workpiece must be properly set up to prevent damage and insure accuracy.

HEADSTOCK CENTER TRUENESS

When machining work between centers, it is very important that the headstock (live) center runs true to produce accurate work. If the live center does not run true, when the work is reversed to machine its entire length, the two cuts will not be concentric (true) with each other, figure 10-1.

The trueness of the live center can be checked by two methods, chalk or dial indicator, depending upon the accuracy required.

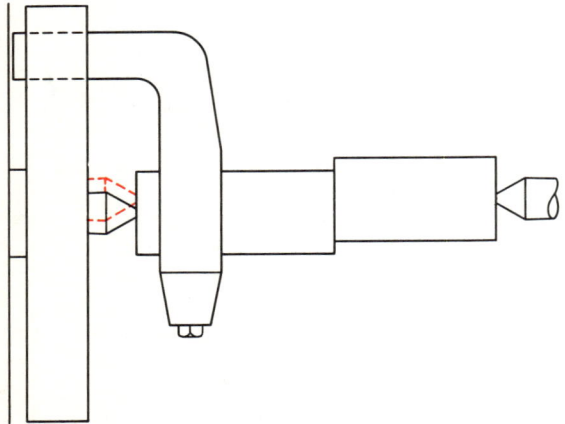

FIG. 10-1 A LIVE CENTER WHICH DOES NOT RUN TRUE WILL PRODUCE ECCENTRIC DIAMETERS WHEN THE WORK IS REVERSED TO MACHINE THE ENTIRE LENGTH.

Procedure: Chalk Method

The chalk method provides a quick means of checking the trueness of the live center. It is an approximate method and should not be used when accurate work must be machined.

1. Start the lathe and bring a piece of chalk close to the revolving center until it marks the piece lightly.
2. The chalk marks only at one point (high spot) on the circumference if the center is not running true.
3. Stop the lathe and remove the center using a knockout bar.
4. Thoroughly clean the headstock spindle taper and the tapered shank of the center.
5. Remove any burrs from both tapers.
6. Replace the center in the headstock spindle with a sharp snap.
7. Recheck the center for trueness.

Procedure: Dial Indicator Method

A dial indicator provides the most accurate method of checking trueness of the live center, and should be used when the workpiece must be machined to close tolerances.

1. Attach a dial indicator to the lathe toolpost or cross-slide.
2. Bring the indicator into contact with the center until it registers approximately one-quarter of a turn.
3. Revolve the lathe spindle by hand and note the indicator pointer.
4. If the center is not running true, the indicator pointer shows a "runout."

Unit 10 Mounting Work Between Centers

FIG. 10-2 USING A DIAL INDICATOR TO CHECK THE TRUENESS OF A LIVE CENTER ACCURATELY.

5. If a runout is indicated, remove, clean, and replace the center as outlined in the chalk method.

6. Recheck the trueness of the live center.

Procedure: Trueing a Live Center

If the workpiece must be held to very close tolerances, the live center should be turned or ground when it is mounted in the headstock spindle. After a center has been trued, it is good practice to mark the spindle, sleeve, and center with matching lines, so that when the center is removed, it can be replaced in the same position, and the accuracy is maintained.

A method commonly used when an accurate live center is required is as follows:

1. Securely hold a piece of tool steel in a three- or four-jaw chuck.

2. Swivel the compound rest 30° to the right of the lathe centerline, figure 10-3.

3. Rough turn the center, feeding the compound rest by hand.

4. Finish turn the center using a high r.p.m. and a slow hand feed.

5. Mount the workpiece between centers.

6. Adjust the lathe dog so that it is driven by the side of one of the chuck jaws.

FIG. 10-3 MACHINING A SPECIAL LIVE CENTER BEING HELD IN A 3-JAW CHUCK

65

Section 2 *Machining Between Centers*

LATHE DOGS

Work which is machined between lathe centers is driven by a lathe dog. The lathe dog is provided with an opening to receive and clamp the workpiece, and a bent tail. The tail of the dog fits into a slot in the lathe driveplate and provides a drive to the workpiece.

Lathe dogs are made in a variety of sizes and styles to suit various operations and workpieces.

The standard bent-tail lathe dog, figure 10-4A, is most commonly used to drive round workpieces on a lathe. They are available in sizes to accommodate diameters ranging from 3/8" to 6". Standard lathe dogs are available with square-head setscrews or headless setscrews to clamp the workpiece. The headless screw is safer than the square-head setscrew since it has no protruding head.

The straight tail dog, figure 10-4B, may be used for the same purpose as the bent-tail type; however, it is driven by a stud attached to the driveplate. The straight tail and the locking screw end are approximately the same weight, thus providing a more balanced lathe dog. Straight-tail dogs are often used in precision work where centrifugal force may cause inaccuracies in the workpiece.

The safety clamp lathe dog, figure 10-4C, can be used to hold a variety of work shapes. It is especially valuable for use on finished work which may be damaged by the setscrew of a common lathe dog. The sliding jaw gives this type of dog a wide range of adjustment and allows it to be mounted on a workpiece while it is between the lathe centers.

The clamp lathe dog, figure 10-4D, can be used to hold round, square, rectangular, hexagonal, and odd-shaped workpieces. The two setscrews allow this type of dog to hold a wider range of workpiece sizes.

Procedure: Setting Up the Cutting Tool and Toolholder

The setup of the cutting tool and toolholder is very important to the accurate and efficient removal of metal. If the setup is incorrect, the cutting tool may not cut too well, or it may move and dig in, ruining the workpiece.

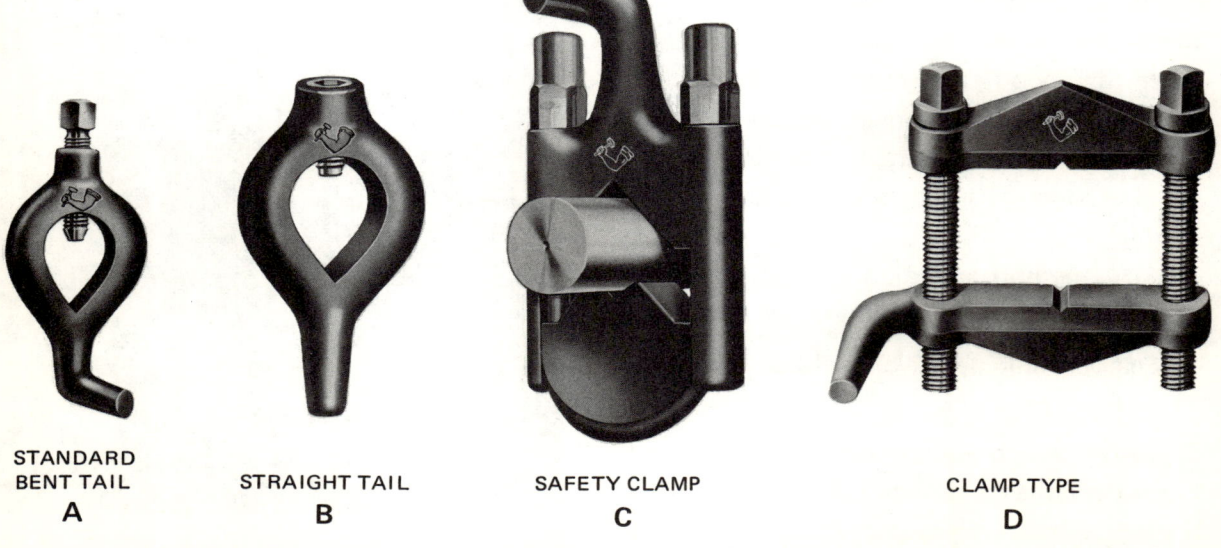

| STANDARD BENT TAIL | STRAIGHT TAIL | SAFETY CLAMP | CLAMP TYPE |
| A | B | C | D |

FIG. 10-4 COMMON TYPES OF LATHE DOGS. (Courtesy Armstrong Bros. Tool Co.)

Unit 10   Mounting Work Between Centers

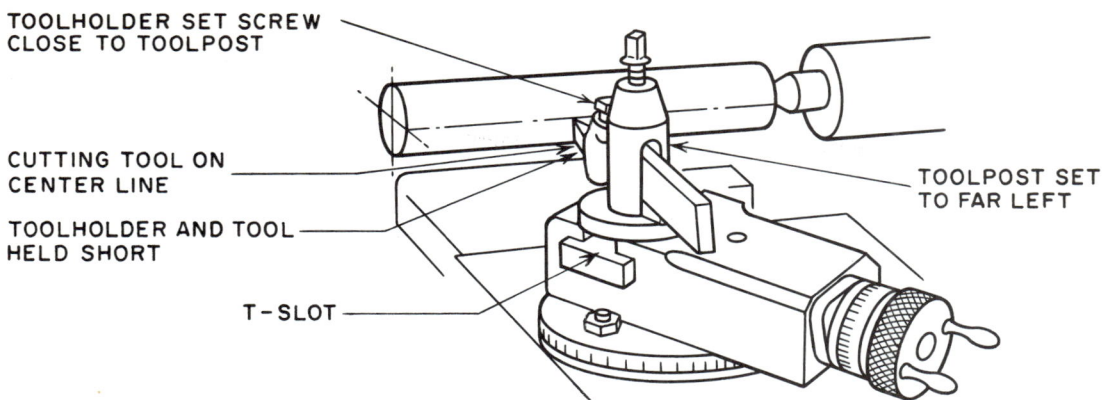

FIG. 10-5   THE TOOLPOST IS SET ON THE LEFT-HAND SIDE OF THE COMPOUND REST WITH THE TOOLHOLDER HELD SHORT.

1. Locate the toolpost on the *left-hand* side of the compound rest T-slot, figure 10-5. *Note:* If the toolpost is set in the middle or on the right-hand side, there is a danger of the lathe dog striking the compound rest, causing damage to the lathe and workpiece.

2. Mount a toolholder so that its setscrew is close to the toolpost (approximately the width of a thumb, figure 10-5). *Note:* More leverage is exerted on the cutting tool the further the point of the cutting tool projects beyond the toolpost. As a result, the tool may be moved under the pressure of the cut. A large overhang allows the tool to spring, causing chattering and inaccurate work due to the tool digging in.

3. Set the toolholder at right angles or pointing slightly towards the tailstock, figure 10-6A.

   If the toolholder moves during the pressure of the cut, the cutting tool does not dig in but swings away from the workpiece causing the diameter to become larger.

   If the toolholder is set as in figure 10-6B, and it swivels due to pressure of the cut, the toolbit digs in, causing the diameter to be cut undersize and thereby ruining the workpiece.

4. Mount the desired toolbit so that it extends only about 1/2" beyond the toolholder, figure 10-5.

5. Tighten the toolholder setscrew sufficiently to hold the cutting tool. If the setscrew is locked too tightly, it may break the toolbit.

6. Set the point of the toolbit to center. Check it for accuracy against the point of the lathe center.

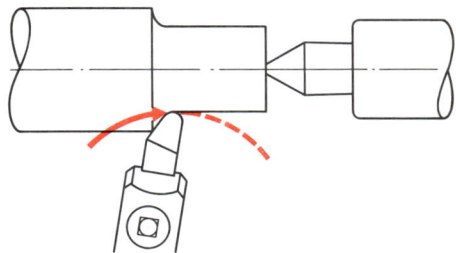

FIG. 10-6A   THE TOOLHOLDER SET SLIGHTLY TO THE RIGHT TO PREVENT THE CUTTING TOOL FROM DIGGING IN

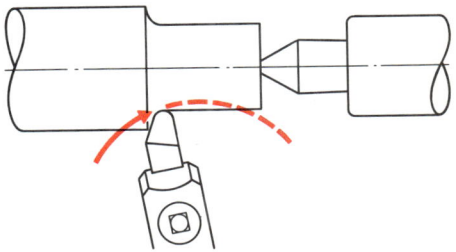

FIG. 10-6B   THE TOOLHOLDER IS INCORRECTLY POSITIONED FOR TAKING HEAVY CUTS.

Section 2   Machining Between Centers

7. *Securely tighten* the toolpost screw to prevent the toolholder from moving under the pressure of a cut.

Procedure: Mounting the Work

Students involved with mounting workpieces between centers on a lathe should be aware of the importance of mounting the work properly to avoid damaging the workpiece and lathe centers.

1. Check the live center for trueness, unit 10.
2. Test the alignment of the lathe centers, unit 9.
3. Set up the cutting tool and toolholder, unit 10.
4. Clean the lathe center points and the workpiece center holes.
5. Adjust the tailstock spindle so that it extends 2 1/2" to 3" beyond the tailstock.
6. Loosen the tailstock lever or nut.
7. Place a suitable lathe dog on the left-hand end of the work, figure 10-7. Do not tighten it onto the workpiece at this time.
8. Apply a suitable center lubricant to the center hole on the right-hand end of the workpiece.

Red and white lead are suitable center lubricants which have been used for many years. Center lubricants, especially developed to withstand the heat and pressures of a cut, have been developed by many manufacturers. They outperform red and white lead as center lubricants.

9. Place the left-hand end of the work on the live center and slide the tailstock until it supports the other end of the workpiece.
10. Lock the tailstock lever or nut.
11. Adjust the tail of the dog so that it fits freely, without binding, in one of the slots of the driveplate, and then tighten the setscrew. *Note:* If a lathe dog binds in the driveplate, the work is forced off the live center, figure 10-8, and the machined diameter is not true (concentric) with the center hole.

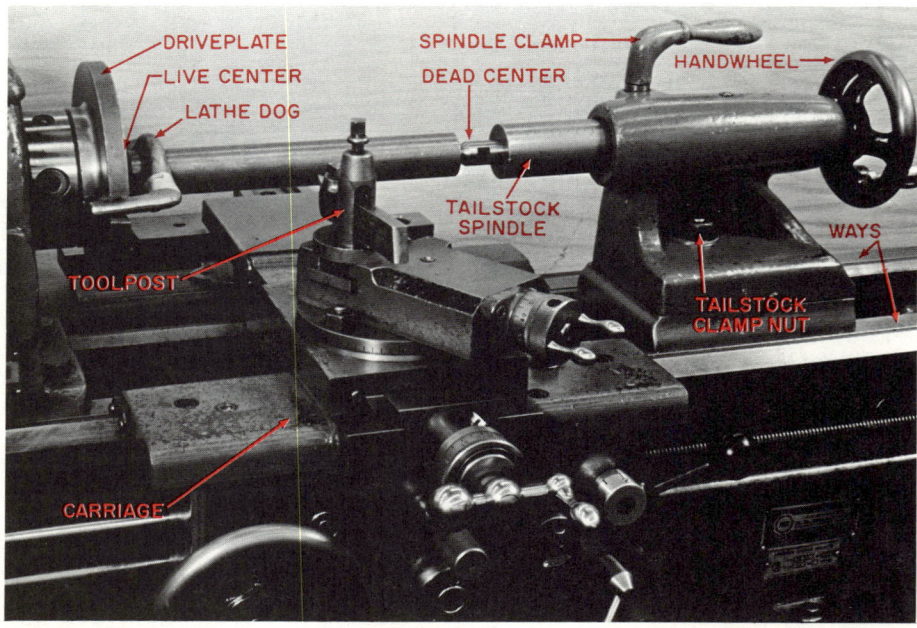

FIG. 10-7   WORK MOUNTED FOR MACHINING BETWEEN CENTERS

Unit 10 Mounting Work Between Centers

12. Revolve the driveplate by hand until the tail of the dog is in a horizontal position, figure 10-7.

13. Hold the *tail of the dog up* in the slot and lightly tighten the tailstock handwheel to hold the dog in the up position.

14. Loosen the tailstock handwheel until the tail of the dog just drops in the driveplate slot. *Note:* If a revolving dead center is used, the tailstock handwheel should be adjusted so that the work fits *snugly* between centers.

15. Hold the tailstock handwheel in this position and tighten the tailstock spindle clamp with the other hand.

16. Check the tension of the work between centers. If the tension is correct, the tail of the dog should drop of its own weight, and there should be no end play between the lathe centers and the work.

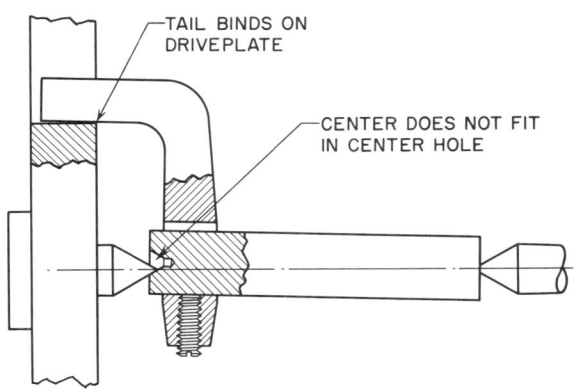

FIG. 10-8 A LATHE DOG BINDING IN THE DRIVEPLATE SLOT CAUSES INACCURACIES SINCE THE WORK IS FORCED OFF THE LATHE CENTER.

17. Move the carriage to the furthest position (left-hand end) of the cut to be taken, bring the cutting tool close to the work diameter, and revolve the lathe spindle by hand to see that the lathe dog does not strike the compound rest.

REVIEW QUESTIONS

1. Why is a large percentage of lathe work in training programs held between centers?

Headstock Center Trueness

2. What occurs if the live center does not run true when machining a workpiece?

3. What is the procedure for *accurately* checking the trueness of a live center?

4. What are two methods of trueing a live center?

Lathe Dogs

5. What type of lathe dog is designed to overcome some of the inaccuracies caused by centrifugal force?

6. What two lathe dogs are best suited to hold a finished diameter without marring its surface?

Cutting Tool Setup

7. Why is it important that the cutting tool be set up properly?

8. Why should the toolpost always be set on the left-hand side of the compound rest?

Section 2   Machining Between Centers

9. Why is a toolholder held "short"?

10. Why should a toolholder be set at right angles or pointing slightly towards the tailstock when taking heavy cuts?

11. At what height should the cutting tool be set?

Work-Mounting Procedure

12. How far should the tailstock spindle extend beyond the tailstock?

13. Why is lubricant used in the center hole supported by the tailstock dead center?

14. What is the result if a lathe dog binds in a drive slot?

15. How is the tension of work adjusted between centers?

16. What is the proper tension for work mounted between lathe centers?

Unit 11 FACING

One of the first turning operations on a workpiece is generally that of facing. The rough stock from which the workpiece is to be machined is usually cut a little longer than the finished size to allow for machining the ends square with the axis of the work. Work may be faced in a chuck or between centers. When it is faced between centers, the lathe centers must first be aligned in order to produce a flat, square end.

Work is usually faced for the following reasons:

- To machine work to the required length
- To smooth and square the ends
- To provide a square, flat surface from which to take measurements

When facing work held in a chuck, a general-purpose toolbit or a facing toolbit may be used. A facing toolbit, figure 11-1, must be used when facing work between centers.

Procedure: Facing Work Between Centers

1. Set the toolpost to the left side of the compound rest.
2. Insert a facing toolbit in the toolholder, extending only about 1/2".
3. Set the point of the toolbit to the height of the lathe center point.

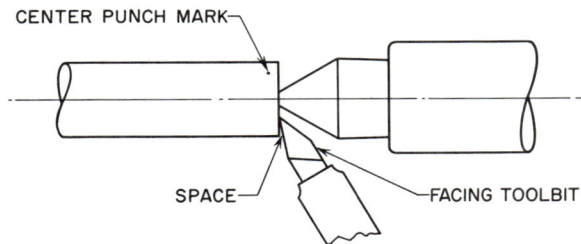

FIG. 11-1 A FACING TOOL IS GROUND TO FIT BETWEEN THE DEAD CENTER AND THE WORK.

FIG. 11-2 A HALF CENTER ALLOWS THE ENTIRE SURFACE TO BE FACED.

4. Mark off the finished length on the workpiece.
5. Clean the lathe centers and the center holes in the workpiece. If the entire surface must be square, a half center should be mounted in the tailstock. The half center, figure 11-2, is ground flat on one side to allow the point of the facing tool to enter the center hole.
6. Mount the work between centers.
7. Set the facing toolbit as shown in figure 11-1. The toolbit should point to the left to permit only the point to cut; otherwise a flat surface is not produced.

Section 2 Machining Between Centers

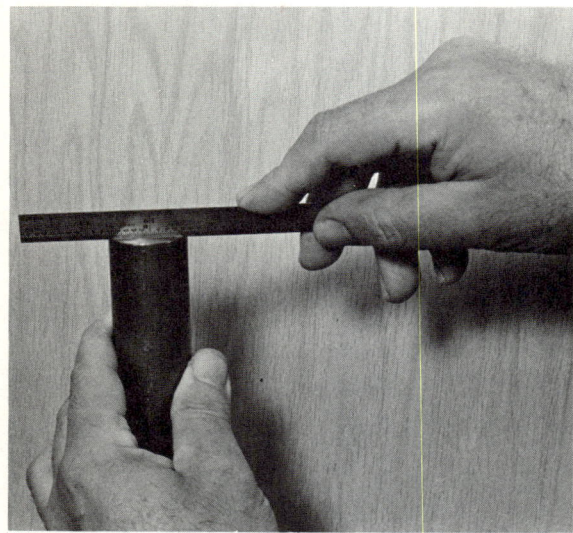

FIG. 11-3 CHECKING THE FLATNESS OF A FACED SURFACE

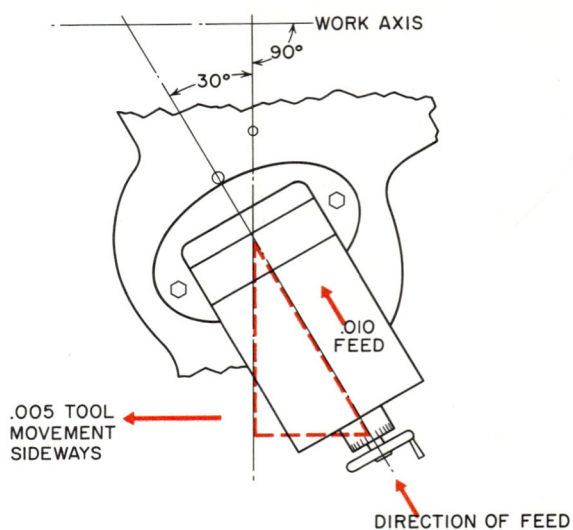

FIG. 11-4 ACCURATE FACING IS POSSIBLE WHEN THE COMPOUND REST IS SET AT 30°

8. Set the lathe to the proper spindle speed for the type and diameter of the work being machined.

9. Start the machine and bring the point of the toolbit close to the lathe center.

10. Move the carriage to the left by hand until a small cut is started.

11. Slowly turn the crossfeed handle to move the tool outwards; at the same time hold the carriage handwheel to prevent any side movement of the carriage. *Note:* If the automatic feed is used to face the workpiece, the carriage should be locked in place.

12. Repeat operations 9, 10, and 11, *always feeding outwards,* until the work is cut to the desired length. *Note:* A center punch mark is often used to mark the length of the work. The work should be faced until the mark is cut in half.

13. Remove the work and test the surface for flatness with the edge of a rule, figure 11-3. The rule must bear against the full width of the faced surface.

Procedure: Facing With the Compound Rest Set at 30°

Work can be accurately faced to length by using the graduated collar of the compound rest.

1. Swivel the compound rest at 30° to the cross-slide, figure 11-4.

2. Move the toolpost to the left-hand side of the compound rest.

3. Set the facing tool to center. Check it against the lathe center point.

4. Bring the toolbit close to the center of the surface to be faced.

5. *Lock the carriage in position* to prevent it from moving during a cut.

6. Feed the compound rest screw in until a light cut is started and face the surface.

7. Measure the work and calculate the amount of material to be removed.

8. Feed the compound rest screw double the amount of material to be removed.

When the compound rest is set at 30°, the amount of side movement of the cutting tool is always one-half the amount fed. For example, if the compound rest is fed .020, the cutting tool moves .010 sideways, figure 11-4.

Procedure: Facing With the Compound Rest at 90°

A series of steps or shoulders can be accurately spaced and faced along the length of a piece of work when the compound rest is set at 90° to the cross-slide, figure 11-5. The amount of toolbit travel is the same as the reading on the graduated collar of the compound rest feed screw.

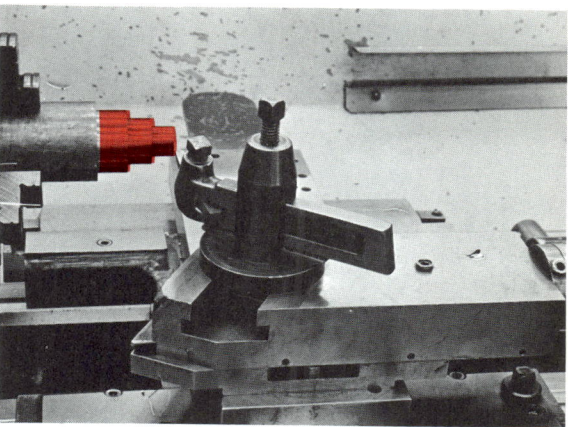

FIG. 11-5 ACCURATE SPACING OF STEPS OR SHOULDERS IS POSSIBLE WHEN THE COMPOUND REST IS SET AT 90°

1. Swivel the compound rest at 90° to the cross-slide, figure 11-5.

2. Set the cutting tool to center.

3. Rough turn all shoulders to within 1/32 of the required length.

4. Set the cutting tool for facing the shoulders.

5. Bring the top slide of the compound rest back as far as possible to insure sufficient travel to cut all shoulders.

6. Set the cutting tool to take a light cut from the end of the workpiece (or a shoulder), and then lock the carriage in position.

7. Machine each shoulder to length by feeding the compound rest the required distance.

REVIEW QUESTIONS

1. What are three reasons for facing a workpiece?

2. What precautions must be taken when setting up a facing toolbit?

3. How should the cut be taken when facing?

4. How may the end of the work be checked for flatness?

Facing With the Compound Rest at 30°

5. Why is the compound rest set at 30° when facing?

6. Why is it important to lock the carriage in position while facing?

Facing With the Compound Rest at 90°

7. What is the advantage of setting the compound rest to 90° for accurately spacing a series of shoulders?

8. What precautions should be taken to insure that the compound rest has sufficient travel to machine all shoulders?

Unit 12 SETTING THE DEPTH OF CUT

The accuracy of any workpiece machined on a lathe is dependent upon the following factors:

- The accuracy and rigidity of the workpiece setup
- The condition of the lathe and the cutting tool
- The care used by the operator in setting the depth of a cut

GRADUATED MICROMETER COLLARS

To assist the lathe operator in setting the cutting tool accurately to remove the required amount of material, lathe crossfeed and compound rest screws are equipped with graduated micrometer collars, figure 12-1. The circumferences of both of these micrometer collars are usually graduated into 100 or 125 equal divisions, each having a value of .001. Therefore, if the *crossfeed screw* is advanced 50 graduations clockwise, the cutting tool will be moved .050 towards the work. Since work in a lathe revolves, the .050 depth of cut is taken from around the entire circumference, thereby reducing the work diameter by .100 (2 x .050), figure 12-2.

The graduated collar indicates only the amount the cutting tool has been moved. Therefore, on any machines where the work revolves (lathes, cylindrical grinders, vertical boring mills, etc.), it is wise to remember that since the cut is taken from the circumference of the workpiece, the cutting tool should be set in only *half the amount of metal to be removed.*

Compound Rest Graduated Collar

The graduated collar on the compound rest feed screw can be used for a number of operations besides setting a cutting tool for a depth of cut. Some of the most common uses are: facing, facing and shoulder turning, making fine cutting-tool settings, and thread cutting.

FIG. 12-1 MICROMETER COLLARS ON THE CROSSFEED AND COMPOUND REST SCREW HANDLES ENABLE SETTING AN ACCURATE DEPTH OF CUT.
(Courtesy South Bend Lathe Co.)

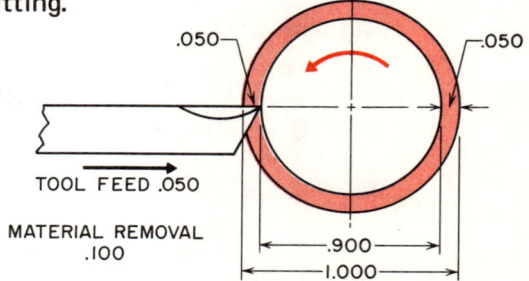

FIG. 12-2 THE WORK IN A LATHE REVOLVES AND, THEREFORE, A .050 DEPTH OF CUT REDUCES THE THE DIAMETER BY .100.

Unit 12   Setting the Depth of Cut

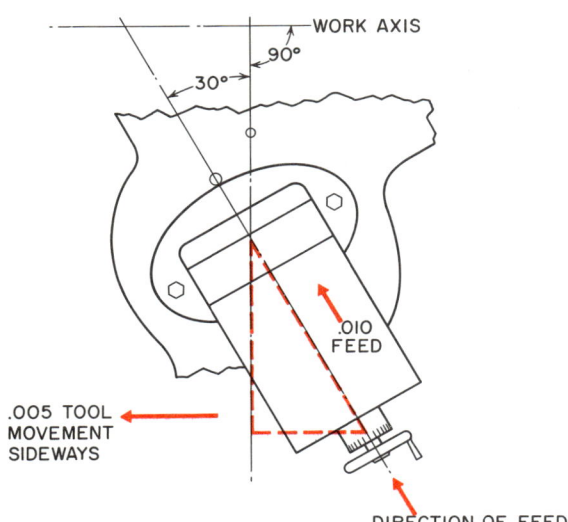

FIG. 12-3 THE CUTTING TOOL MOVES SIDEWAYS ONE-HALF THE AMOUNT FED WITH THE COMPOUND REST AT 30°.

FIG. 12-4 THE COMPOUND REST CAN BE USED FOR ACCURATE FACING AND SHOULDER TURNING WHEN IT IS SET AT 90°.

Facing (compound rest at 30°). The compound rest can be used for facing purposes when it is set at 30° to the cross-slide. In this position, the amount of side movement of the cutting tool is always *one-half the amount of compound rest feed,* figure 12-3. For example, if the compound rest is fed .020, the cutting tool moves sideways .010.

Facing and shoulder turning (compound rest at 90°). Accurate facing and shoulder turning is possible when the compound rest is set parallel to the lathe bed (at 90° to the cross-slide). The amount of toolbit travel is the same as the graduated collar setting since the cut is taken off one surface and not the circumference of the work. For example, if the compound rest is fed .020, the cutting tool advances .020 along the length of the workpiece.

Fine cutting-tool settings (compound rest at 84°16′). Whenever the nature of the workpiece demands that very precise cutting-tool settings be made, the compound rest should be set at 84°16′ to the cross-slide, figure 12-5.

A .001 movement of the compound rest causes the cutting tool to advance towards the workpiece .0001, figure 12-5. This method is generally used when the lathe is used for grinding operations.

Thread cutting. The compound rest is used for most thread-cutting operations. The compound rest is usually set at one-half the included angle of the thread, and the depth of cut is set by feeding the compound rest. In this way, the threading toolbit cuts only on one edge and, therefore, tends to produce a

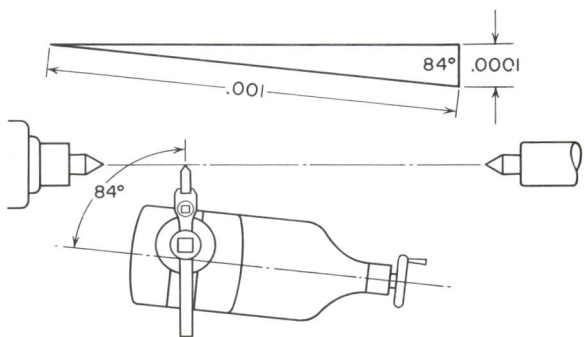

FIG. 12-5 WITH THE COMPOUND REST SET AT 84°16′ IT IS POSSIBLE TO SET .0001 DEPTH OF CUT.

Section 2 Machining Between Centers

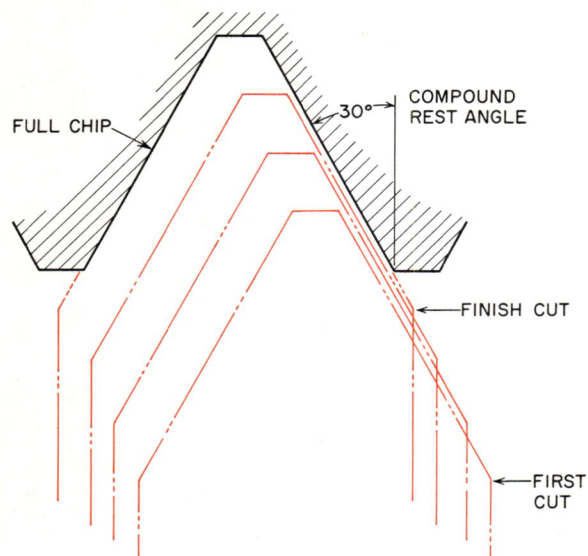

FIG. 12-6 THE TOOLBIT CUTS ONLY ON ONE SIDE WHEN THE COMPOUND REST IS SET AT ONE-HALF THE INCLUDED THREAD ANGLE. (Courtesy Cincinnati Milacron Co.)

smoother thread, figure 12-6. The amount the compound rest should be fed can be easily calculated by trigonometry.

Graduated Collar Hints

If the graduated micrometer collars are used correctly, the operator is able to produce very accurate work. The following points should be carefully observed; otherwise many workpieces may be ruined.

1. If the graduated collars on the compound rest and crossfeed handles are equipped with locking screws, make sure they are tight before making any collar settings.

2. Turn the compound rest feed screw clockwise one-half turn to remove any backlash before setting a cut.

3. Feed the cutting tool towards the workpiece and *never backwards* when setting a depth of cut.

4. If the crossfeed handle is turned past the desired collar setting, turn it *counterclockwise* one-half turn to remove the backlash and then *clockwise* to the proper setting. *Note:* If the handle is turned back only a few thousandths, without removing the backlash (play) between the crossfeed screw and nut, the cutting tool is still set too deep.

5. Many graduated collars are held in position by friction devices and should never be handled when making a setting.

Procedure: Setting a Cutting Tool for an Accurate Depth of Cut

1. Move the toolpost to the left-hand side of the compound rest.

2. Set the cutting tool height to center.

3. Start the lathe and bring the cutting tool so that it overlaps the right-hand end of the work by approximately 1/8".

4. Feed the cutting tool with the *crossfeed handle* until a light cut is made around the circumference of the workpiece.

5. Turn the carriage handwheel until the cutting tool clears the end of the work.

6. Feed the crossfeed screw in (clockwise) approximately .005, and take a cut 1/4" long from the right-hand end of the work.

7. Stop the lathe, but *do not move the crossfeed screw handle.*

8. Turn the carriage handwheel until the toolbit clears the end of the work.

9. Measure the diameter of the turned section and calculate the amount of metal yet to be removed.

10. Turn the graduated collar of the crossfeed screw clockwise one-half the amount of material to be removed. For example, if .020 must be removed, the crossfeed handle should be set in .010 since the cut is taken from the work circumference.

11. Take a cut 1/4" long and recheck the accuracy of the cut. It is good practice to *always check the diameter of a workpiece after the trial cut is made for each setting.*

12. Readjust the cutting tool setting if it is necessary.

REVIEW QUESTIONS

1. What three factors may affect the accuracy of a workpiece machined on a lathe?

Graduated Micrometer Collars

2. What is the purpose of the graduated micrometer collars on the compound rest and crossfeed screws?

3. What is the value of each line on the graduated collar?

4. How much material would be removed from a diameter if a .015 deep cut is taken? Explain why.

5. How far should the graduated collar be set in to turn a .750 diameter to .690?

Compound Rest Graduated Collar

6. Why is accurate facing possible when the compound rest is set at 30°?

7. How can the compound rest be used for accurate facing and shoulder turning?

8. How are .0001 settings made with the compound rest?

9. What are the advantages of feeding the threading toolbit with the compound rest?

Graduated Collar Hints

10. What are two important points to be observed when using graduated collars?

11. What is backlash and how can it be overcome on a lathe?

Setting a Cutting Tool for an Accurate Depth of Cut

12. What is the procedure for setting the cutting tool to the work surface?

13. What precaution should be taken when setting a depth of cut?

Unit 13 PARALLEL TURNING

Turning on a lathe is the process of machining a revolving piece of work with a stationary cutting tool which is fed longitudinally along the workpiece, figure 13-1. The purpose of turning work on a lathe is to produce a true cylindrical diameter, and to cut the work to a specific size. *Parallel turning,* as the term implies, means that the diameter at both ends of the workpiece must be the same size. To machine a parallel diameter on work located between centers, it is important that the *lathe centers be aligned.*

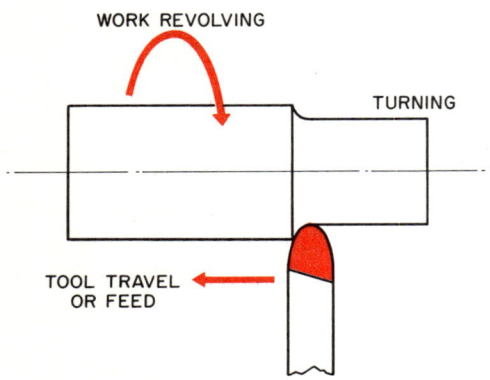

FIG. 13-1 THE PRINCIPLE OF MACHINING A DIAMETER ON A LATHE

Most machining on a lathe should be performed with the cutting tool being fed towards the headstock. This is especially important when turning between centers, since the pressure of the cutting tool should be against the headstock center and not against the stationary dead center.

Work should be turned to size in as few cuts as possible. The general practice is to take two cuts - *one roughing and one finishing cut.* However, the amount of material removed in each cut, or the depth of cut, is dependent upon the following factors:

- The size and condition of the lathe
- The shape and length of the workpiece
- The shape and condition of the cutting tool
- The type of cut being taken (rough or finish)

ROUGH TURNING

The purpose of rough turning is to remove most of the excess material as quickly as possible, true the diameter, and prepare the workpiece for finish turning. The rough cut should be as deep as the machine, workpiece, and the cutting tool will withstand, figure 13-2. Generally, only one rough cut is taken from a workpiece. However, if considerable material (over 1/2") must be removed, it may be necessary to take two roughing cuts.

When machining cast iron or other metals which have a hard outer scale, it is important that the roughing cut be deep enough to cut below the scale. If the toolbit does not pene-

FIG. 13-2 THE ROUGHING CUT SHOULD BE AS DEEP AS THE LATHE, WORKPIECE, AND CUTTING TOOL PERMIT.

trate below the scale, the abrasive action of the scale quickly damages the cutting tool edge.

Since surface finish and accuracy are not important, use as coarse a feed as possible when rough turning to remove excess material quickly. The rough cut should be taken to within 1/32" of the finish size required. This allows sufficient material for the finishing cut.

Procedure: Taking a Roughing Cut

1. Mount a roughing or a general-purpose toolbit into a straight toolholder or a left-hand toolholder.
2. Set the toolpost to the left side of the compound rest.
3. Have the toolholder extend as little as possible beyond the toolpost, about the width of a thumb between the toolpost and the toolholder screw, so that the toolholder is not likely to move under the pressure of a cut.
4. Adjust the toolholder so that if it moves under the pressure exerted during a cut, the cutting tool will swing away from and not into the work, figures 13-3A and B. *Note:* If the cutting tool can swing into the work, the work diameter is cut smaller and the work is probably ruined.
5. Set the cutting tool to center and *securely* tighten the toolpost screw.
6. Properly mount the work between centers.
7. Set the correct lathe r.p.m. for the material being cut.
8. Set the quick-change gearbox for the roughing-cut feed (about .020).
9. Start the lathe and take a *light trial cut* at the right-hand end of the work for 1/4" length.

   The light trial cut is important because:

   a. A true diameter is produced for measuring purposes.

   b. The cutting tool point is set to the true diameter.

   c. The graduated collar reading is now the reference (starting) point for setting the depth of cut.

10. Stop the lathe, but *do not touch the crossfeed handle setting or the graduated collar.*
11. With the carriage handwheel, move the cutting tool until it clears the right-hand end of the work.
12. Measure the diameter and calculate the amount of material to be removed. *Note:* Be sure to leave the diameter .030 to .050 oversize for the rough cut. This provides sufficient material for the finish cut.

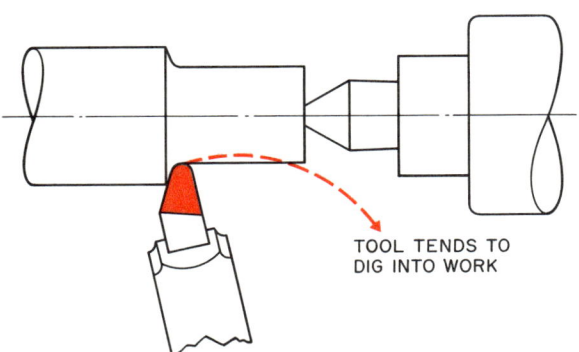

FIG. 13-3A IF AN INCORRECTLY SET TOOLHOLDER MOVES UNDER THE PRESSURE OF A CUT, THE CUTTING TOOL WILL CUT THE WORK UNDERSIZE.

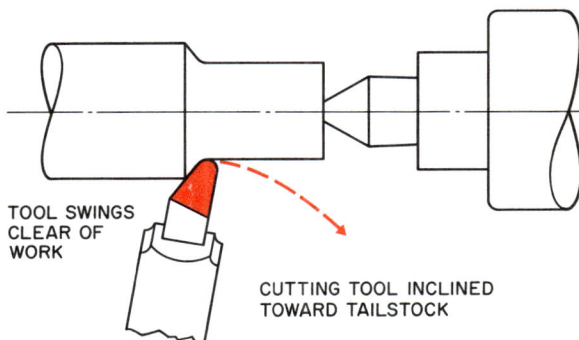

FIG. 13-3B IF A CORRECTLY SET TOOLHOLDER MOVES UNDER THE PRESSURE OF A CUT, THE CUTTING TOOL WILL SWING AWAY FROM THE WORK.

Section 2 Machining Between Centers

13. Turn the crossfeed handle *clockwise* until the desired depth of cut is indicated on the graduated collar.

14. Take a 1/4" long cut, stop the lathe, and measure the diameter for size.

15. Readjust the cutting tool setting, if necessary, and then take the roughing cut for the required length. *Note:* If the entire length of the workpiece must be machined, one of two methods can be employed.

    a. The rough and finish cuts can be taken in one setup, and then the work turned end-for-end and the procedure repeated on the other end. This method is generally used when only one or two parts are required.

    b. After the rough cut has been taken, do not move the crossfeed setting. Reverse the work between centers, and take a rough cut from the other end at the same setting. This method is particularly suitable when many identical parts are required.

FINISH TURNING

After the work has been "roughed down," it must be finish turned to attain the proper size and surface finish required. Generally, only one finish cut is taken since the diameter should be only .030 to .050 oversize after the rough cut. The lathe speed is usually higher than that used for rough turning and a finer feed is used. A sharp toolbit with a slight radius on the point is desirable for finish turning. If the roughing toolbit was used to turn only a few pieces, the cutting edge may still be keen and therefore can be used for the finish cut.

Procedure: Taking a Finishing Cut

1. Remove the work and the roughing toolbit. (If the roughing tool is in good condition, it may be used for finish turning.)

2. Mount a finishing tool in the toolholder and adjust the toolbit to center height.

3. Set the lathe speed and feed for finish turning. The speed should be a little higher, while the feed should be finer than that used for rough turning. A feed of approximately .005 is recommended.

4. Remount the workpiece.

5. Start the machine and take a *light* cut (about 1/4" long) at the right-hand end of the workpiece.

6. Stop the machine and measure the diameter of the trial cut. *Be sure not to move the crossfeed handle or the graduated collar setting.*

7. Move the carriage to the right until the toolbit is clear of the work.

8. By means of the crossfeed handle, feed the cutting tool in one-half the difference between the trial cut diameter and the required finished diameter.

9. Start the machine, engage the automatic feed, and take a cut for about 1/4" long.

10. Stop the machine and recheck the diameter for size.

11. If the diameter is the required size, continue to cut to the desired length.

REVIEW QUESTIONS

1. Define parallel turning.

2. Why must the lathe centers be aligned for parallel turning?

3. Why is it advisable to cut towards the headstock when turning between centers?

4. What factors determine the amount of material which can be removed in one cut?

Rough Turning

5. What is the purpose of rough turning?

6. What factors affect the depth of a rough cut?

7. What precaution should be observed when taking a rough cut from cast iron or metal with a scaly surface?

8. How close to finish size should the rough cut be taken?

9. How is the toolholder set for rough turning?

Finish Turning

10. What is the purpose of finish turning?

11. How does the lathe speed and feed compare with that used for rough turning?

12. Describe the type of cutting tool that should be used for finish turning.

Unit 14 SHOULDER TURNING

When more than one diameter must be turned on a workpiece, the change in diameter, or step, is called a shoulder. Shoulders are generally used to:

- Allow a mating part to fit flatly against the shoulder
- Strengthen the part at the shoulder
- Eliminate sharp corners
- Present a good appearance

Three of the most common shoulders are:

- The *square* shoulder, figure 14-1A
- The *angular* (beveled or tapered) shoulder, figure 14-1B
- The *filleted* shoulder, figure 14-1C

Procedure: Machining a Square Shoulder

Square shoulders are generally used where it is necessary for a mating part to fit right against the shoulder. They may also be used to position parts fitted on a shaft. The square shoulder is used on parts which are not subject to excessive strain at the corner.

1. Face the end of the work to provide a reference point from which to take the measurements.
2. Lay out the position of the shoulder by one of the following methods:
   a. Placing a center punch mark the correct distance from the end of the workpiece, figure 14-2A
   b. Cutting a *light groove* with the point

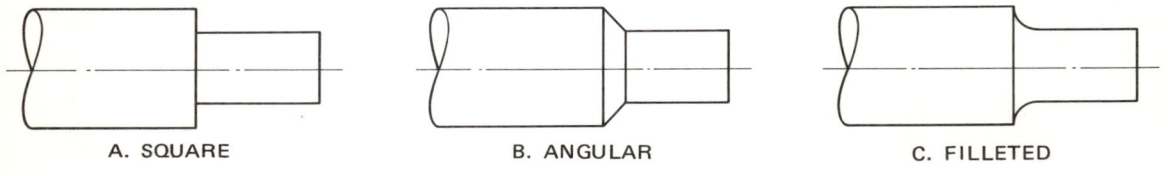

FIG. 14-1 COMMON TYPES OF SHOULDERS

A. SQUARE   B. ANGULAR   C. FILLETED

A. WITH A CENTER PUNCH MARK

B. BY CUTTING A LIGHT GROOVE ON THE WORKPIECE

FIG. 14-2 MARKING THE LENGTH OF A SHOULDER

of a sharp toolbit around the circumference of the work to mark the required length, figure 14-2B.

3. Rough and finish turn the diameter to within 1/32'' of the required length.

4. Mount a facing tool in the toolholder and set it to center. *Note:* Be sure that the facing toolbit is set up as in unit 11, with the point closest to the work and a slight space along the side cutting edge.

5. Apply chalk or layout dye to small diameter, as close to the shoulder as possible.

6. Start the lathe and bring the facing tool in until it just removes the chalk or layout dye. *Note:* Before starting the lathe, the toolbit can be brought fairly close to the diameter by using a piece of paper or shim stock between the toolbit point and the work diameter.

7. Note the reading on the graduated collar of the crossfeed screw.

8. Bring the toolbit towards the shoulder with the carriage handwheel until a cut is started.

9. Face (square) the shoulder by turning the crossfeed handle *counterclockwise,* thus cutting from the center to the outside.

10. For successive cuts, return the crossfeed screw to the same graduated collar setting (as in step 7), and repeat steps 8 and 9 until the shoulder is machined to the correct length.

Procedure: Machining a Beveled (Angular) Shoulder

Beveled or angular shoulders are used primarily to eliminate sharp corners and edges, to make parts easier to handle, and to improve the appearance of the part. They are sometimes used to strengthen a part by eliminating the sharp corner found on a square shoulder. Shoulders are beveled at angles ranging from 30° to 60°; however, the most common is the 45° bevel.

1. Lay out the position of the shoulder along the length of the workpiece. *Note:* This should be the length of the small diameter, figure 14-3B.

2. Rough and finish turn the small diameter to size.

3. Mount a side cutting tool in the toolholder and set it to center.

A. USING A PROTRACTOR TO SET THE TOOLBIT SIDE CUTTING EDGE TO 45°

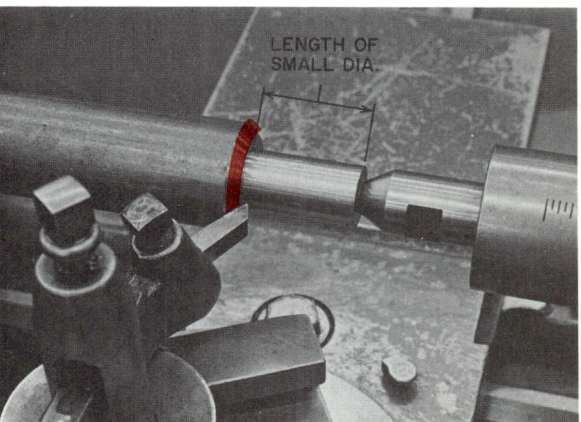

B. MACHINING THE BEVELED SHOULDER

FIG. 14-3 PROCEDURE FOR CUTTING A BEVELED (ANGULAR) SHOULDER

Section 2 Machining Between Centers

4. Use a protractor and set the side cutting edge of the toolbit to the desired angle, figure 14-3A.

5. Apply chalk or layout dye to the small diameter as close as possible to the shoulder location.

6. Set the lathe spindle to approximately one-half of the turning speed.

7. Bring the point of the toolbit in until it just removes the chalk or layout dye.

8. Turn the carriage handwheel by hand to feed the cutting tool into the shoulder.

9. Apply cutting fluid to assist the cutting action and to produce a good surface finish.

10. Machine the beveled shoulder until it is the required size.

If the size of the shoulder is large, and chatter occurs when cutting with the side of the toolbit, it may be necessary to cut the beveled shoulder using the compound rest, figure 14-4.

1. Set the compound rest to the desired angle.

2. Adjust the toolbit so that only the point does the cutting.

3. Machine the bevel by feeding the compound rest by hand.

FIG. 14-4 THE COMPOUND REST SWIVELED TO CUT A LARGE BEVELED SHOULDER

To minimize the possibility of cutting into the small diameter, all cuts should be taken from the center outward.

Procedure: Machining a Filleted Shoulder

Filleted shoulders are generally used on parts which are subject to strain at the shoulder. The rounded corner is pleasing in appearance and also strengthens the shaft at this point without an increase in the diameter of the part.

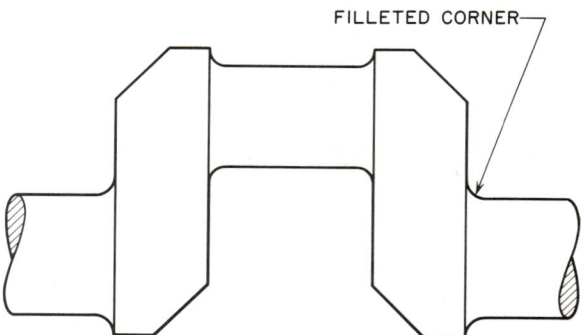

FIG. 14-5A AN APPLICATION OF FILLETED CORNERS ON A CRANKSHAFT

FIG. 14-5B A RADIUS FORM TOOL BEING USED TO PRODUCE A FILLETED SHOULDER

1. Lay out or mark the location of the shoulder on the workpiece. *Note:* When laying out for a filleted shoulder, it is important that allowance be made for the radius to be cut. For example, if a filleted

shoulder having a 1/8" radius is 2 1/2" from end of workpiece, the layout should be 2 3/8" from the end (2 1/2 - 1/8) to leave material for the cutting the radius.

2. Rough and finish turn the small diameter to size.

3. Mount a radius tool in the toolholder and set it to center. Check the toolbit with a radius gage to be sure that it has the correct radius.

4. Apply layout dye or chalk to the small diameter as close as possible to the shoulder location.

5. Set the lathe spindle speed to approximately one-half of the turning speed.

6. Start the lathe and bring the toolbit in until it just removes the layout dye or chalk.

7. *Note the reading* on the graduated collar of the crossfeed screw.

8. Retract the cutting tool by turning the crossfeed handle *counterclockwise one-half turn.*

9. Turn the crossfeed handle clockwise until it is within .030 of the original collar setting. *Note:* The point of the round nose toolbit should now be .030 away from the work diameter. This prevents the cutting tool from undercutting while roughing out the filleted corner.

10. Turn the carriage handwheel *slowly* to start the radius tool cutting the filleted shoulder.

    If chattering occurs while machining the filleted corner, reduce the lathe speed and apply cutting fluid to improve the finish of the fillet.

11. Continue turning the carriage handwheel *slowly* and carefully until the length of the of the shoulder is correct. *Note:* When stopping the lathe to measure the shoulder distance, *do not move the cutting tool setting* by withdrawing it from the diameter.

12. Turn the carriage handwheel to move the cutting tool away from the shoulder slightly.

13. Turn the crossfeed handle clockwise .030 (back to the original collar setting).

14. Finish the filleted corner by carefully advancing the radius toolbit with the carriage handwheel.

    If the radius is too large for a form toolbit, or too much chattering occurs, cut the fillet in steps, using the largest radius tool possible, or use a gooseneck toolholder. Check the accuracy of the fillet with a radius gage.

REVIEW QUESTIONS

1. What is a shoulder?

2. What are the three common shoulders used in machine shop work?

Square Shoulders

3. What is the purpose of a square shoulder?

4. What two methods are used to lay out the position of a shoulder?

5. How should the toolbit be set up for squaring a shoulder?

6. What is one method of setting the cutting tool to the small diameter when starting to square a shoulder?

Section 2   Machining Between Centers

Beveled or Angular Shoulders

7. What is the purpose of a beveled or angular shoulder?

8. Describe two methods of cutting beveled or angular shoulders.

9. What lathe speed should be used when machining a beveled or angular shoulder?

Filleted Shoulders

10. What is the purpose of filleted shoulders and where are they used?

11. What precaution should be observed when laying out the length for a filleted shoulder?

12. How close to the work diameter should a radius tool be set for roughing out a filleted corner?

13. How can chattering be overcome when machining a filleted shoulder?

14. What part of the lathe is used for feeding the toolbit lengthwise?

Unit 15 FILING AND POLISHING

The most *accurate* and *desirable* method for producing a true diameter on a workpiece is to machine it to the desired size using a cutting tool with a sharp, honed cutting edge. The most common industrial practice, when accuracy and surface finish are important, is to turn the diameter close to the required size first, and then to grind it to the size and finish desired.

FILING

Filing on a lathe is performed to remove burrs and sharp edges, reduce the diameter, and to improve the surface finish by removing the cutting tool marks. Filing a diameter on a lathe should be kept to a minimum, since too much filing tends to produce a diameter which is out-of-round. Whenever filing is used to finish work to size, no more than .002 or .003 should be left on the diameter. This should be enough to remove the cutting tool marks without filing the work out-of-round. While filing is permissible and desirable in some cases, the diameter should be turned to size with a keen cutting tool when a true cylindrical object is required.

Types of Lathe Files

Many files can be used on a lathe. Some are especially designed for certain operations and specific metals. The shape of the section to be filed, the type of metal being filed, and the surface finish desired are factors to be considered when selecting a file.

The mill file, figure 15-1A, which tapers in width towards the point, may be used for general-purpose filing on lathe work.

The long-angle lathe file, figure 15-1B, is parallel in both width and thickness, and has two "safe" (uncut) edges. It is designed especially for fine finishing and filing up to a shoulder. The long angle of the teeth tends to eliminate file chatter, provides for the rapid clearing of chips, and reduces the possibility of tearing the metal.

FIG. 15-1A A MILL FILE (Courtesy Nicholson File Co.)

FIG. 15-1B LONG-ANGLE LATHE FILE (Courtesy Nicholson File Co.)

Section 2   Machining Between Centers

The aluminum file, figure 15-1C, has a special tooth construction which is very effective in clearing chips and eliminating clogging. It has a *deep upcut* and a *fine overcut* which tend to break up the filings, allow the file to clear itself, and overcome chatter.

The brass file, figure 15-1D, has a *deep, short upcut angle* with a *fine, long-angle overcut* which tends to produce small scallops, break up the filings, and enables the file to clear itself of chips. It is very effective on tough, ductile metals such as brass and bronze.

The super-shear file, figure 15-1E, has curved teeth cut in an arc that is off-center in relation to the file axis. It has a "double-purpose" milled tooth and is very effective for most nonferrous metals such as aluminum, brass, babbitt, and copper.

Procedure: Preparing the Lathe

As in most other operations, the setup of the lathe is important to efficient filing.

1. Cover the lathe bed with paper. *Note:* This prevents the small chips (filings) from getting into the machine slides and causing excessive wear and damage to the ways of the lathe. A cloth is not advisable since it may become caught in the revolving lathe.

2. Set the lathe spindle at twice the turning speed.

   If the work revolves too slowly, there is a danger of filing the diameter out-of-round. The lathe should be set for top speed when filing small diameters.

3. Disengage the feed rod and lead screw.

   This eliminates the chances of the lathe carriage being engaged accidentally while the lathe is operating at a high speed.

Filing Hints

The following hints are offered so that when it is necessary to file on a lathe, it can be performed efficiently and safely.

FIG. 15-1C   FLAT ALUMINUM FILE (Courtesy Nicholson File Co.)

FIG. 15-1D   FLAT BRASS FILE (Courtesy Nicholson File Co.)

FIG. 15-1E   SUPER-SHEAR FILE (Courtesy Nicholson File Co.)

1. Never use a file without a handle.

   This is especially important in lathe work since the revolving workpiece tends to push the file towards the operator.

2. Make a practice of filing left-handed, figure 15-2.

   This method is recommended by safety councils so that the hands and arms are clear of the revolving driveplate or chuck.

3. Run a *new file* over a flat piece of cast iron before using it on lathe work.

   This removes the extreme sharpness from the top edges of the teeth. This is only necessary when using a new file on work requiring a very smooth finish.

4. Apply enough pressure on the *forward stroke* of the file to make the teeth cut.

   Do not apply pressure or lift the file on the return stroke. If pressure is applied on the return stroke, the file teeth dull very quickly.

5. Use long, slow strokes when filing in a lathe.

   Approximately 30 to 40 strokes per minute are recommended. Fast, short strokes tend to produce out-of-round diameters.

6. Never apply oil or rub the hand over a surface to be filed.

   Oil makes it difficult for the file to cut properly, the oil tends to accumulate in the file teeth, and the filings become stuck in the teeth. These filings cause scratches in the surface being filed.

7. Keep the file teeth clean and free of chips by using a file card. This not only prolongs the life of the file, but also produces a better surface finish. Chalk rubbed into the teeth tends to prevent clogging and facilitates cleaning.

8. Check the lathe center tension occasionally when filing. The work revolving at a high speed causes friction between the work and the dead center. If the center tension is not checked occasionally, the center may be burned and the work ruined.

Procedure: Filing in a Lathe

Before attempting to file in a lathe, be sure to study the sections on "Preparing the Lathe" and "Filing Hints."

1. Mount the work between lathe centers and carefully adjust the center tension. If possible, use a revolving dead center when filing work between centers.

2. Cover the lathe bed with paper.

3. Set the spindle at approximately twice the turning speed.

4. Move the carriage to the right as far as possible.

5. Disengage the feed rod and lead screw.

6. Select the proper file to suit the workpiece material. *Note:* Be sure the file has a properly fitted handle.

7. Start the lathe spindle revolving.

8. Hold the point of the file in the right hand and grasp the handle in the left hand, figure 15-2.

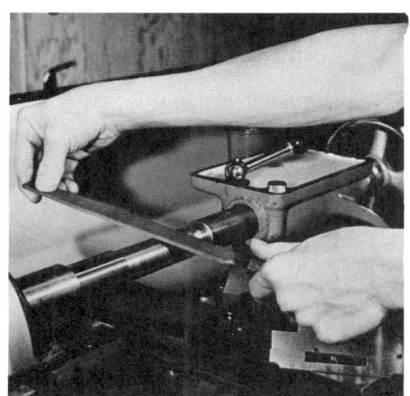

FIG. 15-2 THE LEFT-HAND METHOD OF FILING IN A LATHE IS SAFER.

Section 2 Machining Between Centers

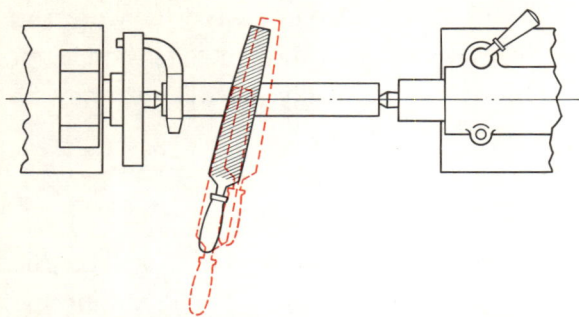

FIG. 15-3 EACH STROKE OF THE FILE SHOULD OVERLAP THE PREVIOUS STROKE BY ONE-HALF THE FILE WIDTH.

9. Use a long forward stroke while applying light pressure on the file. Do not lift the file or apply pressure on the return stroke.
10. Each stroke of the file should overlap the previous stroke by approximately one-half the file width, figure 15-3.
11. Clean the file frequently to keep it from clogging and scratching the work surface.

POLISHING

Polishing is a finishing operation performed with abrasive cloth to improve the appearance of a workpiece. It usually follows the filing operation and produces a very high mirrorlike finish on the diameter. The finish obtained is directly related to the coarseness of the abrasive cloth used while polishing.

Both aluminum oxide and silicon carbide abrasive cloth are available in various grit sizes to suit the finish desired and metal to be polished. *Aluminum oxide* abrasive cloth should be used for polishing steel and most ferrous metals. *Silicon carbide* abrasive cloth should be used for polishing aluminum, brass, and most nonferrous metals. Use 80- to 100-grit abrasive cloth for rough polishing, and 200 grit or finer for finish polishing.

After the work has been finish-polished, a very high gloss can be achieved on the workpiece by using a piece of abrasive cloth which is practically worn out. A few drops of oil on this worn abrasive cloth greatly improves the gloss of the surface finish.

Procedure: Polishing a Diameter

1. Cover the lathe bed with paper to prevent the abrasive dust from entering the lathe slides.
2. Set the lathe at the highest spindle speed.
3. Select a piece of abrasive cloth 1" wide and 10" long to suit the material to be polished and to produce the desired finish.
4. Disengage the feed rod and lead screw.
5. Check the tension of the workpiece between centers. *Note:* Continually check the center tension when polishing in a lathe. Use a revolving center in the tailstock if one is available.
6. Start the lathe and hold the abrasive cloth as shown in figure 15-4.
7. With the fingers of one hand, press the cloth against the diameter while *securely* holding the other end of the abrasive cloth with the other hand, figure 15-4.
8. Slowly move the abrasive cloth back and forth over the surface being polished.
9. Apply a few drops of oil to a piece of worn-out abrasive cloth to obtain a glossy surface.

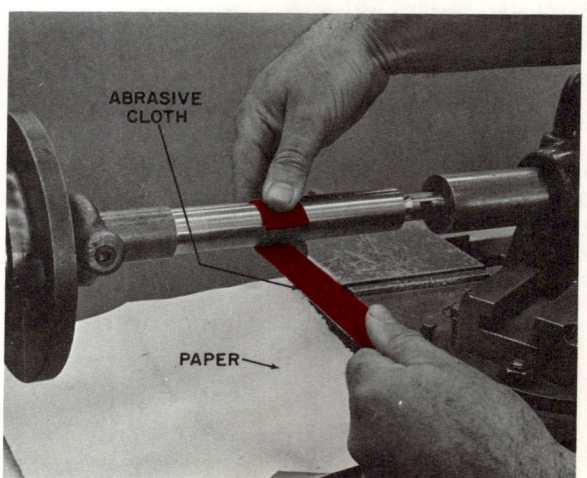

FIG. 15-4 METHOD OF HOLDING ABRASIVE CLOTH WHEN POLISHING A DIAMETER

Polishing With Abrasive Cloth Under a File

Another method of polishing is by holding a piece of abrasive cloth under a file, figure 15-5, and using strokes similar to filing strokes. The abrasive cloth should be slightly wider than the file to insure that the file does not contact the work surface. This method allows more pressure to be applied during polishing, insures parallel diameters, and allows polishing to be performed in a specific area.

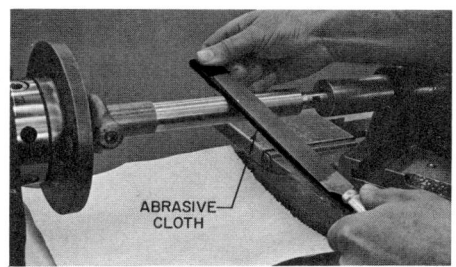

FIG. 15-5 MORE PRESSURE CAN BE APPLIED DURING POLISHING BY HOLDING A PIECE OF ABRASIVE CLOTH UNDER A FILE.

REVIEW QUESTIONS

1. What is the most accurate method of producing a true diameter on a lathe?

Filing

2. Why should filing on a diameter be kept to a minimum?

3. How much material should be left on a diameter for filing?

Types of Lathe Files

4. What factors should be considered when selecting a file?

5. Why is a long-angle lathe file desirable when filing on a lathe?

6. Describe each of the following items. What is the purpose of each item?

   a. The aluminum file b. The super-shear file

Preparing the Lathe

7. Why is it advisable to cover the lathe bed with paper when filing?

8. At what speed should the lathe be set for filing? Explain why.

Filing Hints

9. Why is the left-hand method of filing preferred on a lathe?

10. Why should pressure be applied only on the forward stroke of the file?

11. Why are fast, short strokes not recommended when filing a diameter?

12. Why is it important that oil never be applied to, or the hands rubbed over, a surface being filed?

13. Why should the lathe center tension be checked regularly?

Filing on a Lathe

14. What procedure is followed when using a file on a lathe?

15. How can a file be kept clean and free of chips?

Polishing

16. What is the purpose of polishing?

17. What type of abrasive cloth should be used for polishing

    a. Ferrous metals?   b. Nonferrous metals?

Polishing a Diameter

18. At what speed should a lathe be set for polishing?

19. How can a glossy surface be obtained by polishing?

20. What is the purpose of polishing with a piece of abrasive cloth under a file?

Unit 16 KNURLING

A. DIAMOND PATTERN   B. STRAIGHT LINE PATTERN

FIG. 16-1 TYPES AND PITCHES OF KNURLS. (Courtesy J. H. Williams and Co.)

Knurling is the process of impressing a raised diamond or straight-line pattern on the surface of a cylinder. The diamond pattern is formed by two hardened rolls, one having a right-hand and the other, a left-hand helix. These rolls overlap to create a diamond-shaped pattern, figure 16-1A. Diamond knurling is used to provide a suitable grip and to improve the appearance of the workpiece. Straight-line knurls are formed by two hardened steel rolls with straight grooves cut around the periphery parallel to the axis, figure 16-1B. Straight-line knurls are often used to increase the diameter of a workpiece when a press fit is required. Both types of knurls are available in coarse (14 pitch), medium (21 pitch), and fine (23 pitch) patterns.

TYPES OF KNURLING TOOLS

The knurling tool has a heat-treated body or shank with hardened knurl rolls mounted on hardened steel pins in a movable head.

The knurling tool shown in figure 16-2 has one set of rolls mounted in a self-centering head. The second type, figure 16-3, has three different sets of rolls (coarse, medium, and fine) mounted in a revolving head which pivots on a hardened steel pin. This is also a self-centering unit.

Procedure: Knurling a Diameter

Since knurling is a displacement process and not a cutting process, great pressure is required to make the knurls penetrate the surface of the workpiece; consequently, care

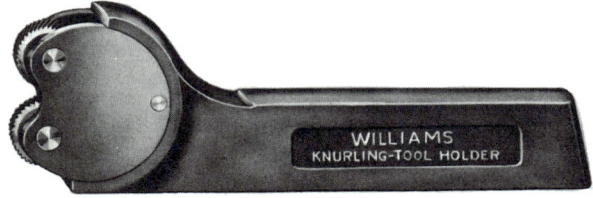

FIG. 16-2 A KNURLING TOOL WITH ONE PAIR OF ROLLERS IN A SELF-CENTERING HEAD. (Courtesy J. H. Williams and Co.)

FIG. 16-3 A KNURLING TOOL WITH THREE SETS OF ROLLS MOUNTED IN A REVOLVING HEAD. (Courtesy J. H. Williams and Co.)

Section 2 Machining Between Centers

FIG. 16-4 SETTING THE KNURLING TOOL ON CENTER

should be taken to set up the knurling tool and the workpiece properly.

1. Check the centers of the work and the machine to see that they are in good condition. The centers of the work should be deep enough to provide good support for the workpiece.

2. Set up the knurling tool in the toolpost with the center of the floating head even with the lathe center point, figure 16-4. *Note:* If the center of the floating head is center punched on the right-hand side, the knurling tool can easily be set on center.

   If a knurling tool with a revolving head is used, the center of the pivot point should be set to center height.

3. Adjust the knurling tool so that it is at right angles (90°) to the axis of the work and tighten it securely, figure 16-5. *Note:* If the knurling rolls become worn, it is advisable to offset the knurling tool slightly so that the edge of the rolls start the pattern

4. Mount the work between centers, being careful to adjust the center tension properly. *Note:* Work held in a chuck should be supported with the dead center whenever possible.

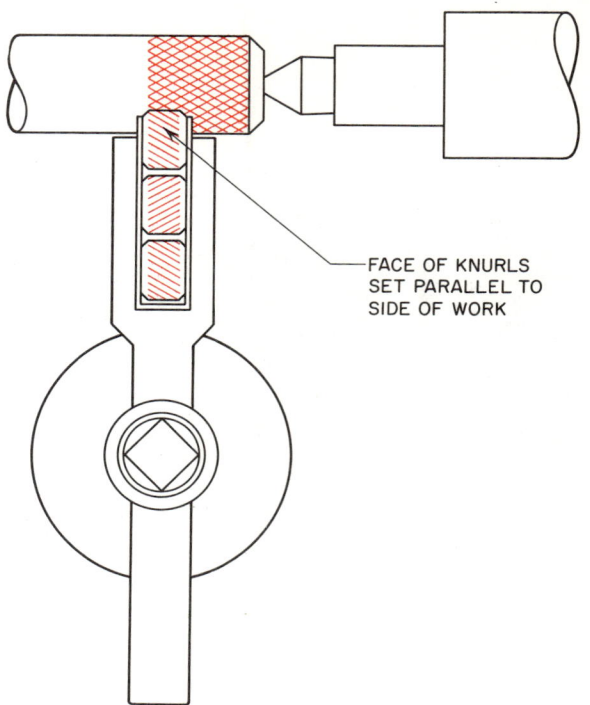

FIG. 16-5 THE KNURLING TOOL SHOULD BE SET AT 90° TO THE CENTERLINE OF THE WORKPIECE.

5. Mark off the length to be knurled.

6. Set the spindle speed to one-quarter of the turning speed.

7. Set the quick-change gearbox for .020-.030 feed.

8. Move the carriage to the right until about one-half the face of the knurling roll overlaps the end of the workpiece, figure 16-6.

FIG. 16-6 THE KNURLING ROLLS SHOULD OVERLAP THE END OF THE WORKPIECE.

This setting allows the knurl to start evenly and produces a true pattern. When the knurling tool is overlapped in this manner, usually only one pass is required.

9. Feed the crossfeed handle in until the knurling rolls touch the work.
10. Start the machine and feed the knurling tool in .025. This should bring the knurled pattern to a point.

or

Feed the knurling tool in .025 and then start the machine.

11. Stop the lathe to see if the knurling pattern is correct, figure 16-7; if necessary, reset the knurling tool.

    If the pattern is incorrect, figure 16-7, it is generally because the knurling tool is not set on center which results in more pressure being applied to one roll.

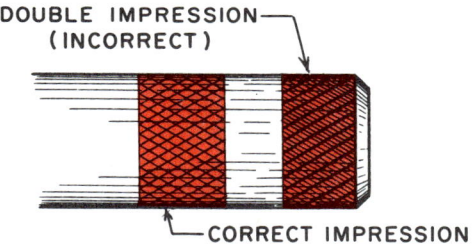

FIG. 16-7 CORRECT AND INCORRECT KNURLING PATTERNS

12. Apply oil to the length to be knurled; this washes away any metal particles, and provides lubrication for the knurling rolls.
13. Engage the automatic feed and knurl for the proper length. *Note:* Do not disengage the automatic feed until the full length of the work has been knurled, otherwise rings are formed on the knurled pattern which are difficult or impossible to remove, figure 16-8.
14. If the knurl is not deep enough, stop the machine, reverse the automatic feed and take another pass across the work until the knurl comes to a point.

FIG. 16-8 KNURLED PATTERN WILL BE DAMAGED IF THE FEED IS SHUT OFF BEFORE THE FULL LENGTH IS KNURLED.

KNURLING CALCULATIONS

When it is important that the diameter of a knurled section be a specific size, the work must be machined to a somewhat smaller diameter to allow for the metal displacement and resultant increase in diameter which occurs during knurling. The diameter of the workpiece, prior to knurling, can be calculated when the number of serrations around the circumference are known.

1. Calculate the number of serrations there will be on the circumference of the workpiece using the formula:

   Pitch of knurl x 3.1416 x (diameter of workpiece − .017)

2. The diameter of the workpiece is then calculated as follows:

   Work diameter = .015 x number of serrations on the workpiece.

   Example: Calculate the diameter of a section which will be 1″ diameter when knurled. A medium knurl (21 pitch) is to be used.

   Number of
   serrations = 21 x 3.1416 x (1″ − .017)
              = 65.973 x .983
              = 64.85 or 65 serrations
   Diameter   = .015 x 65
              = .975 diameter

Section 2   Machining Between Centers

REVIEW QUESTIONS

1. Define knurling.
2. What is the purpose of:
   a. A diamond pattern knurl?
   b. A straight pattern knurl?

Types of Knurling Tools

3. In what pitches are knurling tools available?
4. Describe two types of knurling heads.

Knurling a Diameter

5. What does the statement "knurling is a displacement process" mean?
6. What precaution should be observed regarding centers when knurling?
7. What is the procedure, in outline form, for setting up the knurling tool so that the correct diamond pattern is assured?
8. How should the knurling tool be set if the knurls are worn?
9. What precautions should be observed when knurling work held in a chuck?
10. What speed and feed should be used for knurling?
11. How should the knurl be started so that it is to the proper depth in one pass?
12. To what depth should the knurling tool be fed?
13. Why is it important that the feed not be disengaged while knurling?
14. To what diameter should a piece of work be turned so that a 21-pitch knurl finishes to 7/8 diameter?

Unit 17 GROOVING

Grooving, sometimes called *necking, recessing,* or *undercutting,* is the process of cutting a grooved form or channel on a cylinder. The shape of the cutting tool and the depth to which it is fed determines the shape of the groove. Square, round, or V-shaped grooves are the most common types found in machine shop work.

TYPES OF GROOVES

Square grooves, figure 17-1A, are frequently cut at the end of the section to be threaded in order to provide a channel into which the threading tool may run. If the groove is cut against a shoulder, it allows the mating part to fit squarely against the shoulder. When a diameter is to be finished to size by grinding, a groove is generally cut against the shoulder to provide clearance for the grinding wheel and to insure a square corner. Square grooves may be cut with a cutoff (parting) tool or a toolbit, figure 17-1A, ground to the size required.

Round grooves, figure 17-1B, serve the same purpose as square grooves and are generally used on parts subject to strain. The round groove eliminates the sharpness of a square corner and strengthens the part at the point where it tends to fracture. A round-nose (radius) toolbit, figure 17-1B, ground to the desired radius, is used to cut round grooves.

V-shaped grooves, figure 17-1C, are most commonly found on pulleys driven by V-belts. The V-shaped groove eliminates much of the slippage which occurs in other forms of belt drives. The V-groove can also be cut at the end of a thread to provide a channel into which the threading tool may run. A toolbit ground to the desired angle, figure 17-1C, can be used to cut shallow V-grooves. Larger V-grooves, such as those found on pulleys, should be cut with the lathe compound rest set to form each side individually.

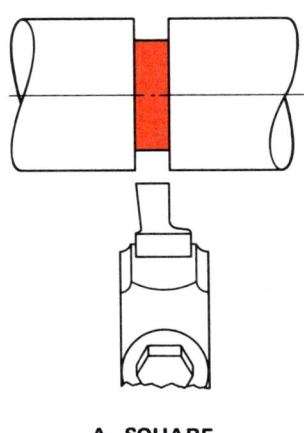
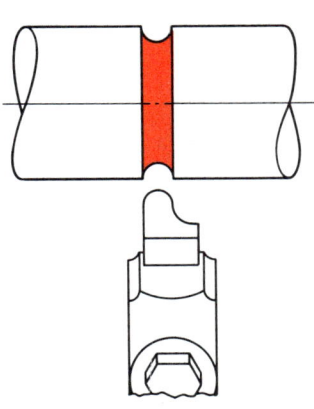
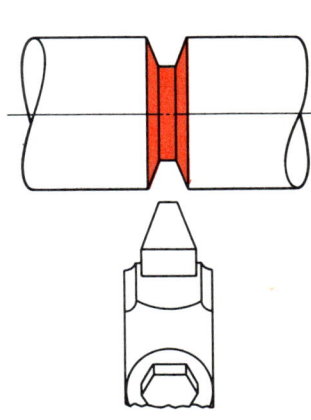

A. SQUARE   **B. ROUND**   **C. V-SHAPED**

FIG. 17-1 COMMON TYPES OF GROOVES

Section 2 Machining Between Centers

FIG. 17-2 EXERCISE CARE WHEN GROOVING BETWEEN CENTERS SINCE THE WORK MAY BEND AND ALSO BREAK THE CUTTING TOOL.

Procedure: Cutting a Groove

The operation of grooving reduces the diameter and tends to weaken the workpiece at that point; therefore, care should be exercised during the grooving operation. This is especially true when grooving work held between centers since the workpiece is weakened and may bend at the groove, ruining the workpiece and damaging the cutting tool, figure 17-2.

Since the shape of the groove is generally governed by the cutting tool shape, the cutting of various grooves is similar and only a general operation of grooving is outlined.

1. Grind the toolbit to the size and shape of the groove required.

2. Mount the toolbit in a toolholder and set the cutting tool height to center.

3. Lay out the location of the groove on the workpiece.

4. Set the lathe to approximately one-half the speed used for turning. *Note:* This is important when grooving since more of the cutting tool edge is in contact with the workpiece and, therefore, chatter tends to occur.

5. Mount the workpiece and carefully adjust the lathe center tension.

6. Start the lathe and carefully bring the cutting tool against the workpiece until it lightly marks the diameter.

7. Note the reading on the graduated collar of the crossfeed screw handle.

8. Calculate the depth of groove by subtracting the groove diameter from the work diameter and dividing this amount by two. This indicates the amount the toolbit must be fed to cut the groove to the required depth.

9. Feed the grooving tool into the work *slowly* by turning the crossfeed handle.

   To insure good surface finish and provide efficient cutting action, apply cutting fluid while the toolbit is being fed into the work.

10. To overcome chatter, it is good practice to move the carriage *slightly* to the left and right as the grooving tool is fed into the work.

    If chattering still occurs, reduce spindle speed until the chatter is eliminated.

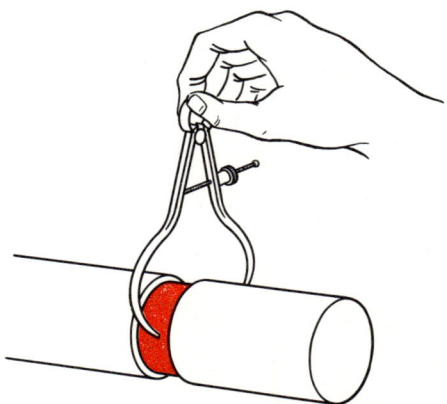

FIG. 17-3 CHECKING THE DIAMETER OF A GROOVE WITH OUTSIDE CALIPERS

Unit 17 Grooving

11. Continue to turn the crossfeed handle until the graduated collar indicates the cutting tool is at the required depth.
12. Stop the machine and check the groove diameter with outside calipers or a knife-edge vernier, figure 17-3. *Note:* It is considered good safety practice to wear safety goggles always when metal is being cut. This is especially true when grooving on a lathe.

REVIEW QUESTIONS

1. Define the operation of grooving.

Types of Grooves

2. What are three purposes of square grooves?
3. What type of cutting tool may be used to cut a square groove?
4. What advantage has the round groove over the square groove?
5. Where are V-shaped grooves generally used?
6. How can large V-grooves be cut on a lathe?

Cutting a Groove

7. Why should care be exercised especially when grooving between centers?
8. At what speed should the lathe be set when grooving? Explain why.
9. Compute how much the toolbit must be fed to cut a 7/8" diameter groove in a 1.115-diameter workpiece.
10. What are two methods of overcoming chatter while grooving?

Unit 18   TAPER TURNING

A *taper* can be defined as a uniform change in the diameter of a workpiece, measured along its axis. Tapers provide a rapid and accurate means of aligning and holding machine parts. Since tapers permit the interchangeability of certain cutting tools or attachments with extreme accuracy, they have contributed to economical industrial practice. Some milling cutters, twist drills, reamers, and lathe centers have tapered shanks which fit into the tapered spindle of a machine tool.

Tapers may be external or internal and are expressed in taper per foot, taper per inch, or degrees. Those used on machines and tools (machine tapers) are classified by the American Standards Association as *self-holding tapers*, and *self-releasing* or *steep tapers*.

FIG. 18-1 THE TANG OF A SELF-HOLDING TAPER PREVENTS SLIPPAGE.

FIG. 18-2 A STEEP TAPER WITH A DRAW BOLT TO HOLD THE ACCESSORY IN ENGAGEMENT

SELF-HOLDING TAPERS

Self-holding tapers, figure 18-1, are those which, when properly seated, remain in position due to the wedging action of the taper. The self-holding taper series is composed of the Brown and Sharpe taper, the Morse taper, and the 3/4" per foot machine tapers, table 18-1. Although the wedging action of the taper provides some frictional drive, the smaller sizes of self-holding tapers have a tang to prevent the tapered adaptor from slipping in the spindle of the machine. Larger tapers generally have keys or draw-in bolts to secure the taper in the spindle.

STEEP OR SELF-RELEASING TAPERS

Steep tapers, figure 18-2, formerly known as milling machine tapers, are used for alignment of machine parts such as milling machine arbors and accessories. The drive to the accessory is provided by means of two diagonally opposite keys, which engage in slots in the collar of the adaptor. Since this taper (3 1/2" per foot) is not a self-holding type, the adaptor is held securely in the spindle by means of a drawbolt.

Other applications of self-releasing tapers may be found on type L and type D-1 lathe spindle noses. The type L spindle nose, figure 18-3A, has a long taper of 3 1/2" per foot. It employs a key drive, and a threaded ring secures the chuck or driveplate to the spindle.

The type D-1 spindle nose, figure 18-3B, has a very short taper (3" taper per foot). The chuck or driveplate is positioned to the spindle by the taper and is secured to the spindle by three cam-locking devices.

Unit 18 Taper Turning

SELF-HOLDING TAPERS (Basic Dimensions)					
Taper Number	Taper per Foot	Gage Line Diameter (A)	Small End Diameter (D)	Length (P)	Origin of Series
.239	.502	.2392	.200	15/16″	Brown & Sharpe Taper Series
.299	.502	.2997	.250	1 3/16″	
.375	.502	.3752	.3125	1 1/2″	
*0	.624	.3561	.252	2″	Morse Taper Series
1	.5986	.475	.369	2 1/8″	
2	.5994	.700	.572	2 9/16″	
3	.6023	.938	.778	3 3/16″	
4	.6233	1.231	1.020	4 1/16″	
4 1/2	.624	1.500	1.266	4 1/2″	
5	.6315	1.748	1.475	5 3/16″	
6	.6256	2.494	2.116	7 1/4″	
7	.624	3.270	2.750	10″	
200	.750	2.000	1.703	4 3/4″	3/4″ Taper per Foot Series
250	.750	2.500	2.156	5 1/2″	
300	.750	3.000	2.609	6 1/4″	
350	.750	3.500	3.063	7″	
400	.750	4.000	3.516	7 3/4″	
450	.750	4.500	3.969	8 1/2″	
500	.750	5.000	4.422	9 1/4″	
600	.750	6.000	5.328	10 3/4″	
800	.750	8.000	7.141	13 3/4″	
1000	.750	10.000	8.953	16 3/4″	
1200	.750	12.000	10.766	19 3/4″	

*Taper No. 0 is not a part of the self-holding taper series. It has been added to complete the Morse Taper Series.

TABLE 18-1

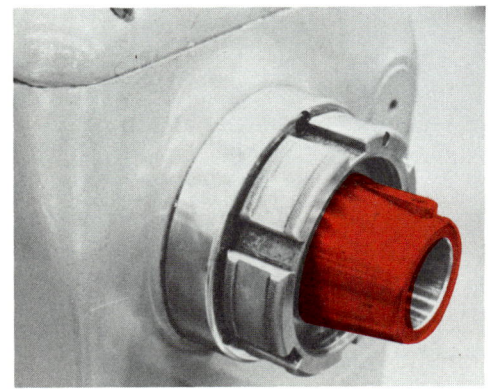

FIG. 18-3A   THE TYPE L SPINDLE NOSE
(Courtesy McGraw-Hill)

FIG. 18-3B   THE TYPE D-1 SPINDLE NOSE
(Courtesy McGraw-Hill)

Section 2 Machining Between Centers

NONSTANDARD TAPERS

Although many of the tapers listed in table 18-1 are taken from the Brown and Sharpe and the Morse taper series, there are several other tapers which are now listed as nonstandard tapers. These include the Jarno taper, the taper pin series, and some of the Brown and Sharpe and Morse tapers.

The Brown and Sharpe taper series, ranging in sizes from No. 4 to No. 18, has .502 taper per foot, excepting No. 10 which has .516 taper per foot. This taper is used on Brown and Sharpe machines and accessories.

The Morse taper, (approximately 5/8" taper per foot), is probably the most widely used taper in the machine industry. It is available in 8 sizes ranging from No. 0 to No. 7. This taper is commonly found on drills, reamers, and lathe center shanks.

The Jarno taper, (.600 taper per foot), is probably the most sensible taper, but has never been accepted to any extent by the machine tool industry, probably due to the high cost of changeover. This taper ranges in size from No. 2 to No. 20. The taper number indicates the large diameter in eighths of an inch and the small diameter in tenths of an inch. The length of the taper is indicated by the taper number divided by two.

Standard taper pins, figure 18-4, having 1/4" taper per foot, are used to assemble parts which must be held and positioned exactly and yet can be easily dismantled. These pins range in size from No. 7/0 to No. 14.

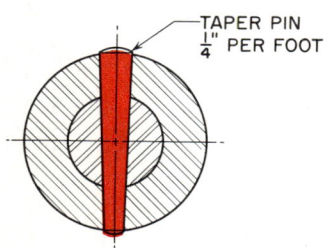

FIG. 18-4 TAPER PINS ALLOW MACHINE PARTS TO BE ASSEMBLED QUICKLY AND ACCURATELY.

TAPER CALCULATIONS

In order to set up the lathe to machine a taper accurately, it is often necessary to first calculate the *taper per foot* or the *taper per inch* of the tapered piece. Taper per foot is the amount of difference between the large and small diameter in 12" of length.

Calculating the Taper per Foot

Since all tapers are not 12" long, if the large diameter, small diameter, and length of taper are known, the taper per foot can be calculated by using the formula:

$$\text{T.P.F.} = \frac{(D - d) \times 12}{\text{T.L.}}$$

Where   D = large diameter
         d = small diameter
         T.L. = taper length

Example: Calculate the taper per foot for the piece of work shown in figure 18-5.

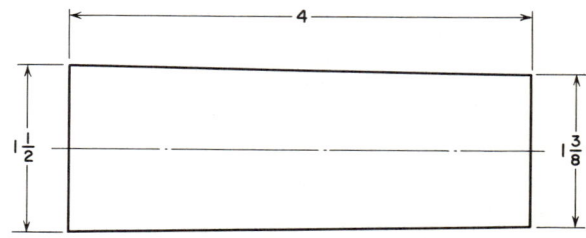

FIG. 18-5   A TAPERED PIECE OF WORK 4" LONG

Solution:
$$\text{T.P.F.} = \frac{(D - d) \times 12}{\text{T.L.}}$$
$$= \frac{(1\,1/2 - 1\,3/8) \times 12}{4}$$
$$= \frac{1/8 \times 12}{4}$$
$$= 3/8''$$

After the taper per foot has been calculated, no further calculations are required if the taper is to be turned by means of the taper attachment. However, if the taper is to be turned by the offset tailstock method, further calculations are required.

Unit 18 Taper Turning

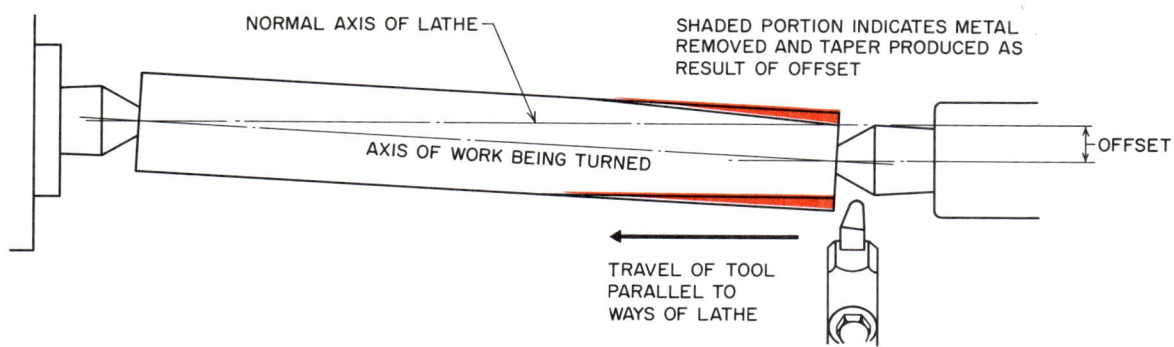

FIG. 18-6 WORK IS OFFSET ONLY ONE-HALF THE AMOUNT OF TAPER IN THE WORKPIECE.

Calculate the Tailstock Offset

To calculate the amount the tailstock must be offset in order to cut a taper, the taper per foot (or the taper per inch) and the overall length of the work must be known. If the T.P.I. is known, the following formula can be applied:

Tailstock offset $= \dfrac{\text{T.P.I.} \times \text{O.L.}}{2}$ (fig. 18-6)

Where T.P.I. = Taper per inch
O.L. = Overall length of workpiece

If T.P.F. is given, convert T.P.F. to taper per inch as follows: T.P.F./12

The formula then will be:

Tailstock offset $= \dfrac{\text{T.P.F.} \times \text{O.L.}}{12} \times \dfrac{1}{2}$

$= \dfrac{\text{T.P.F.} \times \text{O.L.}}{24}$

Problem: Calculate the tailstock offset for the workpiece shown in figure 18-7.

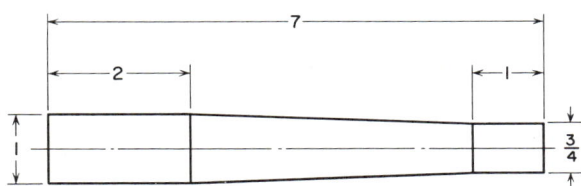

FIG. 18-7 TAPER DIMENSIONS REQUIRED IN ORDER TO CALCULATE THE AMOUNT OF TAILSTOCK OFFSET.

Large diameter = 1″
Small diameter = 3/4″
Length of tapered section = 4″
Overall length = 7″

1. Calculate the taper per foot.

$\text{T.P.F.} = \dfrac{(D - d) \times 12}{\text{T.L.}}$

$= \dfrac{(1 - 3/4) \times 12}{4}$

$= \dfrac{1/4 \times 12}{4}$

$= 3/4″$

2. Calculate the tailstock offset.

Tailstock offset $= \dfrac{\text{T.P.F.} \times \text{O.L.}}{24}$

$= \dfrac{3/4 \times 7}{24}$

$= \dfrac{7″}{32}$

$= .219$

When it is not necessary to know the taper per foot, the following simplified formula can be applied to calculate the amount of tailstock offset.

Tailstock offset $= \dfrac{\text{O.L.} \times (D - d)}{\text{T.L.} \quad 2}$

This formula can be applied to the example in figure 18-7 as follows:

Tailstock offset $= \dfrac{7 \times 1/4}{4 \quad 2}$

$= 7/32$

$= .219$

Section 2 Machining Between Centers

TAPER TURNING

Tapers can be turned on a lathe by three methods:

- By use of the taper attachment which has been set to the desired taper per foot or to the angle required
- By offsetting the tailstock
- By setting the compound rest to the angle of the taper

Taper Attachments

All taper attachments operate on the same principle although the construction and adjustment may vary with the make. Basically, a guide bar A and a sliding block, B, which is connected to the cross-slide, guide the cutting tool as it moves along the work, thus creating a taper on the workpiece, figure 18-8.

The taper attachment provides a quick and accurate method of turning tapers with several advantages over other methods.

- The lathe centers remain in line, eliminating the need to realign the tailstock center for straight-turning operations.
- The center holes in the workpiece are not distorted.
- Tapers can be produced on work held between centers or in a chuck.

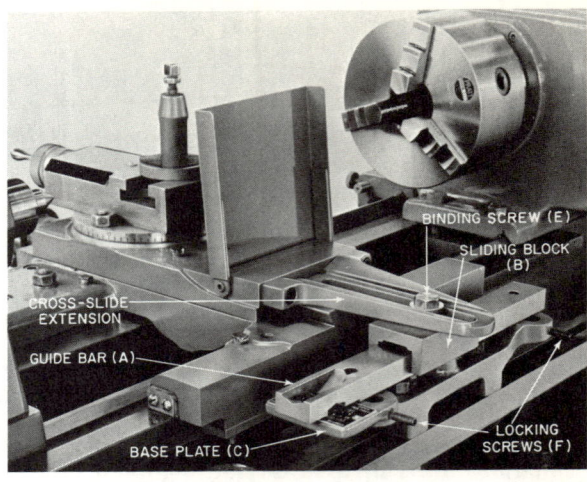

FIG. 18-9 PARTS OF A PLAIN TAPER ATTACHMENT.
(Courtesy Standard Modern Tool Co.)

- A wider range of tapers (up to 3 1/2" per foot) can be produced.
- Matching external and internal tapers can be easily machined with one setting of the taper attachment.
- The length of the work does not have to be considered, since once the taper attachment is set, the same taper per foot (or angle) is turned on any length of workpiece.
- Since the taper attachment is graduated in inches per foot or in degrees of taper, the need for further calculations is eliminated.
- Tapers can be cut on long workpieces.

Taper attachments are supplied in two types:

1. The *plain taper attachment,* figure 18-9
2. The *telescopic taper attachment,* figure 18-10

Procedure: Turning a Taper Using the Taper Attachment

The procedure used to turn a taper using either a telescopic or a plain taper attachment is basically the same with only a few minor changes. When a plain attachment is used, the

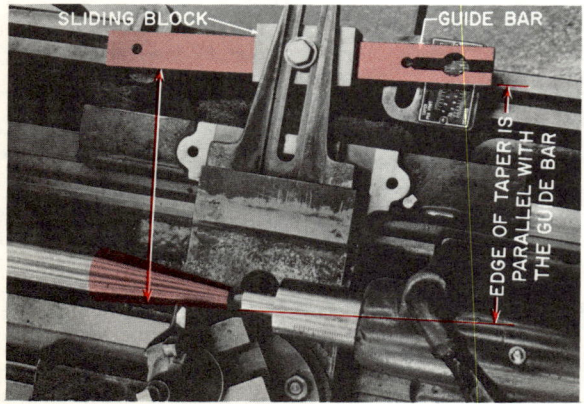

FIG. 18-8 PRINCIPLE OF A TAPER ATTACHMENT

104

Unit 18 Taper Turning

FIG. 18-10 A TELESCOPIC TAPER ATTACHMENT.
(Courtesy Cincinnati Milacron Co.)

binding screw which connects the cross-slide to the crossfeed screw nut must be removed. The screw is then used to connect the sliding block and the slide extension of the cross-slide. When machining a taper with the plain taper attachment, the depth of cut is made by using the compound rest feed handle.

When the telescopic taper attachment is used, the depth of cut is made with the crossfeed handle, since the cross-slide and the crossfeed nut are not disengaged.

1. Check for play between the guide bar and the sliding block and adjust the gibs, if required.
2. Clean and oil the guide bar A, figure 18-10.
3. Loosen the locking screws F_1 and F_2, and swing the guide bar to the required angle or taper as indicated on the ends of the guide bar.
4. Tighten the lock screws.
5. Adjust the baseplate, C, until the ends of the guide bar, A, are equidistant from the cross-slide extension.

6. Set up the cutting tool on center.
7. Mount the workpiece on the lathe and mark the length of the section to be tapered.
8. Adjust the carriage until the cutting tool is opposite the center of the tapered section.
9. Lock the clamping bracket, D, to the lathe bed to secure the taper attachment in this position. *Note:* If a plain taper attachment is used, the following steps should be taken at this point.
   a. Adjust the compound rest so that it is parallel with the cross-slide (90° to the work).
   b. Set up the cutting tool.
   c. Feed the cutting tool in until it is about 1/4" from the work surface.
   d. Remove the binding screw, E, figure 18-9, which connects the cross-slide and the crossfeed screw nut.
   e. Using the binding screw, connect the cross-slide extension and the sliding block, B.
   f. Insert a suitable plug in the hole on top of the cross-slide to protect the crossfeed screw from dirt and metal chips.

   The compound rest feed must now be used to feed the cutting tool into the work.

10. Move the carriage to the right until the cutting tool is about 1/2" past the right-hand end of the workpiece.

    This removes any play in the moving parts of the taper attachment.

11. Take a light cut about 1/16" long and check the end of the taper for size.

Section 2 Machining Between Centers

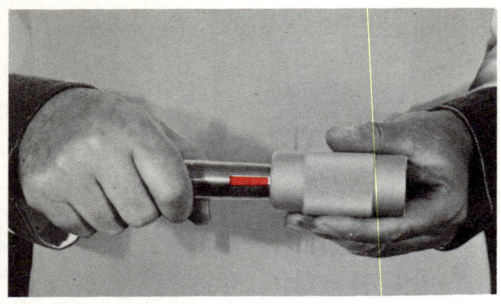

FIG. 18-11 USING A TAPER RING GAGE TO CHECK THE ACCURACY OF A TAPER (Courtesy McGraw-Hill)

12. Set the depth of roughing cut (about .060 over the finish size) and proceed to machine the work as with plain turning.

    Remove the play by moving the cutting tool 1/2" beyond the right-hand end of the work at the beginning of each cut.

13. Check the taper for fit, figure 18-11. See the section on fitting a taper, page 109.

14. Readjust the taper attachment if necessary, take a *light* cut, and recheck the taper.

15. Finish turn the taper to size and fit it to the taper gage. *Note:* If a taper on a workpiece must be duplicated, it is quite convenient to mount the part, or a taper gage having the same taper, on the lathe. Then adjust the guide bar to this taper using a dial indicator mounted in the toolpost, figure 18-12. The guide bar should be adjusted until there is no movement on the needle as the indicator is moved along the tapered surface.

Turning a Taper — Offset Tailstock Method

When the lathe is not equipped with a taper attachment and the part to be turned is held between centers, the tailstock center may be offset from the centerline to produce a taper. Since the amount that the tailstock center can be offset is limited, steep tapers and standard tapers cannot be turned on a long workpiece.

When adjusting a tailstock to taper by the offset tailstock method, certain precautions must be observed.

- Care must be taken to see that the dog does not bind against the driveplate.
- The centers must be adjusted carefully to minimize the distortion on the center holes of the workpiece.
- The tailstock must be accurately set over if accuracy is required in the taper. *Note:* If a number of identical tapered parts are required, it is essential that:

    All workpieces be exactly the same length.

    The center holes are all drilled to the same depth.

Adjusting the Tailstock

The amount of tailstock offset can be checked by three methods.

- By the graduated scale on the end of the tailstock
- By means of a dial indicator mounted on the toolpost
- By using the graduated crossfeed screw collar and a feeler gage

FIG. 18-12 A TAPER ATTACHMENT MAY BE SET UP QUICKLY AND ACCURATELY WHEN A DIAL INDICATOR IS AVAILABLE. (Courtesy McGraw-Hill)

Unit 18   Taper Turning

Procedure: Offsetting the Tailstock by the Visual Method

The graduations on the end of the tailstock provide a quick method of offsetting the tailstock a given amount. This visual method should not be used if accuracy is required.

1. Loosen the tailstock clamp nut.
2. Loosen one setscrew in the base of the tailstock and tighten the other until the upper section of the tailstock has been offset the required amount, figure 18-13.
3. Tighten the opposite setscrew to lock the upper section of the tailstock in place.

Offsetting the Tailstock Accurately

The tailstock can be rapidly and accurately offset by the use of a dial indicator mounted in the toolpost, figure 18-14, or on a magnetic base clamped to the carriage. If the tailstock must be offset accurately to turn a taper, the lathe centers must first be aligned using the procedure given in unit 9.

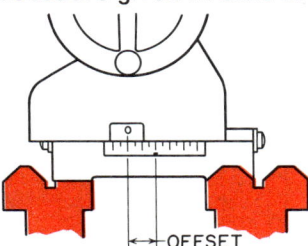

FIG. 18-13 OFFSETTING THE TAILSTOCK USING THE GRADUATED SCALE

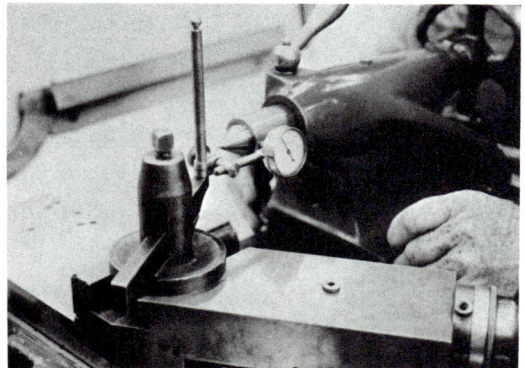

FIG. 18-14 THE TAILSTOCK CAN BE OFFSET ACCURATELY WITH A DIAL INDICATOR.

Procedure: Offsetting the Tailstock with a Dial Indicator

1. Adjust the tailstock spindle to protrude about 2" and *lock the tailstock spindle clamp.*
2. Mount a dial indicator in the toolpost with the plunger in a *horizontal position and on center,* figure 18-14.
3. Bring the point of the indicator into contact with the tailstock spindle until the needle registers about 1/2 revolution.
4. Set the bezel of the indicator and the crossfeed graduated collar to zero.
5. Loosen the tailstock clamp.
6. Offset the tailstock.
   a. If the required offset is less than the range of the dial indicator:
      1) Bring the indicator into contact with the tailstock spindle and set the dial to zero.
      2) Offset the tailstock the required amount.
   b. If the required offset is more than the range of the dial indicator:
      1) Bring the indicator into contact with tailstock spindle until it registers approximately 1/2 revolution.
      2) Turn the crossfeed handle counterclockwise until the indicator moves approximately .010.
      3) Set the indicator dial to zero.
      4) Set the crossfeed collar dial to zero.
      5) Turn the crossfeed handle *counterclockwise* the amount of offset that is required.
      6) Adjust tailstock towards indicator until needle registers zero.
      7) Tighten both setscrews and make sure that indicator still registers zero.

Section 2 Machining Between Centers

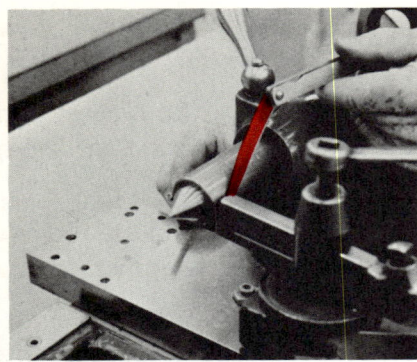

FIG. 18-15A USING A FEELER GAGE BETWEEN THE SPINDLE AND TOOLHOLDER WHEN OFFSETTING THE TAILSTOCK (Courtesy McGraw-Hill)

FIG. 18-15B OFFSETTING A TAILSTOCK USING A GAGE BLOCK BUILDUP BETWEEN THE SPINDLE AND DIAL INDICATOR. (Courtesy McGraw-Hill)

The tailstock can be offset accurately by using a feeler gage between the tailstock spindle and the back of a toolholder, figure 18-15A, or with a gage block buildup or adjustable parallels set to the required size between the spindle and the dial indicator, figure 18-15B.

Turning a Taper — Offset Tailstock Method

Turning a taper with the offset tailstock method is similar to the procedure followed with the taper attachment. However, the following precautions should be observed when using the offset tailstock method.

- Check the accuracy of the tailstock offset and the condition of the lathe centers.
- Make sure that the lathe dog does not bind in the driveplate slot.

Procedure: Cutting a Taper Using the Compound Rest

Short, steep tapers which are given in degrees, can be cut on the lathe by swiveling the compound rest to the required angle, figure 18-16. The cutting tool is fed by means of the compound rest feed handle.

1. Check the drawing for the angle of the taper.
2. Loosen the compound rest locknuts.
3. Swivel the compound rest to the required angle, figure 18-17. Note: If the included angle is given, figure 18-17A, the compound rest is swiveled to half the included angle.
4. Tighten the compound rest locknuts. *Note:* Since the compound rest studs are generally small in diameter, pull the wrench with two fingers when tightening the nuts to avoid stripping the thread, figure 18-18.
5. Set the toolbit on center and position it so *the point and not the side does the cutting.*
6. Bring the toolbit up to the workpiece by means of the crossfeed screw and carriage handwheel.

FIG. 18-16 TURNING AN ANGLE WITH THE COMPOUND REST (Courtesy McGraw-Hill)

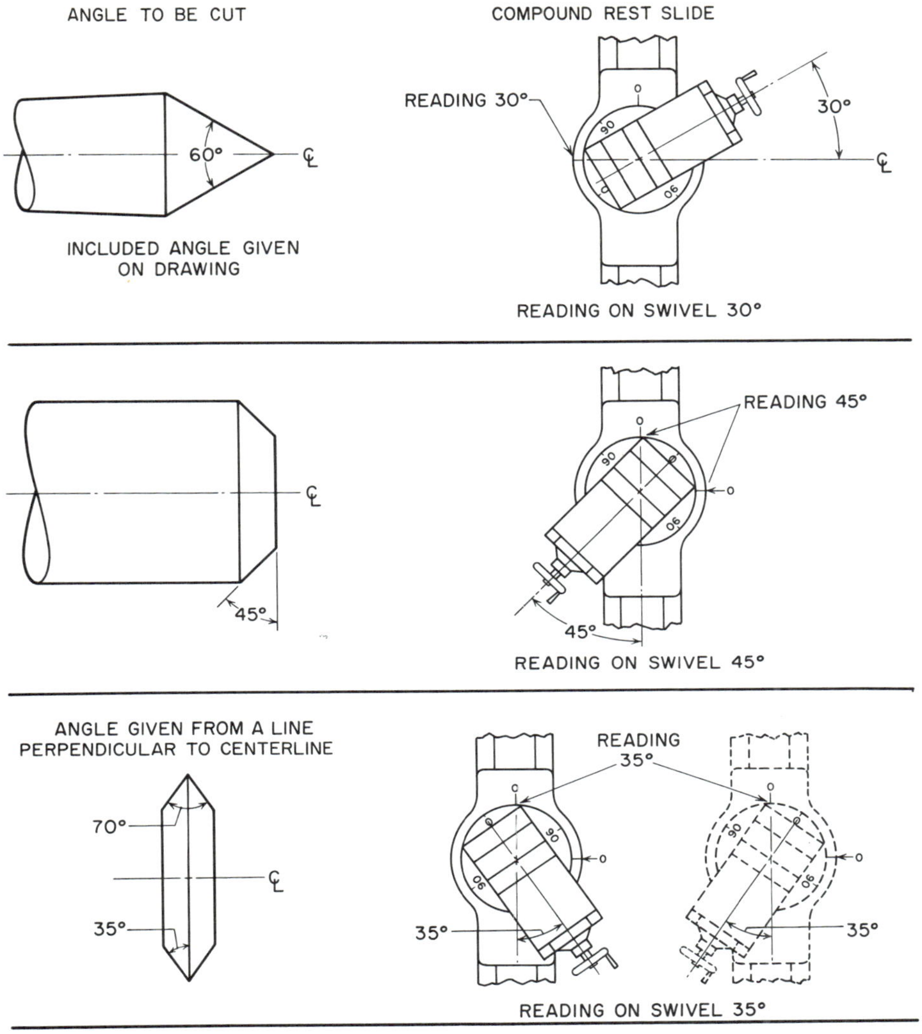

FIG. 18-17 COMPOUND REST SETUP FOR TURNING VARIOUS ANGLES

7. Feed the cutting tool across the work using the *compound rest feed screw.*
8. Check the taper for size and angle.

Procedure: Fitting a Taper

If a tapered part being machined must fit the taper on another part, care must be taken to set up the machine to produce the exact taper. If a tapered plug must fit into a tapered hole, the plug must bear against the hole for the full length, otherwise the purpose of the taper has been defeated. For example, a tapered drill shank does not hold in a drill press

FIG. 18-18 USE TWO-FINGER PRESSURE ON WRENCH WHEN TIGHTENING COMPOUND REST SCREWS.

spindle unless the taper matches perfectly. To insure that the taper fits properly, care must be taken when machining and checking the taper.

1. Rough turn the taper to about .050-.060 oversize.
2. Mark three equally spaced chalk lines around the circumference and extending the full length of the taper.
3. Place the tapered piece into the tapered hole into which it should fit and revolve it about 1/4 turn in a *counterclockwise* direction.
4. Try to move the tapered part radially in the hole. If the taper is not a proper fit, there is a movement at either end between the plug and the hole.
5. Remove the tapered piece and see if the chalk marks have rubbed off evenly for the full length of the taper. This indicates a proper taper.
6. If the chalk marks have not rubbed off evenly, the tapered workpiece should be measured and checked for accuracy of taper.
7. Adjust the taper setup to correct the error and take another *light* cut along the work.
8. Recheck the taper using chalk lines as in step 2.
9. When the taper has the proper bearing, determine the amount to be removed so that the taper fits to the proper depth.
10. Finish machine the workpiece to size.
11. Recheck the taper for the proper fit.

MEASURING TAPERS

External tapers can be checked for accuracy by any of the following methods: ring gage, standard micrometer, sine bar, or taper micrometer.

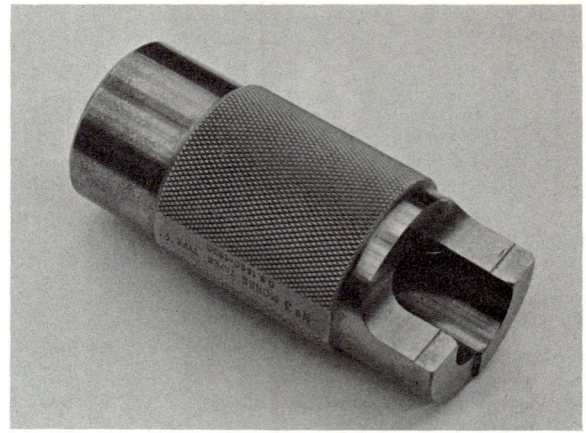

FIG. 18-19   A TAPERED RING GAGE

Checking a Taper Using a Ring Gage

Tapers can be checked for accuracy of fit and for size with a taper ring gage. These gages generally have steps ground on the end to indicate the tolerance on the distance the taper should advance into the ring gage.

When checking the accuracy of a taper with a ring gage, use the same procedure outlined under fitting a taper. The diameter of the taper is correct if the end of the taper comes between the limit steps ground on the gage.

Precautions When Using Tapered Ring Gages

1. Clean the part and the gage thoroughly before checking the taper.
2. Do not force or twist the gage excessively as this causes the gage to wear.

Procedure: Checking a Taper Using the Standard Micrometer

If a ring gage is not available, the accuracy of the taper can be checked by using a standard micrometer.

1. Clean the tapered workpiece and apply bluing to it.
2. Mark off 2 lines exactly 1" apart, figure 18-20.

Unit 18 Taper Turning

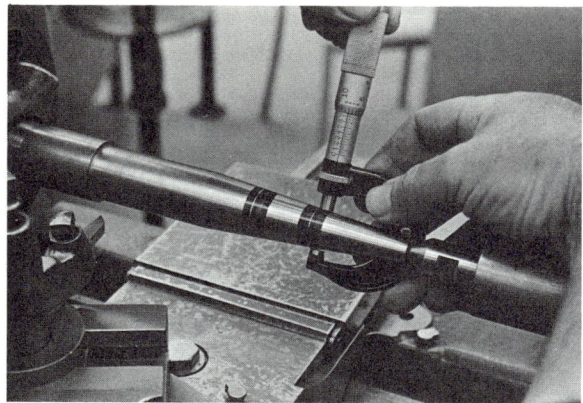

FIG. 18-20 CHECKING THE ACCURACY OF A TAPER WITH A MICROMETER

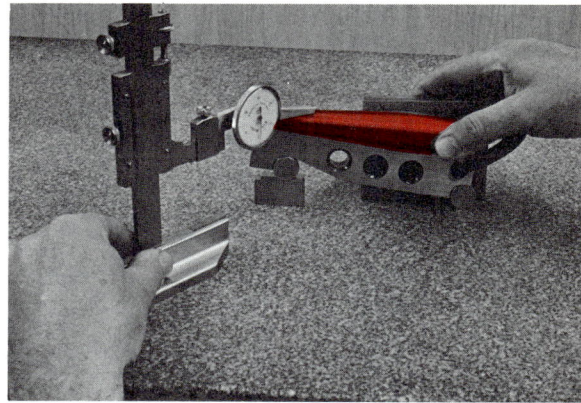

FIG. 18-21 TAPERS MAY BE ACCURATELY CHECKED WITH A SINE BAR.

3. Carefully measure the taper with a micrometer so that the edge of the anvil and spindle just touch the lines. Record both readings; the difference between these two is the taper per inch of the workpiece.

4. Convert the taper per inch to taper per foot or angle of taper as required, see table 18-2.

5. Compare the T.P.F. of the workpiece to the taper required; correct if necessary.

Procedure: Checking a Taper Using the Sine Bar

The angle of a taper can be accurately checked with a sine bar, gage blocks, and a dial indicator, figure 18-21.

1. Check the blueprint dimensions.

2. Calculate the *included* angle of taper of the workpiece. See table 18-2.

FORMULAS USED IN CALCULATING DIMENSIONS FOR MEASURING TAPERS		
Given	Required	Formula
T.P.F.	T.P.I.	$\dfrac{T.P.F.}{12}$
T.P.I.	T.P.F.	T.P.I. × 12
Diameters for 1" length	T.P.I.	D − d
End diameters and length of taper	T.P.I.	$\dfrac{D-d}{T.L.}$
Included angle of taper	T.P.F.	Tan 1/2∠ × 24
T.P.F.	Included angle of taper	Tan 1/2∠ = $\dfrac{T.P.F.}{24}$
End diameters and length of taper	Included angle of taper	Tan ∠ = $\dfrac{(D-d)}{T.L.}$

TABLE 18-2

Section 2 Machining Between Centers

3. Calculate the gage block buildup (5 x sine angle) required for the angle of the taper.
4. Place a 5" sine bar on a clean surface plate with the proper gage block buildup under the one end.
5. Place the plug on top of the sine bar, figure 18-21, with the centerline of the taper parallel to the centerline of the sine bar.
6. Move a dial indicator mounted on a surface gage across the top of each end of the tapered section. The angle of the taper is correct if the indicator reads the same at both ends.

Example: Calculate the gage block buildup required to check a No. 3 Morse taper using a 5" sine bar.

Taper per foot of No. 3 Morse taper = .602 (table 18-1).

Angle of taper = Tan 1/2 angle = $\frac{T.P.F.}{24}$

= Tan 1/2 angle = $\frac{.602}{24}$

= Tan 1/2 angle = .0251

1/2 angle = 1° 26'

angle = 2° 52'

Gage block buildup = 5 x sine 2° 52'

= 5 x .05001

= .25005

Procedure: Checking a Taper With a Taper Micrometer

Tapers can be checked quickly and accurately with a taper micrometer, figures 18-22A and B. This relatively new measuring instrument permits tapers to be checked while they are still on the lathe, thus saving the time required to reset the workpiece in the machine if the taper is not correct. This device consists of an adjustable anvil and a 1" sine bar which is attached to the frame and is adjusted by the micrometer spindle.

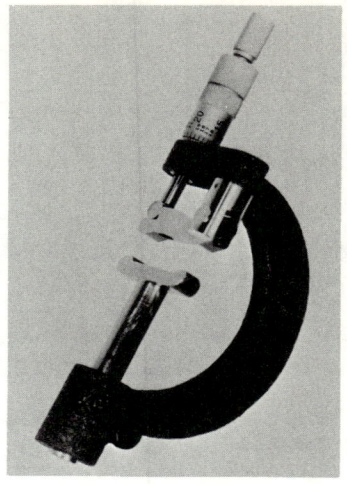

FIG. 18-22A A TAPER MICROMETER

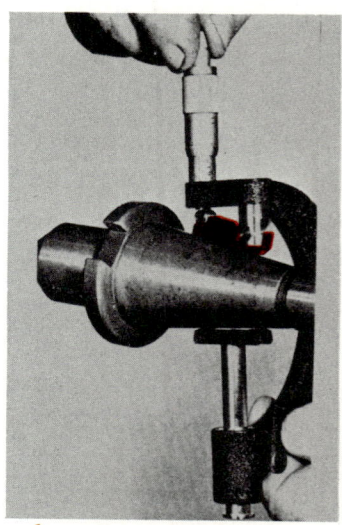

FIG. 18-22B CHECKING A TAPER WITH A TAPER MICROMETER (Courtesy Taper Micrometer Corp.)

1. Clean the surface of the taper and the faces of the taper micrometer.
2. Place the taper micrometer over the workpiece and tighten the thimble until the jaws bear on the tapered section.
3. Note the reading on the micrometer. This indicates the taper per inch of the tapered section.
4. From the tables supplied with the instrument, convert this reading to taper per foot or angle of taper as required.

REVIEW QUESTIONS

1. Define a taper.

2. What is the purpose of a taper?

3. How may tapers be expressed?

Self-Holding Tapers

4. Describe a self-holding taper.

5. What tapers make up the self-holding taper series?

Self-Releasing Tapers

6. How does the drive of a self-releasing taper used on a milling machine compare to the drive of a self-holding taper?

7. How do the type L and type D-1 spindle noses differ?

Nonstandard Tapers

8. What are two nonstandard tapers and where is each one used?

Taper Calculations

9. For the following workpieces, calculate

   a. The taper per foot.   b. The tailstock offset.

	Large Dia.	Small Dia	Taper Length	Overall Length
1)	2	1 3/4	4	10
2)	1 3/64	15/16	7	18
3)	1 5/8	1 1/2	2 1/2	12 1/2
4)	7/8	7/16	4 1/2	9
5)	19/32	11/32	3	8 3/4

10. Using the simplified formula, what is the tailstock offset for workpieces No. 2 and 3 in question 9?

Taper Turning

11. What are the three methods by which tapers can be turned on a lathe?

12. Describe the principle of the taper attachment.

13. What are six advantages of using the taper attachment to turn a taper?

14. What are the two types of taper attachments?

Turning a Taper Using a Taper Attachment

15. How is the toolbit fed to depth when using

   a. A plain taper attachment?   b. A telescopic taper attachment?

Section 2   Machining Between Centers

16. What are the steps required to set up the plain taper attachment to turn a taper of 1/4" T.P.F.?

17. How is the play removed from a taper attachment before starting each cut?

Turning a Taper by the Offset Tailstock Method

18. What precautions should be observed when turning a taper by the offset tailstock method?

19. What is one method of offsetting the tailstock *accurately?* Describe the steps of this method.

Cutting a Taper Using the Compound Rest

20. What steps are required to machine a taper having a 22° included angle on a workpiece?

21. How should the cutting tool be set when machining a taper with the compound rest?

Fitting a Taper

22. What are the steps required to fit a taper to a ring gage?

Measuring Tapers

23. How may a taper be checked using a standard micrometer?

24. Why should the dial indicator be moved across each end when checking a taper with a sine bar and gage blocks?

25. Calculate the buildup required to check a No. 200 taper, table 18-1, using a 5" sine bar?

26. Describe the construction of a taper micrometer.

27. What is the advantage of using a taper micrometer to check a taper?

28. How is the accuracy of the taper determined with a taper micrometer?

Unit 19 FORM TURNING

It is often necessary to form curved or irregular shapes on a workpiece. This can be accomplished by various methods, such as:

- Coordinating the carriage and crossfeed movements to produce the form manually
- Using a form tool which has the cutting edge ground to the contour of the form desired
- Manually operating the lathe to follow a template of the form desired.
- Using a hydraulic or electronic tracer attachment to produce the form automatically

The use of templates and tracer attachments is covered in detail in the section "Special Operations."

FREEHAND FORM TURNING

Freehand turning of curved forms involves the simultaneous operation of the carriage and crossfeed controls to produce the desired shape on the workpiece. Freehand turning is generally used when it is uneconomical to provide a template and set up a tracer attachment if only a few parts are required. The beginning lathe operator generally finds this a difficult operation since it involves the coordination of both hands. However, with practice and by following the basic steps which follow, an operator can soon produce curved forms with reasonable accuracy.

Procedure: Turning a Concave Form Freehand

1. Mount the workpiece on the lathe.
2. With a sharp, pointed toolbit, cut a light groove at each end and also in the center of the concave form.

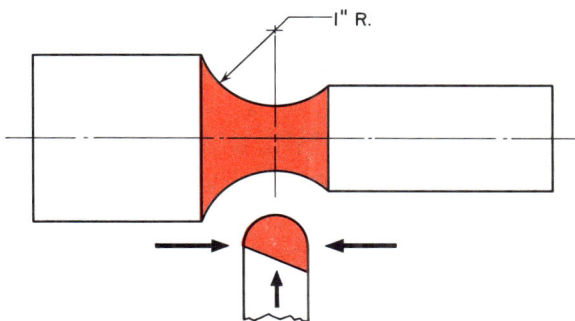

FIG. 19-1 MACHINING A CONCAVE FORM ON A WORKPIECE

3. Mount a round-nose (radius) toolbit on center. *Note:* A concave form can be produced easily and accurately if a toolbit having as large a radius as possible is used.
4. Set the lathe speed to approximately three-quarters of the turning speed.

   This is important to avoid chatter, since a greater amount of the toolbit cutting edge is in contact with the work.
5. Move the carriage to bring the edge of the radius toolbit close to the start of the concave form.
6. Place one hand on the carriage handwheel *(not the handle)* and the other hand on the crossfeed handle.
7. Slowly turn the carriage handwheel, feeding the toolbit towards the center of the form, while the other hand turns the crossfeed handle clockwise, moving the tool into the work. *Note:* Coordinating the amount of carriage movement in relation to the crossfeed to cut the desired form requires practice.
8. Take successive cuts from each side, figure 19-1, always *starting at the large end of* the form and working *towards the center.*

Section 2 Machining Between Centers

It is wise to take light cuts until the proper coordination between the carriage and crossfeed is attained.

9. Check the small diameter of the form with a caliper and the accuracy of the form with a radius gage.

   If a radius gage is not available, obtain a piece of steel having the correct diameter and use its circumference to check the accuracy of the radius. For example, a 2" diameter (1" radius) could be used to check the accuracy of the form in figure 19-1.

10. Cut the form as close to the size and contour as possible.

11. Use a half-round file to remove the machine marks and bring the radius to the required shape.

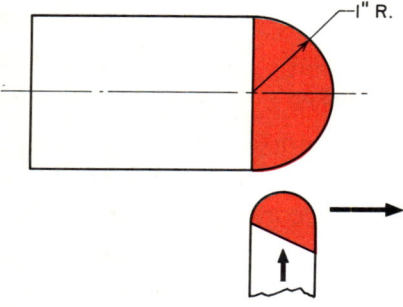

FIG. 19-2 MACHINING A 1" RADIUS ON THE END OF A WORKPIECE

Procedure: Turning a Convex Radius Freehand

Basically, the same procedure used to machine a concave form can be used to produce a convex form. The following steps outline the procedure for machining the 1" radius on the work illustrated in figure 19-2.

1. With a pointed toolbit, cut a light groove 1" from the end.

2. Using a round-nose toolbit, take cuts starting at the 1" line and feeding towards the end of the work, figure 19-2. *Note:* For the first half of the radius (1/2"), the carriage must be moved faster than the crossfeed handle. However, from this point to the end of the work, the crossfeed handle must be moved faster than the carriage handwheel.

3. Take successive cuts, *always* starting at the outside diameter and cutting towards the end.

4. Check the accuracy of the form with a radius gage.

   If a 1" radius gage is not available, drill or bore a 2" diameter hole in a piece of thin material, and then cut the piece through the hole center. This produces two 1" radius gages.

5. Machine the radius as close to size and shape as possible.

6. With a file, remove the machine marks and finish the form.

FORM TURNING TOOLS

When it is required to produce intricate forms, radii, contours, etc., on a number of workpieces, they can be reproduced conveniently and accurately with form tools. Form tools are ground so that the contour of the cutting edge corresponds to the desired shape, figure 19-3. If the toolbit is ground accurately, an accurate form is reproduced on the workpiece. If the form must be held to close tolerances, it is wise to check the accuracy of the cutting edge on an optical comparator.

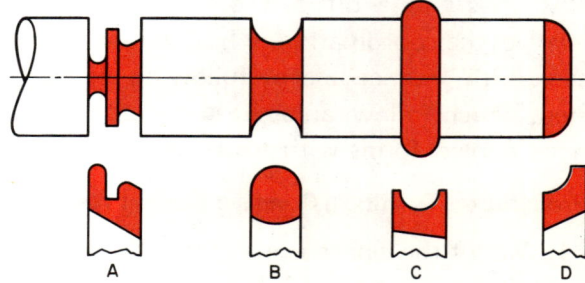

FIG. 19-3 FORM TOOLS ARE USED TO REPRODUCE SPECIAL SHAPES AND CONTOURS ACCURATELY AND EASILY.

The need for constant checking and inspection of parts with a master template, which is necessary when forms are produced manually, is eliminated when using an accurately ground form tool. The contour or shape of each part is identical if the same form toolbit is used. When a form tool requires sharpening, it is important that the grinding occurs only on the top of the cutting edge, otherwise the shape and accuracy of the form is altered.

When a convex radius, as shown in figure 19-3C, is required, a raised collar must be left on the workpiece to insure that there is enough material to produce the desired form.

Form Turning Hints

1. The lathe speed should be approximately one-half of the turning speed. This is necessary since so much of the toolbit cutting edge is in contact with the workpiece at one time.

If chatter occurs during the form turning operation, reduce the lathe speed until the chatter is eliminated.

2. A gooseneck tool is recommended when turning a large radius since it tends to reduce chatter and produce a smoother surface.

3. Apply cutting fluid while form turning to produce a good cutting action and eliminate chatter.

4. Feed the toolbit into the work slowly while applying cutting fluid.

To eliminate chatter, the lathe carriage should be moved *slightly* to the left and right (longitudinally) while feeding the toolbit.

5. Form turning tends to cause the work to spring, therefore, wherever possible, support the workpiece with a steady rest.

6. Abrasive cloth should be used to improve the surface finish of a contour form. To avoid altering the shape of the contour, filing should be kept to a minimum.

REVIEW QUESTIONS

1. What are four methods of producing curved or irregular shapes on a round piece of work?

Freehand Form Turning

2. When is freehand form turning used?

Turning a Concave Form Freehand

3. What shape cutting tool should be used when machining a concave form freehand?

4. How should successive cuts be taken on a concave form?

5. How may the accuracy of a large radius be checked if a radius gage is not available?

Turning a Convex Form Freehand

6. How is a convex radius turned freehand? Describe each step.

Section 2 *Machining Between Centers*

Form Turning Tools

7. What is a form tool and why is it used?

8. How should a form tool be sharpened?

Form Turning Hints

9. At what speed should a lathe be operated when using form turning tools?

10. What are three methods of eliminating chatter when using form turning tools?

11. How should the surface finish of contour forms be improved?

Unit 20 THREADS

One of the most important mechanical units to the manufacturing industry is that of the screw thread. There are few assemblies or components which do not use threads for fastening, adjusting, or transmitting motion.

A thread can be defined as a helical ridge of uniform section formed on the outside of a cylinder or cone, or on the inside of a round hole. Threads can be either external as in the case of a bolt, or internal as found on nuts. In order to cut a thread properly, the machinist should have a knowledge of thread types, shapes, proportions, and the calculations required.

Screw threads can be either right- or left-hand, figure 20-1. A bolt or nut having a *right-hand thread* is advanced or tightened by turning the part *clockwise*. Right-hand threads can be identified visually by holding the threaded part in a horizontal position (with the axis pointing from left to right). The helix angle slopes down and to the right, figure 20-1A. *Left-hand* threads turn *counterclockwise* to advance, and the helix angle slopes down and to the left when held in a horizontal position, figure 20-1B.

Screw threads are used for four general purposes:

- *For fastening devices* such as **screws**, studs, bolts, and nuts
- *To transmit motion.* When thread cutting on a lathe, the threaded lead screw advances the carriage (and threading tool) along the work.
- *To provide accurate measurement,* as in the case of the micrometer
- *To increase torque.* Screw jacks may be used to raise heavy loads.

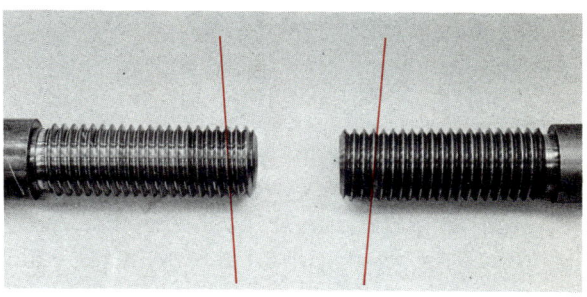

A. RIGHT-HAND THREAD   B. LEFT-HAND THREAD
FIG. 20-1 SCREW THREADS

THREAD TERMINOLOGY

Before the machinist can calculate thread proportions or cut a thread, the following terminology which applies to screw threads should be understood, figure 20-2.

Screw thread — a helical ridge of uniform section formed on the inside or outside of a cylinder or cone.

External thread — a helical ridge cut on the outside of a cylinder, such as a bolt, or on a cone (wood screw).

Internal thread — a helical ridge cut on the inside of a round hole such as on the inside of a nut.

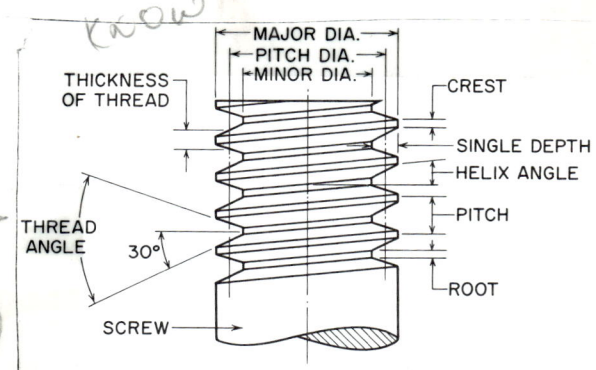

FIG. 20-2 PARTS OF A SCREW THREAD

Section 2 Machining Between Centers

Major diameter — the largest diameter of an external or internal thread. The major or the outside diameter (O.D.) is also the nominal diameter of the thread, e.g., a 1/2″ – 13NC thread has an outside or nominal diameter of 1/2″.

Minor diameter (formerly root diameter) — the smallest diameter of an external or internal thread. It is equal to the major diameter minus two depths.

Pitch diameter — the diameter of an imaginary cylinder which passes through the thread at a point where the width of the thread is equal to the width of the groove. The pitch diameter is equal to the major diameter minus one depth.

Number of threads — the number of roots or crests per inch of a threaded section.

Pitch — the distance from a point on one thread to the corresponding point on the next thread, measured parallel to the axis. The pitch is always one over the number of threads per inch (1/N).

Lead — the distance a thread advances axially in one revolution. For a single-start thread, the lead is equal to the pitch.

Crest — the top surface joining the sides of a thread. On an external thread, the crest is on the major diameter; on an internal thread it is on the minor diameter.

Root — the bottom surface joining the sides of two adjacent threads. The minor diameter of an external thread is formed by the root of the thread. On an internal thread, the root forms the major diameter.

Flank — the side of the thread connecting the crest and the root

Depth of thread — the distance between the crest and the root, measured perpendicular to the axis of the thread.

Angle of thread — the included angle between the sides or flanks of a thread measured in an axial plane.

Helix angle — the angle which the thread makes with a plane perpendicular to the axis of the thread. This is often called the lead angle.

TYPES OF THREADS

There are many kinds of threads used in industry, each having a special purpose; however, only the more common types will be discussed. The following symbols apply to all the threads described in this unit.

C — Width of flat at root
D — Single depth of thread
F — Width of flat at crest
N — Number of threads per inch
P — Pitch
R — Radius at crest or root
W — Width of groove or ridge

The American National Thread

The American National thread, figure 20-3A, is divided into four series; National Coarse (NC), National Fine (NF), National Special (NS), and National Pipe Thread (NPT). All of these threads have the same shape and proportions, i.e., a 60° angle with a root and crest truncated to 1/8 the pitch of the thread. This is a general-purpose thread used in machine construction, assembly, and fabrication.

The fine thread series was devised for use with parts subject to considerable vibration, such as aircraft and automobile parts. The finer pitch tends to resist loosening and provides for finer adjustment than coarse pitch threads. The shallower depth of thread also provides for greater strength in the bolt. American National Threads have been superseded by the Unified Thread series.

$$D = .6495 \times P \text{ or } \frac{.6495}{N}$$

$$C \& F = .125 \times P \text{ or } \frac{.125}{N}$$

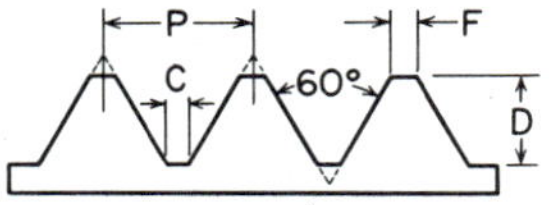

FIG. 20-3A AMERICAN NATIONAL THREAD FORM

The British Standard Whitworth Thread

The British Standard Whitworth thread (BSW), figure 20-3B, is commonly used in Britain and is being superseded by the Unified thread. It has a 55° angle with rounded crests and roots. The rounded root tends to minimize cracking at the bottom of the thread. It is used in machine assembly, construction, and fabrication.

$$D = .6403 \times P \text{ or } \frac{.6403}{N}$$

$$R = .1373 \times P \text{ or } \frac{.1373}{N}$$

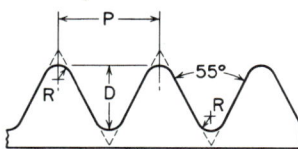

FIG. 20-3B WHITWORTH THREAD FORM

The Unified Thread

The Unified thread (UN), figure 20-3C, was developed by the United States, Britain, and Canada to become a standardized thread in these countries. It incorporates features of the American National Form and the British Standard Whitworth threads, and was designed because of the lack of interchangeability of these two thread systems, particularly during World War II. This system was agreed upon in 1948, and since then there has been a gradual transition to Unified threads.

The thread has a 60° angle with rounded roots and rounded or flat crests. It is available in the coarse and fine series (UNC and UNF). Threads in the Unified system are interchangeable with threads of the same pitch and diameter in the American National series

$$D \text{ (external thread)} = .6134 \times P \text{ or } \frac{.6134}{N}$$

$$\text{(internal thread)} = .5413 \times P \text{ or } \frac{.5413}{N}$$

$$F \text{ (external thread)} = .125 \times P \text{ or } \frac{.125}{N}$$

$$\text{(internal thread)} = .250 \times P \text{ or } \frac{.250}{N}$$

FIG. 20-3C UNIFIED THREAD FORM

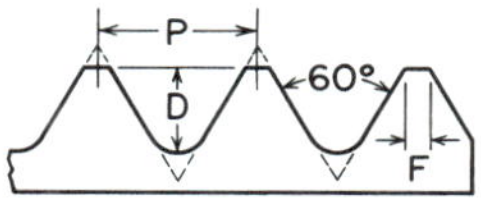

The American National Acme Thread

This thread, figure 20-3D, has a 29° angle and is a modification of the square thread. Because the Acme thread can be cut more readily with taps, dies, single-point lathe tools, and milling cutters, it has generally replaced the square thread. The angular face of the thread makes it particularly suitable for lead screws on lathes, since it allows the split nut to be engaged easily. Other common uses include screws in jacks and vises.

$$D \text{ (min.)} = .500 \times P \text{ or } \frac{.500}{N}$$

$$\text{(max.)} = .500 \times P + .010 \text{ or } \frac{.500}{N} + .010$$

$$F = .3707 \times P \text{ or } \frac{.3707}{N}$$

$$C = .3707 \times P - .0052 \text{ or } \frac{.3707}{N} - .0052$$

(for maximum depth)

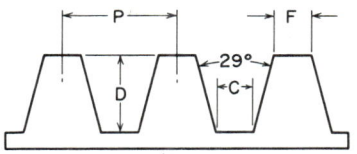

FIG. 20-3D ACME THREAD FORM

Section 2   Machining Between Centers

The Brown and Sharpe Worm Thread

This thread, figure 20-3E, has an included angle of 29° and resembles the Acme thread; however, the depth is greater and the width of the crests and roots differ from those of the Acme. This thread is used in conjunction with worm gears to transmit motion between two shafts at right angles to each other but not in the same plane as in the dividing head. The self-locking feature of a worm and gear drive makes it particularly suitable for winches and steering mechanisms in cars.

$$D = .6866 \times P \text{ or } \frac{.6866}{N}$$

$$F = .335 \times P \text{ or } \frac{.335}{N}$$

$$C = .310 \times P \text{ or } \frac{.310}{N}$$

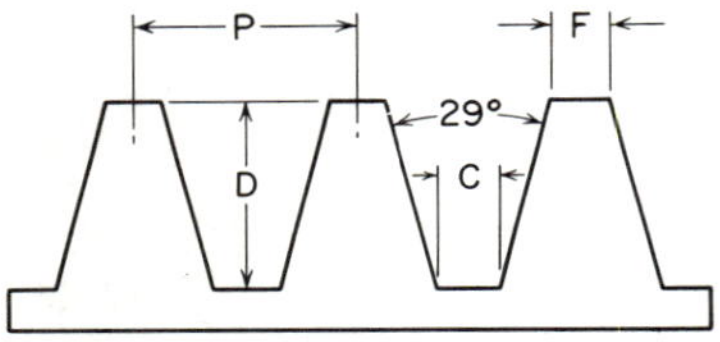

FIG. 20-3E   BROWN AND SHARPE WORM THREAD FORM

The Square Thread

This thread, figure 20-3F, formerly used for vises and screw jacks, is being replaced by the Acme thread. Square threads are more difficult to cut particularly with taps and dies.

$$D = .500 \times P \text{ or } \frac{.500}{N}$$

$$F = .500 \times P \text{ or } \frac{.500}{N}$$

$$C = .500 \times P + .002 \text{ or } \frac{.500}{N} + .002$$

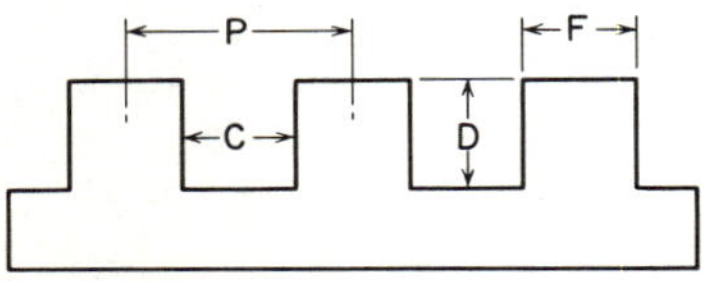

FIG. 20-3F   SQUARE THREAD FORM

The International Metric Thread

This thread, figure 20-3G, has a 60° angle and a flat crest and root truncated to 1/8 the depth of the thread. It is the common thread standard used in continental Europe and is used in North America mainly on instruments and spark plugs.

$$D \text{ (max.)} = .7035 \times P \text{ or } \frac{.7035}{N}$$

$$\text{(min.)} = .6855 \times P \text{ or } \frac{.6855}{N}$$

$$C \& F = .125 \times P \text{ or } \frac{.125}{N}$$

$$R \text{ (max.)} = .0633 \times P \text{ or } \frac{.0633}{N}$$

$$\text{(min.)} = .054 \times P \text{ or } \frac{.054}{N}$$

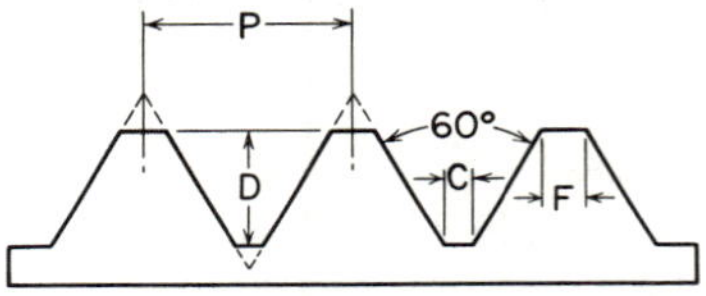

FIG. 20-3G   INTERNATIONAL METRIC THREAD FORM

THREAD FITS AND CLASSIFICATION

In order to understand any thread system, one should become familiar with the terminology relating to thread fits and classifications.

Fit — The range of tightness between two mating parts. It is determined by the clearance or interference between parts when assembled. For example, the class of fit for National Form threads is specified as Class 1 (loosest) to Class 4 (tightest).

Allowance — the *intentional difference* in the size of mating parts. It is the minimum clearance (*positive allowance*) or the maximum interference (*negative allowance*) intended between the parts, figure 20-4. With screw threads it is the allowable difference between the largest external thread and the smallest

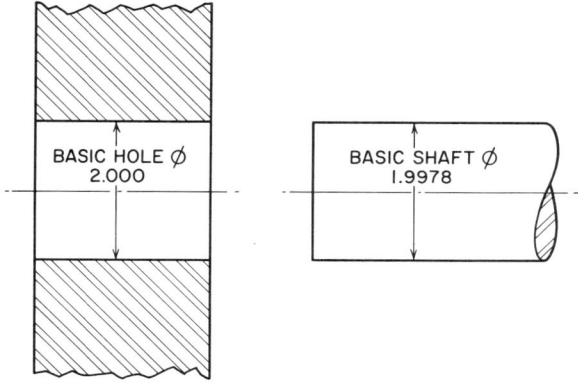

FIG. 20-4 THE ALLOWANCE (INTENTIONAL DIFFERENCE) BETWEEN THE SHAFT AND HOLE IS .0022.

internal mating thread. This produces the tightest fit acceptable.

Tolerance — the *variation permitted,* or the tolerated error, in a part size. Because it is impractical, from a cost standpoint, to machine work to the exact size, a slight variation is generally permitted in the size of the part. This tolerance can be expressed as a plus or minus dimension, or both. If the tolerance is expressed in one direction only, it is called *unilateral* tolerance. For example, a part may be made to 2.000 + .002. This means that the part may vary from 2.000 to 2.002. The tolerance is *bilateral* when it is given as a plus and minus dimension. In this case, the total tolerance is the sum of the plus and minus tolerances. For example, if the 2.000 part is made to a tolerance of ± .002 (2.000 ± .002), the size of the part may vary from 2.002 (2.000 + .002) to 1.998 (2.000 − .002) and still be the correct size for the job.

The tolerances and allowances for threads having angular sides, are given at the pitch diameter. Accurate thread measurements are also taken at the pitch diameter. The blank diameter of a rolled thread is also the pitch diameter of the finish-rolled thread. Since thread rolling is a displacement operation, the metal displaced to form the groove below the pitch line is forced up to form the thread above the pitch line.

The allowance and tolerance for a 1 1/4″ − 7 UNC thread, class 2A and 2B fit are as follows:

- *Allowance*
  Minimum pitch diameter of the internal thread (2B) = 1.1572
  Maximum pitch diameter
  of the external thread (2A) = 1.1550
  Allowance (intentional dif.) = .0022
- *Tolerance*
  Maximum pitch diameter of the external thread (2A) = 1.1550
  Minimum pitch diameter of
  the external thread (2A) = 1.1476
  Tolerance or permissible = .0074
  variation

Note: The allowance (intentional difference) for a 1 1/4″ − 7 UNC thread is .0022; however, a difference of .0074 is permissible.

Limits — The maximum and minimum size of a part. The upper limit is the nominal size of the part plus the tolerance, while the lower limit is the nominal size of the part minus the tolerance. For example, the specifications for a shaft diameter are 2.500 ± .002. The upper limit of the shaft is 2.502 while the lower limit is 2.498. Any size between these two limits is acceptable.

The pitch diameter limits for a 1 1/4″ − 7 UNC Class 2A thread are:

- Maximum pitch diameter of external thread (2A) = 1.1550
- Minimum pitch diameter of external thread (2A) = 1.1476

This thread is acceptable only if its pitch diameter measures between 1.1550 (upper limit) and 1.1476 (lower limit).

Nominal Size — the size designation of the part. For example, if a thread is designated 1 1/4" – 7 UNC, the nominal diameter is 1 1/4" (major diameter).

Actual Size — the measured size of a part. Diameters to be threaded are generally turned slightly under the nominal size. For example, the nominal or major diameter of a 1 1/4" – 7 UNC 2A thread is 1.250. The actual size of this diameter may vary from 1.248 (maximum diameter) to 1.232 (minimum diameter) and still be within the acceptable limits.

Classification of Unified Thread Fits

The fits for American National Form or American Standard threads were formerly designated by Class 1, 2, 3, or 4. These fits were as follows:

- Class 1 – loose fit, suitable for rapid assembly
- Class 2 – free fit, commercial standard and general use
- Class 3 – medium fit, where there is little or no play between the mating parts
- Class 4 – tight fit, where a wrench must be used to force the nut onto the bolt

With the adoption of the Unified Thread series, the method of designating thread fits was changed. The "class" now refers to tolerance, or tolerance and allowance; it no longer refers to fit or "class of fit."

Unified thread tolerances and allowances have been designated by the Screw Thread Committee as *1A, 2A, and 3A for external threads and 1B, 2B, and 3B, for internal threads.*

Classes 1A and 1B replace Class 1 of the former American Standard threads. They have about 1.5 more tolerance than does a Class 2A and 2B. Class 1A and 1B threads are used on parts that must be assembled quickly and easily and where the threads may become dirty or bruised.

Classes 2A and 2B are the recognized standard for the majority of commercial nuts and bolts. These provide a medium or free fit. The fit produced by Class 2A and 2B threads corresponds to the Class 2 fit of the American Standard threads.

Classes 3A and 3B are used where a very close fit is required. No allowance is provided and tolerances are about 75% of those used in Classes 2A and 2B.

This system of thread classification permits the use of any combination of mating threads. It is possible to use a Class 3B nut with a Class 2A bolt to provide a closer fit than is achieved with a 2B nut and a 2A bolt. This interchangeability reduces the cost of threaded parts. The dimensions and tolerances for Unified threads can be found in any handbook.

REVIEW QUESTIONS

1. Define a thread.
2. How can a right-hand thread be identified?
3. In which direction should a right-hand nut be turned to tighten it?
4. What are four general purposes of screw threads and what is one use for each?

Thread Terminology

5. What is the definition of each of the following thread terms:
   a. Major diameter?
   b. Minor diameter?
   c. Pitch diameter?
   d. Number of threads?
   e. Root?
   f. Angle of thread?

6. Define and compare the lead and pitch of a thread.

Types of Threads

7. Describe an American National thread.

8. What are the four American National threads and where is each used?

9. Describe a Unified thread form. Why was it developed?

10. How does the American National Acme thread compare with the Brown and Sharpe Worm thread and where is each used?

11. Describe the International Metric thread. Where is this thread used?

Thread Fits and Classification

12. What is the definition for each of the following as they relate to screw threads:
    a. Fit?
    b. Allowance?
    c. Tolerance?
    d. Limits?

13. What is the difference between bilateral and unilateral tolerance?

14. For the following thread pitch diameters, what is the basic pitch diameter, upper limit, lower limit, and tolerance:
    a. $.8425 \begin{array}{l} +.0015 \\ -.000 \end{array}$?
    b. $.9675 \pm .002$?
    c. $1.2139 \begin{array}{l} +.000 \\ -.002 \end{array}$?

15. For each of the above problems, is the tolerance bilateral or unilateral?

Classification of Unified Thread Fits

16. How were the American National Form thread fits revised in the Unified thread system?

17. What type of fit should be specified for:
    a. The loosest fit?
    b. The tightest fit?

18. What advantages does the Unified system of thread fits have over the system previously used for American National Form threads?

Unit 21 THREAD CUTTING

Before the lathe is set up for thread cutting, the functions of two important parts of the lathe, the quick-change gearbox and the thread-chasing dial, should be understood.

THE QUICK-CHANGE GEARBOX

The quick-change gearbox, figure 21-1, provides a means of quickly setting the lathe to cut the desired number of threads per inch. This box contains several different size gears, any of which can be quickly engaged to give the proper ratio between the rotation of the headstock spindle and the lead screw. For example, when cutting eight threads per inch, the quick-change gearbox must be set so that the carriage travels one inch while the work revolves eight times. The feed and thread plate on the front of the gearbox indicates the proper setting for the levers to produce the desired number of threads per inch. The top numbers (generally whole numbers) in each block refer to the number of threads per inch that are cut when threading. The *decimals* in the lower part of each block indicate the feed or the distance the carriage travels for each revolution of the work when the feed rod is engaged.

Procedure: Setting the Quick-Change Gearbox for Thread Cutting

1. Check the blueprint for the number of threads per inch to be cut.

2. Check the feed and thread plate and locate the *whole number* which represents the desired threads per inch.

3. With the lathe stopped, engage the tumbler lever pin in the hole at the bottom of vertical column in which number is located.

4. Set the top lever in the position indicated in the horizontal column in which the desired number is located.

FIG. 21-1 THE QUICK-CHANGE GEARBOX. (Courtesy Standard Modern Tool Co.)

5. Adjust the sliding gear in or out as required. *Note:* Different makes of lathes vary as to the method of setting the quick-change gearbox. However, all the information that is required will be found on the feed and thread plate.

6. Turn the lathe spindle by hand to see that the lead screw revolves. *Note:* Some lathes may be equipped with a sliding collar and others with a lead screw lever on the gearbox which must be engaged so that the lead screw revolves.

7. Recheck the lever settings.

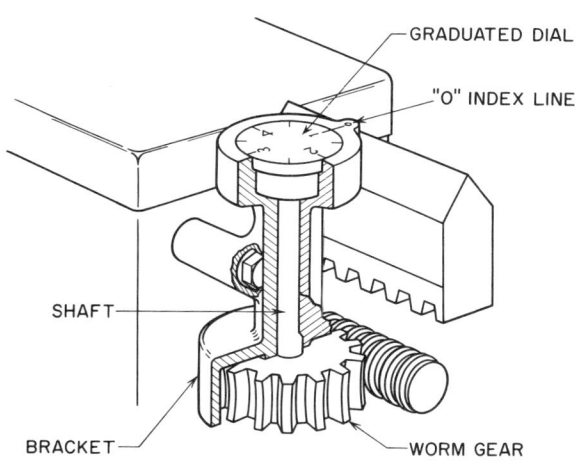

FIG. 21-2 THE THREAD-CHASING DIAL MECHANISM

THE THREAD-CHASING DIAL

When thread cutting on a lathe, the lathe spindle and the lead screw must be in the same relative position for each successive cut. On lathes not equipped with a thread-chasing dial, this is achieved by leaving the split-nut lever engaged throughout the whole thread-cutting operation. At the end of each cut, the lathe spindle is reversed to return the carriage to the starting position. This long and tedious process is seldom used now, since most modern lathes are equipped with a thread-chasing dial.

The thread-chasing dial mechanism, figure 21-2, which is fastened to the carriage of the lathe or built into it, indicates when the split-nut lever should be engaged so that the threading tool follows the previously cut groove. This mechanism consists of a graduated dial which is connected to a worm gear by means of a shaft. On some lathes, this unit is mounted in a bracket attached to the carriage. When thread cutting, the bracket is adjusted until the worm gear meshes with the threads of the lead screw. As the lead screw revolves, the worm gear and graduated dial revolve. By means of the graduations on the dial and the *O* index line, the operator can determine where to engage the split-nut lever so that the threading tool enters the previously cut groove.

The face of the dial is graduated into eight divisions, four numbered and four unnumbered divisions. The position of these lines as the dial revolves indicates the relationship of the lead screw and cutting tool to the work and the groove when threading. Figure 21-3, page 128, indicates when the split-nut lever should be engaged to cut various numbers of threads per inch.

LATHE SETUP FOR CUTTING 60° THREADS

Threads may be produced by several methods such as rolling, grinding, milling, and by the use of dies. However, workpieces that are produced on the engine lathe are generally threaded on a lathe. Work may be held in a chuck or between centers for threading. Work that is held in a chuck should be finish-turned and threaded before it is removed from the chuck.

Procedure: Setting Up a Lathe for Cutting a Right-Hand Thread (National Form or Unified)

1. Check the diameter of the part to be threaded with the blueprint and determine the number of threads per inch to be cut.

Section 2 Machining Between Centers

THREADS PER INCH TO BE CUT	WHEN TO ENGAGE SPLIT NUT		READING ON DIAL
EVEN NUMBER OF THREADS	ENGAGE AT ANY GRADUATION ON THE DIAL	1 1 ½ 2 2 ½ 3 3 ½ 4 4 ½	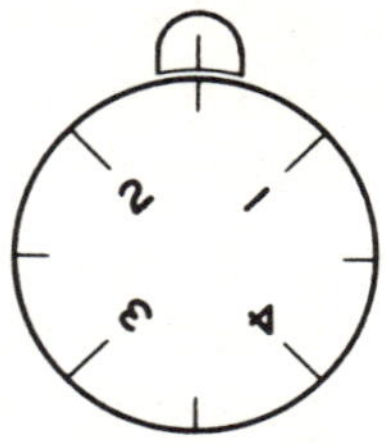
ODD NUMBER OF THREADS	ENGAGE AT ANY MAIN DIVISION	1 2 3 4	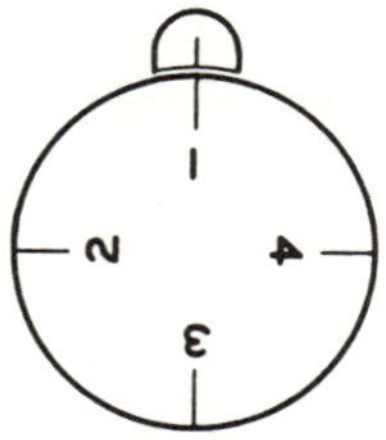
FRACTIONAL NUMBER OF THREADS	1/2 THREADS, E.G. 11 1/2 ENGAGE AT EVERY OTHER MAIN DIVISION 1 & 3, OR 2 & 4 OTHER FRACTIONAL THREADS ENGAGE AT SAME DIVISION EVERY TIME		
THREADS WHICH ARE A MULTIPLE OF THE NUMBER OF THREADS PER INCH IN THE LEAD SCREW	ENGAGE AT ANY TIME THAT SPLIT NUT MESHES		USE OF DIAL UNNECESSARY

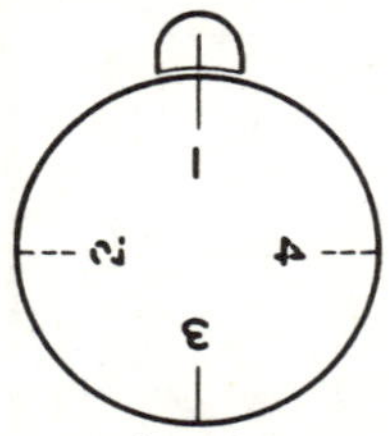

FIG. 21-3 RULES FOR ENGAGING THE SPLIT-NUT FOR THREAD CUTTING

Note: It is good practice to have the diameter .002 – .003 undersize.

2. Set the spindle speed to approximately 1/4 of the turning speed.
3. Set the quick-change gearbox to the required number of threads per inch.
4. Engage the lead screw.
5. Set the compound rest to 29° to the right, figure 21-4. Note: With the compound rest set to 29°, a slight shaving action occurs on the following edge of the thread and tends to produce a better finish.

FIG. 21-4 THE COMPOUND REST IS SWUNG 29° TO RIGHT WHEN CUTTING 60° RIGHT-HAND THREADS.

Unit 21 Thread Cutting

6. Select a 60° threading tool and check it for accuracy with a center gage, figure 21-5.

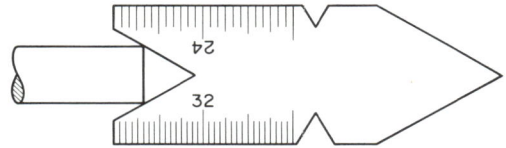

FIG. 21-5 A CENTER GAGE IS USED TO CHECK THE ANGLE OF A 60° THREADING TOOL.

FIG. 21-6A A THREADING TOOLBIT GROUND FOR USE IN A LEFT-HAND OFFSET TOOLHOLDER

FIG. 21-6B A THREADING TOOLBIT GROUND FOR USE WITH A STRAIGHT TOOLHOLDER

7. Set the threading toolbit in a toolholder, having it extend only about 1/2", figure 21-6. *Note:* Sixty-degree threading tools can be ground in two different ways, figure 21-6. Figure 21-6A shows a threading tool ground for use with a left-hand offset toolholder. The toolbit shown in figure 21-6B is used with a straight toolholder and is quite easy to sharpen. When the tool blank is ground, only side clearance should be ground on the following edge for a length of about 1/2". The leading edge is then ground to form an angle of 60° with the following side. Whenever the toolbit requires resharpening, only the leading edge should be ground.

8. Set the point of the toolbit on center.

9. Mount the work between centers.

10. *Mark the drive plate slot* in which the tail of the dog is engaged. This insures that the work is replaced in the same position when it is removed from the lathe for testing a thread.

11. Set the toolbit square with the work using a center gage, figure 21-7.

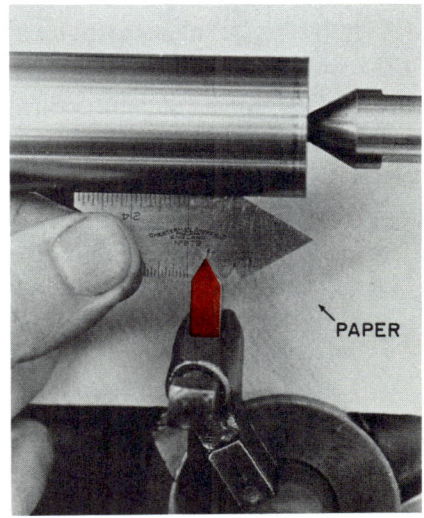

FIG. 21-7 METHOD OF SETTING THREADING TOOL SQUARE WITH THE WORK. NOTE THE PAPER.

Section 2 Machining Between Centers

A piece of paper placed on the cross-slide under the work makes it easier to check the tool alignment.

Note: Never jam the threading tool into the center gage. Align left side of cutting tool with left side of *V* in the center gage.

12. Arrange the apron controls as required for thread cutting.

CUTTING 60° THREADS

Threading probably requires more attention and skill than any other machining operation performed on a lathe. After the lathe has been set up for threading, a few trial passes without actual cutting should be taken by engaging the split-nut lever in order to get the "feel" of the lathe.

Procedure: Cutting a 60° Right-hand Thread

1. Start the lathe and, with the leading edge of the threading tool, chamfer the end of the workpiece to just below the minor diameter.

2. Move the toolbit in with the crossfeed handle until it just touches the diameter to be threaded.

OR

   a. Bring the toolbit close to the work and stop when the handle of the crossfeed screw is at the 3 o'clock position, figure 21-8.

FIG. 21-8 THE HANDLE OF THE CROSSFEED SCREW SET AT APPROXIMATELY 3 O'CLOCK POSITION FOR THREAD CUTTING

   b. With the compound rest handle, move the toolbit in until it lightly marks the the diameter of the revolving work.

Note: If the crossfeed handle is in the same relative position for every thread-cutting operation, through repetition, an operator very quickly finds thread cutting an easy, routine operation.

3. Set the crossfeed and compound rest graduated collars to zero.

4. Move the carriage to the right until the toolbit clears the end of the work.

5. Using the compound rest feed handle, feed the tool in .003 – .005.

6. Engage the split-nut lever at the proper line on the thread-chasing dial, figure 21-3, and take a trial cut along the work.

7. At the end of the cut, first back out the toolbit, using the crossfeed handle, and then disengage the split nut.

8. Stop the lathe and check the number of threads per inch with a thread pitch gage, figure 21-9A. If a thread pitch gage is not available, a ruler or a center gage can be used, figures 21-9B and C.

9. Move the carriage to the right until the cutting tool clears the end of the work and turn the crossfeed handle back to zero.

10. Set the depth of cut with the compound rest feed handle.

    a. The first two cuts should be about .015 – .020 deep.

    b. Succeeding cuts should be governed by the depth of the thread and should be about .010.

    c. Reduce the depth of the cuts as the thread becomes deeper.

    d. The last few cuts should be about .001 deep.

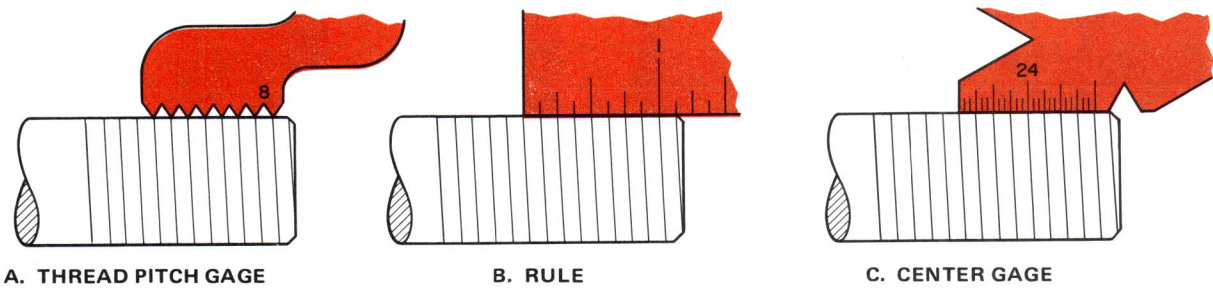

A. THREAD PITCH GAGE **B. RULE** **C. CENTER GAGE**

FIG. 21-9 CHECKING THE NUMBER OF THREADS PER INCH

11. Take successive cuts until the thread is the desired depth, table 21-1, or until the thread fits a gage nut freely with no play. *Note:* To improve the finish of the thread, cutting fluid should be used when thread cutting.

RESETTING A THREADING TOOL

Whenever the relative position of the work and the threading tool is changed, the tool must be reset in order to finish the thread. The threading tool must be reset for the following reasons, otherwise the thread may be ruined.

- The threading tool has been removed for grinding.
- The toolholder has moved due to the pressure of the cut.
- The work has been removed from the machine before the threads have been completed.
- The dog has slipped on the work.

DEPTH SETTINGS FOR NATIONAL FORM THREADS

T.P.I.	Compound Rest Setting		
	0°	30°	29°
24	.027	.031	.0308
20	.0325	.0375	.037
18	.036	.0417	.041
16	.0405	.0468	.046
14	.0465	.0537	.0525
13	.050	.0577	.057
11	.059	.068	.0674
10	.065	.075	.074
9	.072	.083	.082
8	.081	.0935	.092
7	.093	.1074	.106
6	.108	.1247	.1235
4	.1625	.1876	.1858

TABLE 21-1

Note: When using this table for cutting National Form threads, the correct width of flat (.125P) must be ground on the toolbit, otherwise the thread will not be the correct width.

FIG. 21-10 USE ONLY THE CROSSFEED AND COMPOUND REST HANDLES WHEN RESETTING A THREADING TOOL.

Unit 21 Thread Cutting

Procedure: Resetting a Threading Tool

1. Set up the lathe and the work as for thread cutting.

2. Start the lathe; with the toolbit clear of the work, engage the split-nut lever at the proper line on the thread-chasing dial.

3. Allow the carriage to move along until the toolbit is opposite any portion of the threaded section.

4. Stop the machine, *leaving the split-nut lever engaged.*

5. Using only the *crossfeed* and the *compound rest feed handles,* figure 21-10, feed the toolbit into the groove until the following edge of the toolbit touches the following edge of the thread.

6. Set the crossfeed graduated collar to zero.

7. Back out the threading tool, using the crossfeed handle, and move the carriage to the right until the toolbit clears the end of the work.

8. Turn the crossfeed handle in to zero and take a trial cut.

9. For successive cuts, feed the compound rest the desired amount.

10. Continue cutting until the thread is to the required depth.

Procedure: Cutting a Left-Hand Thread (National Form or Unified)

When there is a tendency for a right-hand thread to loosen due to the rotation of the spindle, a left-hand thread should be used. The left-hand thread tends to tighten under the same condition as previously mentioned. The pedestal grinder is a classic example of right- and left-hand thread applications on the same spindle. The thread on the right-hand end of the spindle (when standing in front of the grinder) has a right-hand thread. The thread on the left end is left-

FIG. 21-11 THE COMPOUND REST IS SET 29° TO THE LEFT WHEN CUTTING LEFT-HAND THREADS.

hand. If these were reversed, the grinding wheels would loosen in use.

Although the basic procedure for cutting right- and left-hand threads is the same, there are a few changes in the machine setup.

1. Check the diameter of the part to be threaded with the blueprint and determine the number of threads per inch to be cut.

2. Set the spindle speed and the quick-change gearbox as for right-hand threads.

3. Engage the lead screw to revolve in the opposite direction as for a right-hand thread.

4. Set the compound rest at 29° to the left, figure 21-11.

5. Set up the left-hand threading tool and square it with the work.

6. When cutting the left-hand thread, the threading tool is set to the left end of the section to be threaded. *Note:* A groove, generally to the minor diameter, is cut at the starting point of the thread, figure 21-12.

7. Proceed to cut the thread to the same dimensions as for a right-hand thread.

Unit 21 Thread Cutting

FIG. 21-12 A GROOVE IS USED AT THE BEGINNING OF A LEFT-HAND THREAD.

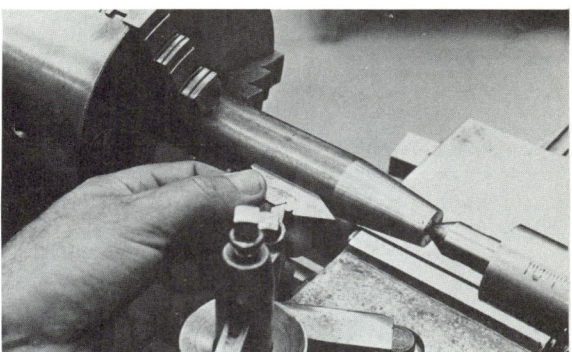

FIG. 21-13 THE TOOLBIT MUST BE SET SQUARE WITH THE AXIS OF THE WORK WHEN CUTTING THREADS ON A TAPERED SECTION.

Cutting a Thread on a Tapered Section

When a thread such as a pipe thread is required on a workpiece, either the taper attachment or the offset tailstock method may be used for cutting the taper. The threads are then cut on the tapered section, using the same tapering setup and the regular setup for thread cutting. *Note:* It is most important that the threading tool be set at 90° to the axis of the work and not square with the tapered section, figure 21-13.

ACME THREADS

The Acme Thread form is generally replacing the square thread because it is stronger, easier to cut on the lathe, and can be made with taps and dies. This thread has a clearance of .010 at the crest and the root of mating threads. The hole for an internal Acme thread is cut .020 larger than the minor diameter of the screw, while the major diameter of the internal thread is .020 larger than the screw. Taps for Acme threads are .020 larger than the major diameter of the screw.

Procedure: Cutting a Right-Hand Acme Thread

1. Grind the toolbit to fit the 29° angle in the end of the Acme thread gage, figure 21-14. Allow sufficient side clearance so that the sides of the tool do not rub on the workpiece when cutting the thread.

2. Grind the point of the toolbit flat until it fits into the proper slot on the gage. The width of the toolbit point is then correct for the pitch indicated on the gage. If a gage is not available, the width of the toolbit point can be calculated as follows and then measured with a micrometer.

   width of point = $\dfrac{.3707}{N}$ - .0052

3. Set the proper spindle speed.

4. Set the quick-change gearbox to the required number of threads per inch.

5. Set the compound rest 14 ½° to the right.

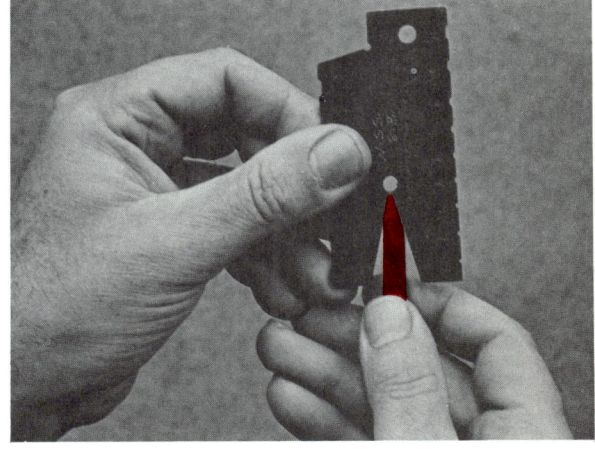

FIG. 21-14 AN ACME THREAD GAGE IS USED TO CHECK THE ANGLE OF THE THREADING TOOL.

133

Section 2 Machining Between Centers

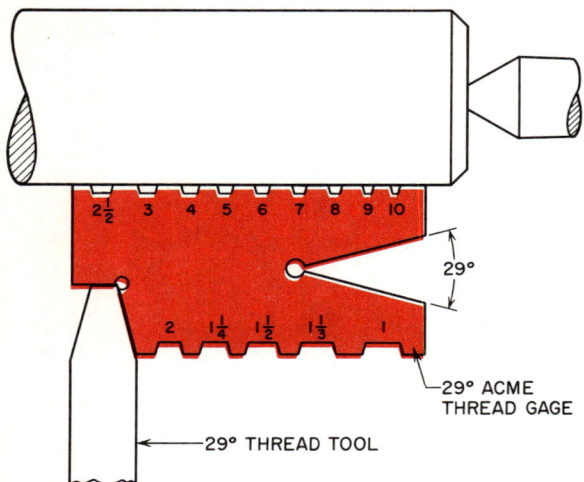

FIG. 21-15 THE TOOLBIT IS SQUARED WITH WORK BY AN ACME THREAD GAGE.

6. Set the point of the threading tool to center height.

7. Mount the work on the lathe.

8. Set the toolbit square with the work using the Acme thread gage, figure 21-15.

9. Cut a section 1/16" long on the right end of the work to the thread minor diameter. This indicates when the thread is cut to the proper depth.

10. Cut the thread to depth using the compound rest handle. *Note:* Some machinists prefer to rough out coarse pitch threads first with a square tool, the width of which is slightly smaller than the root of the Acme thread. The Acme toolbit is then required only to form the angular sides.

SQUARE THREADS

Although the square thread has been superseded by the Acme thread, it may be necessary for the machinist to cut a square thread when replacing parts in certain machines and equipment.

If a person has mastered the cutting of American Form and Unified threads, no problems should be encountered in cutting square threads. One of the biggest problems in producing a square thread is that of properly grinding the toolbit.

Square Threading Tool Calculations

A square threading tool must be ground so that it fits into the helical groove of the thread to be cut. It must have the same slope as the helix angle of the thread and clearance must be provided so that the sides do not rub against the work, figure 21-16A. It should be understood that the helix angle of a thread varies with the *lead and the diameter of the thread*. Because of this, it is first necessary to calculate the leading and following angles of the thread. The leading and following edge angles of the thread can be represented by a right-angle triangle, figure 21-16B. From this

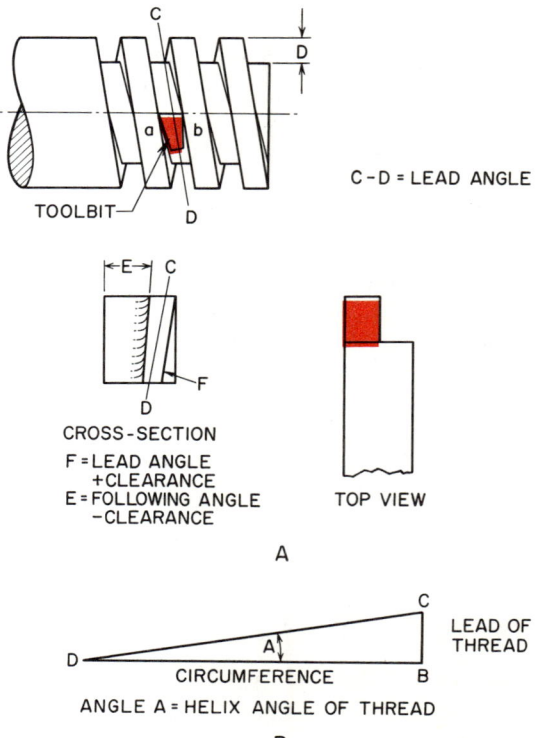

FIG. 21-16 SHAPE AND CALCULATIONS REQUIRED FOR A SQUARE THREADING TOOL

triangle it can easily be seen that the helix angle changes if the lead or the diameter (and circumference) is altered.

By referring to figure 21-16B, the leading and following edge angles of the thread can be calculated as follows:

$$\text{Tan leading angle} = \frac{\text{lead of thread}}{\text{circum. of minor dia.}}$$

$$\text{Tan following angle} = \frac{\text{lead of thread}}{\text{circum. of major dia.}}$$

After these angles have been calculated, side clearance must be provided for the toolbit.

Leading angle   add 1°

Following angle   subtract 1°

Example: Calculate the leading and following angles of a threading tool required to cut a 1 1/8" − 6 square thread.

Solution: Lead = 1/6

             = .1667

Single depth = $\frac{.500}{6}$

             = .083

Double depth = .166

Major diameter = 1.125

Minor diameter = 1.125 − .166

             = .959

$$\text{Tan leading angle} = \frac{\text{lead}}{\text{circum. of minor dia.}}$$

$$= \frac{.1667}{.959 \times 3.1416}$$

$$= \frac{.1667}{3.0128}$$

$$= .0553$$

∴ The leading angle of the thread = 3° 10′

Angle to which to grind the leading edge of the toolbit

$$= 3° \ 10' + 1° \text{(clearance)}$$

$$= 4° \ 10'$$

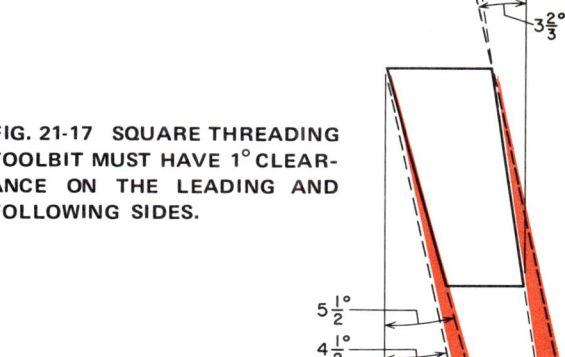

FIG. 21-17 SQUARE THREADING TOOLBIT MUST HAVE 1° CLEARANCE ON THE LEADING AND FOLLOWING SIDES.

$$\text{Tan following angle} = \frac{\text{lead}}{\text{circum of major dia.}}$$

$$= \frac{.1667}{1.125 \times 3.1416}$$

$$= \frac{.1667}{3.5343}$$

$$= .04716$$

∴ The following angle of the thread = 2° 42′

Angle to which to grind the following edge of the toolbit

$$= 2° \ 42' - 1° \text{ (clearance)}$$

$$= 1° \ 42'$$

Procedure: Cutting a Square Thread

Although cutting a square thread involves procedures similar to other thread cutting methods, it probably requires more care and attention.

1. Calculate the leading and following angles for the toolbit.

2. Grind the toolbit .002 wider than one half the thread pitch. This allows sufficient clearance for proper assembly of mating parts. *Note:* If a coarse pitch thread is to be cut, it is advisable to first cut the thread to depth using a roughing or stocking tool about .015 smaller than the thread groove. This should be followed with the finishing tool .002 wider than one half the thread pitch.

Section 2 Machining Between Centers

3. Check and align the lathe centers.
4. Set the spindle speed.
5. Set the quick-change gearbox for the required number of threads per inch.
6. Set the compound rest at 30° to the right. This moves the compound rest clear of the crossfeed handle and provides longitudinal movement if it is necessary to pick up the thread.
7. Set the cutting tool on center.
8. Mount the work.
9. Set the threading tool square with the work.
10. Cut the right-hand end of the work to the minor diameter for about 1/16″ long. This indicates when the thread is at the proper depth.
11. If possible, cut a groove to the minor diameter at the other end of the threaded section. This permits the cutting tool to "run out" at the end of the thread.
12. Calculate the depth of the thread (1/2P or .500/N).
13. Start the machine and touch the toolbit to the diameter of the workpiece.
14. Set the *crossfeed graduated collar* to zero.
15. Move the cutting tool in .002 – .003 and take a trial cut.
16. Check the thread with a screw pitch gage.
17. Using the *crossfeed handle,* move the tool in .005 – .007 for each cut. The depth depends on the size of the thread, the shape of the workpiece, and the condition of the machine. Long threads should be supported with a follower rest. *Note: All cuts are made using the crossfeed handle.*
18. Cut the thread to depth, using cutting fluid on each pass.

MULTIPLE-START THREADS

When it is necessary to increase the rate of travel along a screw or threaded shaft, the lead of the thread must be increased. If the lead is increased on a single-start thread, the depth also has to be increased which often weakens the thread. To overcome this problem and to create a stronger, more attractive thread, multiple-start threads are used.

The *pitch* of a thread is the distance from a point on one thread to a corresponding point on the next thread.

The *lead* is the distance the thread advances in one revolution. On a single-start thread the lead is equal to the pitch. On multiple-start threads, the lead is equal to the pitch times the number of starts on the threaded section. For instance, the lead is twice the pitch on a two-start thread; on a triple-start thread, the lead is three times the pitch, figure 21-18.

When cutting multiple threads, the depth is always calculated in relation to the pitch and

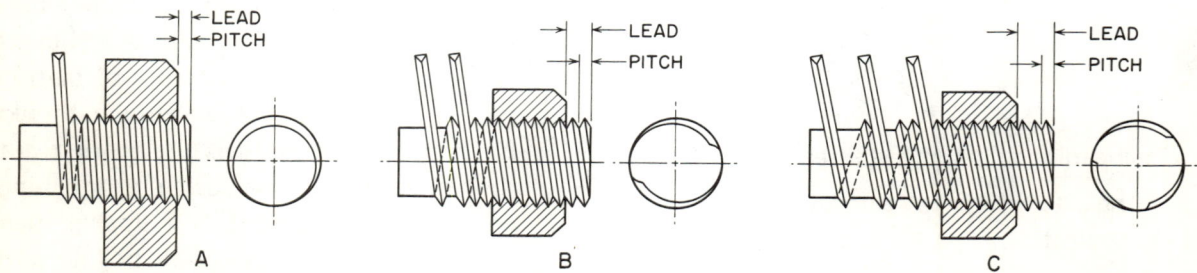

FIG. 21-18 THE RELATIONSHIP OF PITCH TO LEAD IN SINGLE- AND MULTIPLE-START THREADS. (A) SINGLE THREAD, (B) DOUBLE THREAD, (C) TRIPLE THREAD. (Courtesy Cincinnati Milacron Co.)

not the lead. Although there does not appear to be any standard way to designate multiple-start threads, it is important that there be no confusion in the thread identification; the pitch, the lead, and the number of starts should always be specified. For example, a 3/4" - 2 start - 16 pitch thread is fully identified as follows: 3/4" - 16 UNC - 1/8 lead - 2 start. This method leaves no doubt as to the setting of the quick-change gearbox.

Multiple-start threads can be cut on the lathe by several methods:

- By means of an accurately indexed or slotted driving plate. This permits rotation of the work the desired amount to cut the required number of thread starts. When cutting a double-start thread, the dog is rotated to a diagonally opposite slot or hole in the driveplate.

- By disengaging the intermediate gear on the end gear train and rotating the spindle the desired amount.

- By using the thread-chasing dial. This method applies only to certain threads, for example, a double-start thread with *an odd number or a fractional lead*.

- By setting the compound rest parallel to the centerline of the lathe. This method permits the use of the compound rest graduated collar to position the threading tool for the next start when cutting multiple threads. When this method is used, the threading tool must be fed into the work with the crossfeed handle.

Procedure: Cutting a Multiple Thread

A double-start 60° thread, having 12 threads per inch and a 1/6 lead is required.

1. Set up the lathe for cutting a single-start thread.
2. Set the quick-change gearbox to the *lead* of the thread (6 tpi).
3. Calculate the depth of thread and the amount of compound rest feed required for 12 tpi (6 tpi to one-half depth).
4. Cut the first thread to within .001 of the calculated depth, using the compound rest feed. Note the reading on the compound rest graduated collar.
5. Back out the threading tool using the compound rest handle. *Do not move the crossfeed handle setting.*
6. Revolve the work one-half turn by either of the following methods:

    a. *Driveplate Method*

    1) Remove the work from the lathe centers, leaving the dog attached to the work.

    2) Rotate the work one-half turn so that the tail of the dog engages in the diagonally opposite slot and set the work between centers.

    OR

    b. *Gear Train Method* (figure 21-19)

    1) Rotate the spindle by hand until one tooth in the spindle gear *A* is oppo-

FIG. 21-19 GEAR TEETH IN THE GEAR TRAIN MUST BE MARKED PRIOR TO INDEXING FOR MULTIPLE THREADS.

site a groove in the intermediate gear *B* and mark both of these with chalk, figure 21-19. Also mark a corresponding tooth and groove on the intermediate gear *B* and the lead screw gear *C*.

2) Count the number of teeth in the spindle gear *A* (for instance, 24) and divide this in half. Starting with the tooth *next to the marked one,* count 12 teeth (half the number in the spindle gear) and mark the twelfth tooth with chalk.

3) Disengage the intermediate gear by sliding it out of engagement with the spindle gear or both gears if necessary.

4) Revolve the spindle by hand until the marked upper tooth has revolved to the approximate lower position.

5) Re-engage the intermediate gear with the other gears, being sure that all chalk marks line up.

7. Since the crossfeed handle was not moved, feed the compound rest in until the toolbit lightly marks the revolving work.

8. Use the compound rest and cut the second thread to the calculated depth (.001 more than for the first thread).

9. Leave the crossfeed and compound rest handles at the final depth setting, index the work one-half turn, and take one pass through the first thread. This insures that both threads are cut to exactly the same depth.

The Thread-Chasing Dial Method of Cutting Multiple Threads

When a double-start thread has an odd-numbered lead (1/3, 1/5, etc.), or a fractional lead (1/1 1/2, 1/1 1/3, etc.), both threads may be cut in the same setup.

Procedure for Odd-Numbered Leads

1. Set up the lathe for thread cutting.
2. Take one cut along the thread by engaging the split-nut lever on *a numbered line.*
3. Return the cutting tool to the start of the thread.
4. Return the crossfeed handle to zero and engage the split-nut lever *on an unnumbered line.* This locates the second thread in the center of the first thread.
5. Continue cutting the threads to depth by alternately engaging on a numbered line and an unnumbered line for each compound rest setting. *Note:* Threads having fractional leads can be cut in a similar manner as those with odd-numbered leads. In this case, the split-nut lever must be engaged at opposite numbered divisions for one cut (e.g., 1 or 3). For the second cut, it should be engaged at either of the two other numbered lines (2 or 4).

Procedure: Cutting a Multiple Thread Using the Compound Rest Method

The compound rest may be set parallel to the ways of the machine and used to index the threading tool to cut multiple threads, figure 21-20.

To cut a two-start NC thread having 1/8 lead (1/16 pitch):

1. Convert the pitch of the thread to thousandths, e.g., 1/16 = .062.
2. Set the compound rest parallel to the ways of the machine (90° to the cross-slide).
3. Set up the lathe for thread cutting.
4. Set the quick-change gearbox to cut 8 tpi.
5. Turn the compound rest feed handle *1/4 turn clockwise* to remove the backlash and set the graduated collar to zero.
6. Using the crossfeed handle, move the threading tool in until it lightly touches

Unit 21   Thread Cutting

the work. Set the crossfeed graduated collar to zero.

7. Calculate the depth of thread (.6495/16 = .041).

8. Cut the first thread to depth (.041), using the crossfeed handle for setting the depth for each cut.

9. With the cutting tool clear of the work, advance the compound rest handle clockwise .062. This positions the threading toolbit for cutting the second thread.

10. Cut the second thread to depth, making all settings by feeding the crossfeed handle.

FIG. 21-20 THE COMPOUND REST MAY BE SET AT 90° FOR CUTTING CERTAIN MULTIPLE THREADS.

METRIC THREADS

Although the International Metric standard screw thread has similar proportions to that of the National Form thread, the method of identifying the pitch differs. In all threads except Metric, the pitch is generally referred to as threads per inch; with the Metric thread the pitch is expressed in millimeters (mm.).

Metric threads can be cut on any lathe that is equipped with two change gears having 50 and 127 teeth respectively.

Since most lathes in this country are equipped for the English system of measurement, it is necessary that the machinist understand the relationship of the English and the Metric systems of measurement.

1″ = 2.54 centimeters

∴ The ratio of inches to centimeters is 1 : 2.54 or 1/2.54.

To cut Metric threads on a standard quick-change lathe, it is necessary to incorporate certain gears in the gear train to produce a ratio of 1/2.54.

The required gears are $\frac{1}{2.54} \times \frac{50}{50} = \frac{50}{127}$

In order to cut Metric threads, these gears are placed in the gear train of the lathe; the 50-tooth gear is placed on the spindle and the 127-tooth gear is placed on the lead screw.

Procedure: Cutting a 2mm. Metric Thread on a Standard Quick-Change Lathe

1. Mount the 50-tooth gear on the spindle and the 127-tooth gear on the lead screw.

2. Convert the 2-millimeter pitch to the number of threads per centimeter.

   10 millimeters = 1 centimeter

   ∴ Pitch = $\frac{10}{2}$ = 5 threads per centimeter

3. Set the quick-change gearbox to 5 threads per inch. By incorporating the 50- and 127-tooth gears into the gear train, the lathe now cuts 5 threads per centimeter (2 mm.) instead of 5 threads per inch.

4. Set up the lathe for thread cutting.

Section 2 Machining Between Centers

5. Take a light trial cut; at the end of the cut *do not disengage the split-nut lever.*
6. Reverse the spindle rotation until the cutting tool is just past the end of the threaded section.
7. Check the thread with a Metric thread pitch gage.
8. Cut the thread to the required depth, being careful *not to disengage the split nut* until the thread is complete.

REVIEW QUESTIONS

1. Describe the construction of the quick-change gearbox. What is its purpose?
2. What is the procedure for setting a quick-change gearbox for thread cutting?
3. Describe the construction of a thread-chasing dial.
4. How is the thread-chasing dial used in the thread-cutting operation?
5. What do the lines on a thread-chasing dial indicate?

Lathe Setup for Cutting 60° Threads

6. What are five methods of producing a thread?
7. At what angle should the compound rest be set to cut a 60° right-hand thread? Explain why.
8. What is the purpose of a thread center gage?
9. How do the two styles of 60° threading tools compare? Discuss their use, regrinding, and advantages.
10. How is a threading toolbit set square with the workpiece?

Cutting 60° Threads

11. What is the procedure for setting a threading toolbit to the diameter to be threaded?
12. What three methods can be used to check the number of threads per inch?
13. Why should the depth of cut become progressively smaller as the thread is cut deeper?

Resetting a Threading Tool

14. What are three reasons why a threading tool may have to be reset?
15. What are the steps in the procedure used to reset a toolbit into a thread groove?

Cutting a Left-Hand Thread

16. What purpose do left-hand threads serve?

17. What are three differences in the lathe setup for left-hand threads as compared with right-hand threads?

Cutting a Thread on a Tapered Section

18. How is a threading tool set up when cutting threads on a tapered section?

Acme Threads

19. Compare the sizes of an internal and external Acme thread.

20. Calculate the width of the point of an Acme threading tool used to cut four threads per inch.

21. How does the compound rest setting for an Acme thread differ from that for an American National thread? Explain why.

22. What procedure is recommended when cutting coarse pitch Acme threads?

Square Threads

23. Illustrate with a neat sketch the shape of a square threading tool in a thread groove.

24. What two factors affect the helix angle of a thread?

25. Calculate the leading and following angles for a square threading tool used to cut a 1 1/2" – 4 square thread.

26. How is a threading tool ground so that its sides do not rub against the sides of the thread?

Cutting a Square Thread

27. What procedure should be followed when cutting a coarse pitch square thread?

28. How should the compound rest be positioned for cutting square threads? Explain why.

29. How can the workpiece be prepared to indicate when a thread is cut to the proper depth?

30. How are successive cuts taken when cutting square threads?

Multiple-Start Threads

31. What are three purposes of multiple threads?

32. How does the pitch and lead of a multiple-start thread compare with that of a single-start thread?

33. How is the depth of a multiple-start thread calculated?

Section 2   Machining Between Centers

34. What are three methods of indexing the work when cutting multiple threads? Briefly describe each method.

Cutting a Multiple Thread

35. What determines the setting of the quick-change gearbox for multiple threads?

36. How is the second thread cut to depth in a double-start thread?

37. For what threads can the thread-chasing dial be used? Explain the procedure.

38. How can the compound rest be used to space multiple threads accurately?

Metric Threads

39. What two gears are essential in order to cut Metric threads on a standard lathe?

40. How are these two gears derived and where is each placed in the gear train?

41. Outline the procedure for *cutting* a Metric thread.

Unit 22   THREAD MEASUREMENT

Modern industrial practices demand that components be manufactured to certain standards so that they are interchangeable with other parts on assembly. Since threaded components are widely used, it is important that threads be cut to an accurate size to assure interchangeability. There are many methods of checking the accuracy of a thread, depending on the accuracy required and the equipment available.

The more common methods of measuring a thread are:

- *Gages,* such as the thread ring, thread plug, and thread snap gage
- *Micrometers,* such as the screw thread micrometer, and the thread comparator micrometer
- *The three-wire method*
- *Optical comparators*

Whenever a thread must be accurate, its size should be measured on the pitch diameter. The size at the pitch diameter is more important than the size of the thread at the root or crest since the contact with a mating part should occur on the angular sides of a thread for a proper fit.

GAGES

Thread ring gages, figure 22-1, consisting of a "go" and "no-go" gage, are used to check the accuracy of external threads. Both gages have a knurled outside diameter; however, an annular groove cut on the knurled surface identifies the "no-go" gage. Each gage has a threaded hole in the center and three radial slots, one of which is equipped with a setscrew to permit a small adjustment. The "no-go" gage only determines whether the pitch diameter is below minimum limits.

Both gages should be thoroughly cleaned and lightly oiled to prolong the life of the gage before being used. When checking a thread, the threaded components should enter the "go" gage freely but should not enter the "no-go" gage for more than 1 1/2 turns.

Thread plug gages, figure 22-2, are used to check the accuracy of internal threads. Thread plug gages consist of a "go" and "no-go" gage mounted either in individual holders, figures 22-2A and B, or in the same holder, figure 22-2C. The "go" gage, which always has a longer thread, has a chip groove for cleaning the thread in a hole.

FIG. 22-1 A "GO" AND "NO-GO" THREAD RING GAGE HELD IN A HOLDER (Courtesy Greenfield Tap and Die)

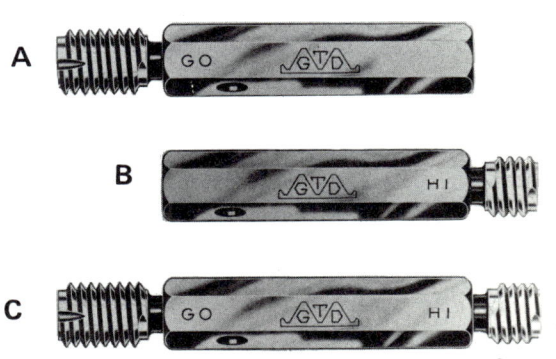

FIG. 22-2 THREAD PLUG GAGES, A. "GO" GAGE, B. "NO-GO" GAGE, C. DOUBLE-END GAGE (Courtesy Greenfield Tap and Die)

Section 2 Machining Between Centers

FIG. 22-3 (A) AN ADJUSTABLE ANVIL THREAD SNAP GAGE, (B) A THREAD ROLL SNAP GAGE
(Courtesy Taft-Peirce Mfg. Co.)

Thread plug gages are used to determine whether the pitch diameter of the internal threads is within the acceptable limits. The "go" gage is manufactured to the minimum pitch diameter, while the "no-go" gage is made to the maximum pitch diameter. If an internal thread is within limits, the "go" gage should enter the threaded hole for the full length of the gage, but the "no-go" gage should not enter a hole for more than three threads.

Thread snap gages, figure 22-3, which may have either threaded rolls or grooved anvils can be used to check the accuracy of external threads. The *outer* or *"go" rolls (anvils)* are set to the maximum pitch diameter of the thread and check all thread elements simultaneously. The *inner* or *"no-go" rolls* are set to check only the minimum pitch diameter of the thread. External threads which are the correct diameter pass through the outer "go" rolls, but are stopped by the inner "no-go" rolls.

MICROMETERS

A *screw thread micrometer,* figure 22-4, can be used to check accurately the pitch diameter of 60° threads such as the American National and Unified threads. This micrometer is equipped with a V-shaped swivel anvil and a spindle which is cone-shaped. When the 60° micrometer spindle point is brought into contact with the anvil, the zero on the micrometer represents a line drawn through the plane A-B, figure 22-4. Since this line, or point of contact is midway between the pointed spindle and the V-shaped anvil, the micrometer reading represents the pitch diameter of a thread.

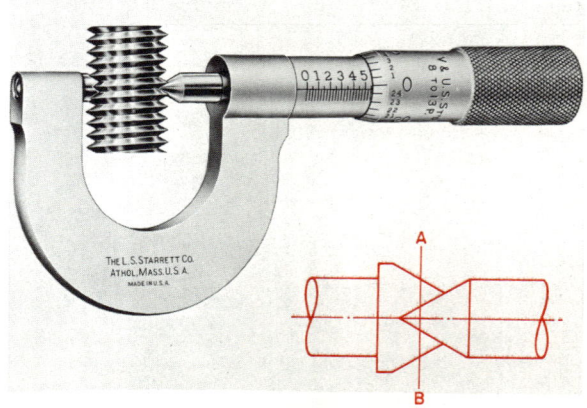

FIG. 22-4 THE SCREW THREAD MICROMETER MEASURES THE PITCH DIAMETER OF A THREAD. (Courtesy L.S. Starrett Co.)

Screw thread micrometers are limited to measuring threads within a certain range depending upon the V-shaped anvil. The range of each micrometer is stamped on its frame. Screw thread micrometers having a capacity of 1" are available to cover four ranges of threads per inch.

8 to 13 tpi	22 to 30 tpi
14 to 20 tpi	32 to 40 tpi

The screw thread micrometer does not give an absolutely accurate pitch diameter measurement because of the helix angle of the thread. Usually this error is minute and, therefore, of no practical importance. However, if extreme accuracy is required, it is recommended that the micrometer be set to a master thread plug gage which has the same pitch and diameter as the thread required.

FIG. 22-5 A THREAD COMPARATOR MICROMETER COMPARES THE ACCURACY OF A THREAD WITH A KNOWN STANDARD. (Courtesy L.S. Starrett Co.)

A *thread comparator micrometer*, figure 22-5, has two conical points, and can be used to compare quickly the accuracy of a thread with a known standard. This micrometer does not measure the pitch diameter of a thread and, therefore, must first be set to a standard thread plug gage. The threaded workpiece must be measured, and its reading compared with the micrometer reading of the thread plug gage.

THREE-WIRE THREAD MEASUREMENT

The three-wire method of thread measurement is recommended by the National Screw Thread Commission and the Bureau of Standards as a reliable method of measuring threads and securing uniformity. It is considered one of the best methods of measuring the pitch diameter, since it is least affected by any error which may be present in the included angle of the thread.

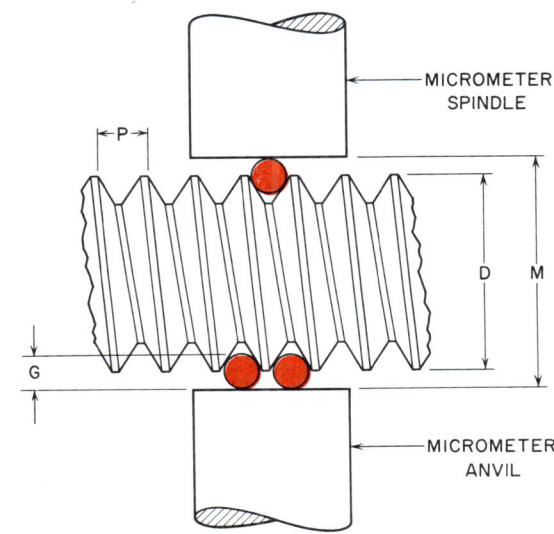

FIG. 22-6 THREE WIRES ARRANGED CORRECTLY FOR MEASURING 60° THREADS

Three hardened and lapped wires of equal diameter are placed in the thread groove; two on one side and one on the other side, figure 22-6. If the thread must be accurate to within a thousandth of an inch, a standard micrometer can be used to measure the distance over the wires M. A dial gage setup should be used if the thread accuracy is required in ten-thousandths of an inch. Electronic comparators should be employed if the accuracy is required in hundreds of thousandths of an inch.

Different size wires are used in the measurement of different size threads; the larger the thread, the larger the wire size that should be used. The *best size wire* is recommended when the greatest accuracy is required, since it contacts the side of the thread exactly at the pitch line, figure 22-7. However, since the best size wire is not always available, larger or smaller wire sizes may be used for measuring a thread. The wires should not be so large that they contact the edge of the crest nor so small that they are below the major diameter of the thread.

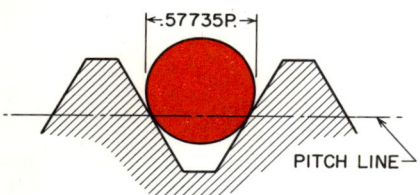

FIG. 22-7 THE BEST SIZE WIRE CONTACTS THREAD AT PITCH LINE.

Calculating the Measurement Over the Wires

The following formula can be applied to calculate the measurement over the wires M for American National threads:

$$M = D + 3G - \frac{1.5155}{N}$$

where: M = Measurement over the wires
D = Major diameter of the thread
G = Wire size diameter
N = Number of threads per inch

Since the best size wire is not always available, any of the following formulas can be used to calculate the wire size G.

- Largest size wire $= \dfrac{1.010}{N}$ or 1.010 P

- Best size wire $= \dfrac{.57735}{N}$ or .57735 P

- Smallest size wire $= \dfrac{.505}{N}$ or .505 P

Any size wire between the calculated minimum and maximum size can be used to measure thread. However, for the greatest accuracy, the best size wire (.57735 P) is recommended since it contacts the thread at the pitch line.

Example: Calculate the measurement over the wires M for a 7/8 – 9NC thread using the best size wire.

1. Calculate the wire size G.

$$G = \frac{.57735}{9}$$

$$= .06415$$

2. Calculate the measurement over the wires M.

$$M = D + 3G - \frac{1.5155}{N}$$

$$= .875 + (3 \times .06415) - \frac{1.5155}{9}$$

$$= .875 + .1925 - .1684$$

$$= 1.0675 - .1684$$

$$= .8991$$

Simplified Three-Wire Formula

The following simplified three-wire thread formula can be applied *only if the best size wire is used.*

$$M = D + \frac{.2165}{N}$$

Using the same thread as in previous example

$$M = .875 + \frac{.2165}{9}$$

$$= .875 + .0241$$

$$= .8991$$

Table 22-1 contains the best size wire and the measurement over the wires for a number of American Standard threads. Consult a handbook for formulas and best size wire for other threads which can be measured by the three-wire method.

OPTICAL COMPARATORS

An *optical comparator,* figure 22-8, that is equipped with a screw thread accessory, can be used to check accurately the thread angle, helix angle, depth, and width of a thread. The part is first mounted between the centers of the screw thread accessory, and its image is projected onto the screen. The frame is set to the helix angle of the thread in order to bring the thread parallel to the light beam. With this setup, various thread characteristics can be measured accurately, depending upon the magnification of the optical comparator.

Measuring Acme Threads

A three-wire system can also be employed to measure the accuracy of Acme threads. Consult a handbook for the formula and also

Thread Diameter (D)	Number of Threads per Inch (N)	Diameter of Best Wire $G = \frac{.57735}{N}$	Measurement Over Best Wire Diameter (M)
1/4	28	.0206	.2577
1/4	20	.0288	.2608
5/16	24	.0240	.3215
5/16	18	.0320	.3245
3/8	24	.0240	.384
3/8	16	.0360	.3885
7/16	20	.0288	.4483
7/16	14	.0412	.4529
1/2	20	.0288	.5108
1/2	13	.0444	.5166
9/16	18	.0320	.5745
9/16	12	.0481	.5805
5/8	18	.0320	.637
5/8	11	.0525	.6447
3/4	16	.0360	.7635
3/4	10	.0577	.7716
7/8	14	.0412	.8904
7/8	9	.0641	.899
1	14	.0412	1.0154
1	8	.0721	1.027

TABLE 22-1 THREAD MEASUREMENT DIMENSIONS
(American Standard Threads)

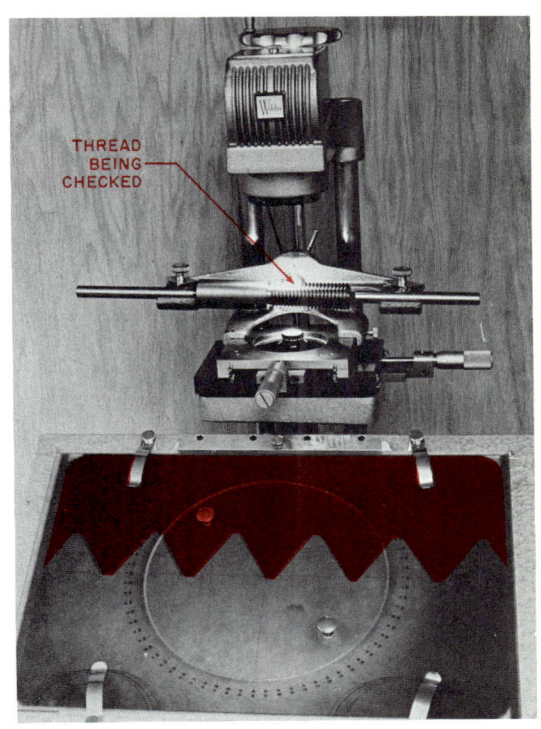

FIG. 22-8 AN OPTICAL COMPARATOR WITH A SCREW-THREAD ACCESSORY BEING USED TO MEASURE A THREAD

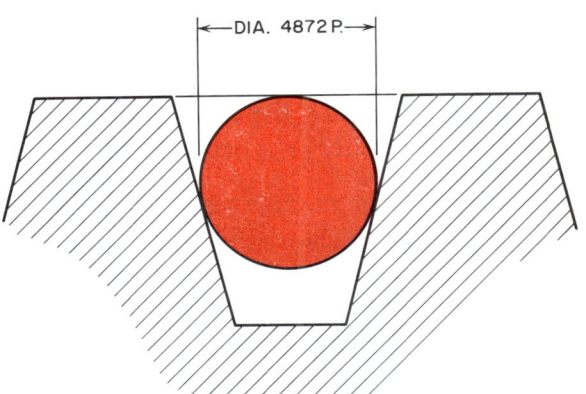

FIG. 22-9 THE ONE-WIRE METHOD OF MEASURING ACME THREADS

the best wire size for each thread pitch. However, for most 29° Acme threads and especially those with a small helix angle (small pitch), the *one-wire method* is accurate enough. Figure 22-9 shows how a single wire or pin can be used in checking the accuracy of an Acme thread. *Note:* When using the one-wire

Section 2 Machining Between Centers

method, it is *important that the burrs are first removed* from the diameter of the thread. A micrometer is used over the wire and the thread is correct when the micrometer reading is the same as the major diameter of the thread, and *the wire cannot be pulled out* of the thread.

Since the diameter of the wire changes with each thread pitch, it is important that the correct size wire is used. This can be calculated by applying the following formula:

Wire size = .4872 x pitch or $\frac{.4872}{N}$

For example, if an Acme thread having 4 threads per inch is required, the diameter of the wire which should be used is:

$$\text{Wire size} = \frac{.4872}{4}$$

$$= .1218$$

REVIEW QUESTIONS

1. What are four common methods of measuring threads?

Gages

2. Compare a "go" and a "no-go" thread ring gage.
3. How should a thread be checked for accuracy with a thread ring gage?
4. Describe a "go" and a "no-go" thread plug gage.
5. To what dimensions are the outer and inner rolls of a thread snap gage set?

Micrometers

6. Describe the construction of a screw thread micrometer. How is it used to check the accuracy of a thread?
7. How should a screw thread micrometer be used if a thread with extreme accuracy is required? Explain why.
8. How can a thread comparator micrometer be used when measuring threads?

Three-Wire Thread Measurement

9. How is the three-wire system used to measure thread? Illustrate by a neat sketch.
10. Describe the wires used in the three-wire system of thread measurement.
11. Why should the best size wire be used whenever possible?
12. Using the largest size wire, calculate the measurement over the wires for the following threads:

    a.   3/4" – 10 NC   b.   1" – 14 NF

13. Using the best size wire, calculate the measurement over the wires for the following threads.

     a.   5/8" – 18 NF   b.   1 ¼" – 7 NC

Measuring Acme Threads

14. What precaution should be taken before measuring an Acme thread with the one-wire system?

15. Calculate the one-wire size required to measure an Acme thread having eight threads per inch.

Unit 23   STEADY AND FOLLOWER REST

STEADY REST

A *steady rest,* figure 23-1, is used to support long slender work to prevent it from springing and bending while it is being machined. Long shafts, which are stiff enough to be machined without bending, can also be supported by a steady rest to prevent deflection caused by the cutting action. In this way, heavy cuts can be taken.

A steady rest is composed of a frame, usually containing three adjustable jaws; however, some rests used on large lathes may have as many as five jaws. The base of the frame is machined with one or more V-shaped grooves to fit the inside ways of the lathe bed, figure 23-1, while the top portion is hinged to allow work to be mounted or removed. The steady rest jaws are usually made of a soft material, such as brass or a fiber, so that they do not mar the surface of the revolving work. Rollers are generally attached to steady rest jaws used to support heavy workpieces. The jaws can be moved in or out radially by means of adjusting screws to suit various work diameters and are held in position by the lock screws.

Steady rests can be used to support work being machined between lathe centers and also work being held in a chuck. This unit deals only with the use of a steady rest on work between centers.

Units 28 and 30 cover the use of a steady rest to support work being held in a lathe chuck for center truing and turning operations.

Hints on Preparing the Workpiece and Steady Rest

The following hints are offered in order to overcome some of the problems associated with the use of steady rests. If these hints are carefully observed, damage to the steady rest jaws is eliminated and accurate work produced.

1. Never set a steady rest on a rough, untrue diameter.

2. Before a steady rest is mounted, a true diameter should be turned on the workpiece with a sharp pointed toolbit. *Note:* Take light cuts to produce a smooth, true diameter. This section should be slightly wider than the steady rest jaws and slightly past the center of the work. This enables the workpiece to be reversed and turned between centers without resetting the position of the steady rest. If possible, turn this section to some mandrel size.

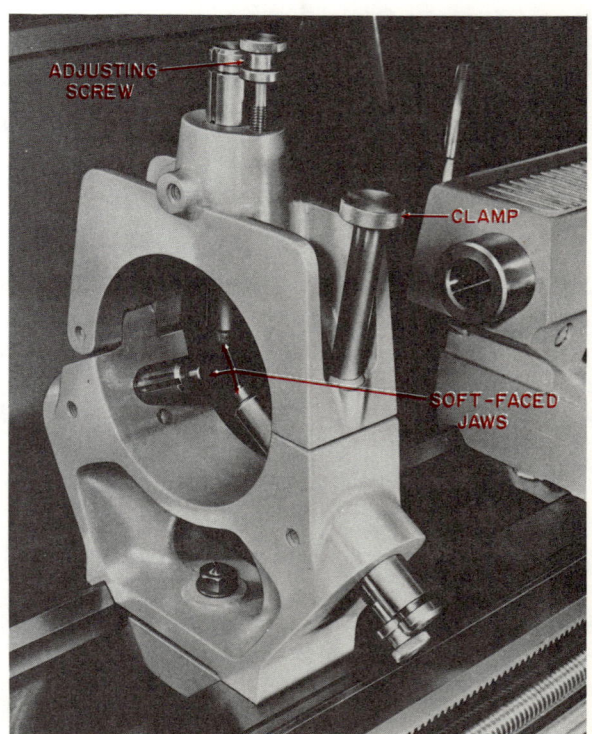

FIG. 23-1   A STEADY REST IS USED TO PREVENT SLENDER WORK FROM SPRINGING DURING A MACHINING OPERATION. (Courtesy Cincinnati Milacron Co.)

Then mount the mandrel between centers and set the steady rest jaws to the shorter and sturdier mandrel diameter. This overcomes the difficulty of making a true setting on a long slender workpiece which can spring easily. Once the jaws are set, the steady rest may be moved along the lathe bed to suit the position of the turned section.

3. If it is impossible, due to the nature of the workpiece, to turn a true section, mount and adjust a *cat-head,* figure 23-2, on the work. The cat-head consists of a sleeve, which is placed over the workpiece, and four adjusting screws on each end which are used to true the cat-head diameter with the axis of the workpiece. The steady rest is then set to the cat-head diameter.

4. Never set the jaws too tightly against the surface of the work as this tends to spring the work and also causes overheating and marring of the work surface.

5. To avoid marring the surface of a finished diameter, place a strip of leather or abrasive cloth (abrasive side against the jaws) between the steady rest jaws and the work surface.

The abrasive cloth can be held in a fixed position (especially when the work is revolving) by looping it around the work and clamping both ends between the hinged and bottom section of the steady rest.

6. Keep the jaws well lubricated to prevent seizing and marring the work surface.

Procedure: Setting Up a Steady Rest

The proper setup of a steady rest is important since an inaccurate setup can result in tapered work, marring of the work surface, chattering, and springing.

1. Mount the work between the lathe centers.

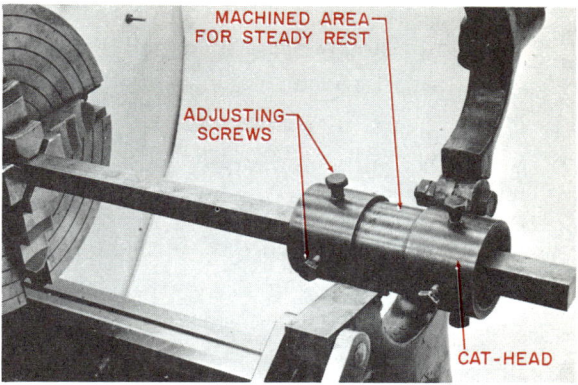

FIG. 23-2 THE SCREWS OF A CAT-HEAD CAN BE ADJUSTED TO PROVIDE A TRUE SURFACE FOR THE STEADY REST EVEN IF SQUARE WORK IS TO BE TURNED.

2. Turn a true diameter on the workpiece as outlined in step 2 of "Hints on Preparing the Workpiece and the Steady Rest," page 150.

If it is impossible to turn a true diameter, mount a cat-head as outlined in step 3 of "Hints on Preparing the Workpiece and the Steady Rest."

3. Mount a steady rest on the lathe and locate it along the bed so that it does not interfere with the machining operation.

4. Clamp the steady rest to the lathe bed, and then close and lock the top hinged section, figure 23-3.

FIG. 23-3 MOUNTING A STEADY REST TO PREVENT LONG, SMALL-DIAMETER WORKPIECE FROM SPRINGING DURING THE MACHINING OPERATION

Section 2 Machining Between Centers

5. Adjust the steady rest jaws close to the work by using a piece of paper or a .002 feeler gage between the diameter and the jaws.

6. Start the lathe and apply red or white lead to the work at the steady rest jaws.

7. Carefully adjust each jaw until the red or white lead just smears. This indicates that the jaw is in contact with the work surface.

8. Tighten the lock screw on each jaw to hold it in position.

9. Keep the jaws well lubricated to prevent seizing and marring the work surface.

10. Machine the diameter to size, coming as close as possible to the steady rest.

11. Reverse the work in the lathe, position and reset the steady rest jaws to the machined diameter.

12. Machine the other half of the shaft to the required size.

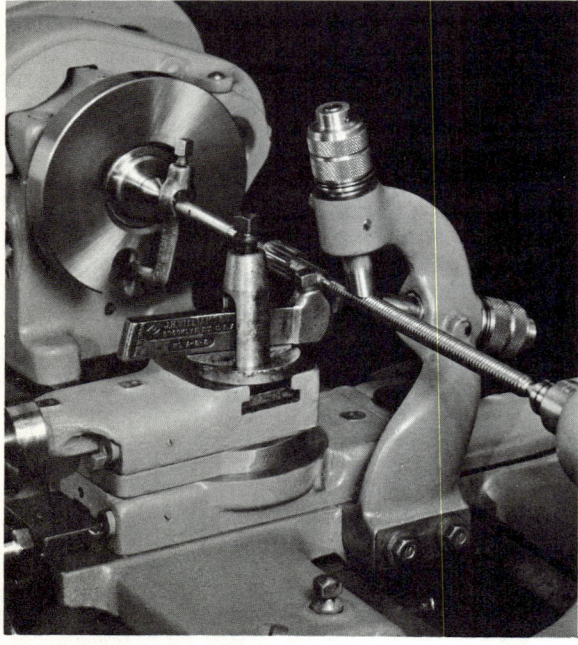

FIG. 23-4 A FOLLOWER REST IMMEDIATELY FOLLOWING THE CUTTING TOOL PREVENTS THE WORK FROM SPRINGING FOR THE ENTIRE LENGTH OF THE CUT.
(Courtesy South Bend Lathe Inc.)

FOLLOWER REST

The *follower rest,* figure 23-4, is used when it is necessary to have support close to and following the cutting tool to prevent long slender work from springing. The follower rest differs from the steady rest in that it has only two adjustable jaws, and is attached to and moves with the lathe carriage. One jaw is located opposite the cutting tool, while the other jaw bears on the top of the workpiece to prevent the work from springing up and away from the cutting tool. Since the follower rest is positioned immediately to the right of and following the cutting tool, the jaws support the work for the whole length of the cut.

Some follower rests use a bushing bored to suit the diameter to be cut, to provide a bearing surface for the work, and to hold it rigidly. Different-sized bushings are required for each diameter being cut. If the bushing is split through one wall, it can be adjusted slightly to compensate for wear by the follower rest jaws. This type of follower rest is generally used when long shafts are being machined.

Procedure: Setting Up a Follower Rest

1. Mount the workpiece between the centers.

2. Turn a section at the tailstock end to the diameter of the cut to be taken. *Note:* This diameter should be 1/4'' longer than the width of the follower rest jaws.

3. Mount the follower rest to the saddle of the lathe.

4. Set the jaws close to the turned diameter by using paper or a .002 feeler gage between the diameter and follower rest jaws.

5. Start the lathe and apply red or white lead to the diameter at the jaws.

6. Carefully adjust each jaw until the red or white lead just smears.

7. Tighten the lock screw on each jaw to hold it in position.

8. Lubricate the diameter and follower rest jaws to reduce friction and prevent marring of the work surface.

9. When taking successive cuts on a workpiece, steps 4 to 8 should be repeated to adjust the jaws to the smaller diameter.

Turning Precautions

When using a follower rest, it is important that the operator pay close attention, especially for *any change in diameter*. A change in diameter, generally due to cutting tool wear, breakage, or movement due to the pressure of the cutting action, causes:

- The work to be forced into the cutting tool by the follower rest jaws, resulting in inaccurate work

- Excessive friction and undue wear on the follower rest jaws

REVIEW QUESTIONS

Steady Rest

1. What purpose does a steady rest serve?

2. Describe the construction of a steady rest.

3. How does a steady rest used to support heavy workpieces differ from a standard steady rest?

Hints on Preparing the Workpiece and the Steady Rest

4. Why should a steady rest never be set on an untrue surface?

5. What is the procedure for producing a true diameter on the workpiece for the steady rest?

6. To what size should the true diameter be cut, if possible?

7. Describe the construction of a cat-head. What is its purpose?

8. What precaution can be taken to prevent the steady rest jaws from marring the surface of a finished diameter?

Setting Up a Steady Rest

9. How is the steady rest placed on the lathe bed?

10. What is the procedure for accurately setting the steady rest jaws to the work surface?

Follower Rest

11. Compare a follower and a steady rest as to construction, purpose, and method of attaching to the lathe.

12. Why does the follower rest provide support for the entire length of the cut?

13. Explain the use of a bushing on a follower rest. Where is it generally used?

Section 2 Machining Between Centers

Setting Up a Follower Rest

14. What is the procedure for turning a section on the workpiece and the setting of the follower rest jaws?

15. Why is it important that the follower rest jaws be lubricated?

Turning Precautions

16. How can a change in diameter cause problems when using a follower rest?

17. What may cause a change in diameter during any turning operation?

Unit 24   MANDRELS

Many times it is necessary to machine the outside diameter of thin parts such as gears, flanges, and small pulleys, concentric with the hole in the center of the part. Since it is impossible to hold and machine the full outside diameter in a chuck in one setup, a *mandrel* is used. The mandrel, when pressed into a hole, permits the work to be mounted between centers and also provides a means of driving the workpiece. The names *mandrel* and *arbor* are often confused and used interchangeably. A mandrel is used to hold workpieces, while an arbor (such as found on a milling machine) is used to hold cutting tools.

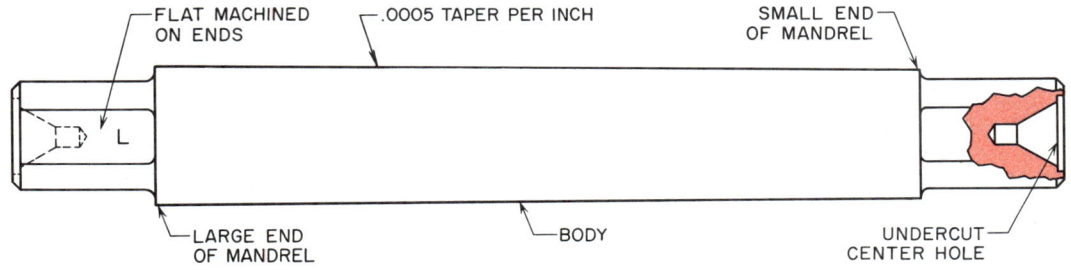

FIG. 24-1   CHARACTERISTICS OF A STANDARD SOLID MANDREL

TYPES OF MANDRELS

Many types and styles of mandrels are available to suit a variety of applications and workpieces. The most common ones used on a lathe are:

- The standard solid mandrel
- The expansion mandrel
- The gang mandrel
- The threaded mandrel
- The taper shank mandrel

Standard Solid Mandrel

The *standard solid mandrel,* shown in figure 24-1, is generally made of tool steel which has been hardened and then ground to a specific size. It is pressed or driven into a bored or reamed hole of a workpiece so that it can be mounted in a lathe for further machining operations.

The following are the characteristics of a standard solid mandrel.

1. The body of the mandrel is tapered slightly; generally .006 taper per foot (.0005 taper per inch).

2. The nominal size of a mandrel is generally closer to the small end.

3. The small ends of mandrels up to 1" diameter are ground .0005 under the nominal size, while those over 1" are usually .001 smaller. This permits the easy starting of a mandrel in a hole.

4. The large end of the mandrel is usually .004 over the nominal size; however, this does vary, depending upon the length of the mandrel.

5. Both ends of the mandrel are machined somewhat smaller than the body size and are provided with a flat for the clamping

Section 2 Machining Between Centers

screw of the lathe dog. This prevents the mandrel from being damaged when the lathe dog is clamped on, and also preserves the mandrel accuracy.

6. The size of the mandrel is always stamped on the large end.

7. The center holes, drilled into slightly recessed ends, are large enough to provide a good bearing surface and withstand the strains during machining. The recessed ends protect the center holes from being damaged.

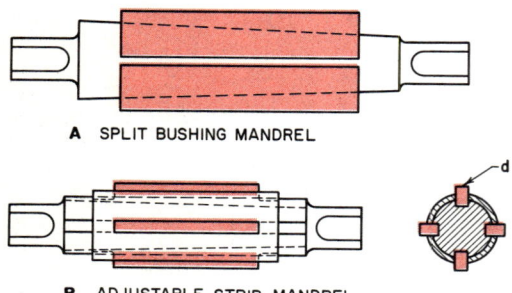

A SPLIT BUSHING MANDREL

B ADJUSTABLE STRIP MANDREL

FIG. 24-2 EXPANSION MANDRELS ARE USED WHEN WORKPIECES HAVING ODD-SIZED HOLES MUST BE MOUNTED ON A MANDREL.

Expansion Mandrels

Expansion mandrels, figure 24-2, can be used to advantage especially when workpieces with odd-sized holes must be mounted on a mandrel. There are a great variety of expanding-type mandrels and, therefore, only the most common types are discussed in this unit.

The *expansion mandrel* shown in figure 24-2A consists of a solid tapered mandrel and a split bushing (sleeve) which expands when forced on the mandrel. A number of different-sized bushings, all fitting the same solid mandrel, greatly increase the range and, therefore, only a relatively few mandrels are required. This type of mandrel is usually available in sets to suit a variety of hole sizes. Sets for small diameters can be expanded approximately 1/16", while the larger diameter sets can be expanded a proportionately larger amount.

The *adjustable strip mandrel,* figure 24-2B, consists of a straight body with four tapering grooves cut along its length and a sleeve which is slotted to correspond to the tapering grooves. Four strips (marked *d*) are fitted into the slots, and when the arbor is driven in, the strips are forced out by the tapering grooves and contact the hole. Sets of different-sized strips greatly increase the range of each mandrel. This type of mandrel is not suited to thin work since the pressure which is applied in four places may distort the workpiece.

Gang Mandrel

The *gang (flanged) mandrel,* figure 24-3, is used to hold a number of parts, all having the same size hole, for a machining operation. It consists of a parallel body (no taper) with a flange at one end and a threaded portion at the opposite end. A number of parts whose internal diameter is not more than .001 larger than the mandrel body size can be mounted and held securely when the nut is tightened against the *U* washer. A gang mandrel is especially useful when it is necessary to perform a machining operation on a number of thin parts which might be easily distorted if held by any other method.

Threaded Mandrel

A *threaded mandrel,* figure 24-4, is used when it is necessary to hold and machine workpieces having a threaded hole. This mandrel has a threaded portion which corresponds to

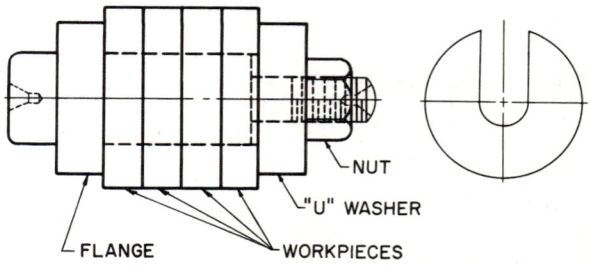

NUT
"U" WASHER
FLANGE
WORKPIECES

FIG. 24-3 A GANG (FLANGED) MANDREL IS USED TO HOLD A NUMBER OF IDENTICAL WORKPIECES FOR A MACHINING OPERATION.

the internal threads of the part to be machined. An undercut at the shoulder insures that the part can be threaded snugly against the flat shoulder. If the threaded part does not have one face which is squared with the axis of the thread, an equalizing washer W should be used between the square shoulder and the part to insure that the part is true and not canted on the threads.

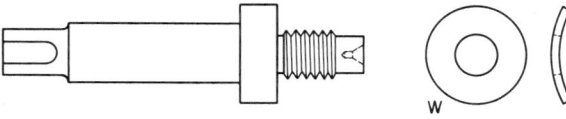

W – AN EQUALIZING WASHER

FIG. 24-4 THREADED MANDRELS ARE USED TO HOLD PARTS HAVING A THREADED HOLE.

Taper-Shank Mandrels

Taper-shank mandrels, figure 24-5, are not used between centers but are fitted to the internal taper of the headstock spindle. The portion which extends can be machined to suit the workpiece to be turned. Taper-shank mandrels are generally used to hold small workpieces.

The *expansion-stud mandrel,* figure 24-5A, has an internal thread and is slotted to allow it to be expanded slightly. When the tapered plug is tightened, the outside diameter of the stud expands against the inside of the workpiece. This type of mandrel is especially useful when machining a number of similar parts whose internal diameters may vary slightly.

The *threaded-stud mandrel,* figure 24-5B, has a projecting portion which is threaded to suit the internal thread of the part to be machined. This type of mandrel is especially valuable for holding workpieces which have blind holes.

Procedure: Mounting Work on a Mandrel

The following steps should be closely observed when mounting work on a *standard solid mandrel* to prevent damage to the mandrel or workpiece, and also to insure the accuracy of the finished part.

1. Select the proper size mandrel to suit the hole in the workpiece.
2. Clean and apply a light film of oil or grease on the diameter of the mandrel (see [1] figure 24-6).
3. Remove any burrs from the edge of the hole in the workpiece (see [2] figure 24-6).
4. Thoroughly clean and lubricate the hole in the part to be mounted (see [1] figure 24-6).

This prevents seizing or scoring as the mandrel is pressed into or out of the part.

5. Insert the small end of the mandrel squarely into the hole by *hand* (see [3] figure 24-6). It should enter the hole easily for approximately 1", and square itself. *Note:* The large end of the mandrel always has the size stamped on it.

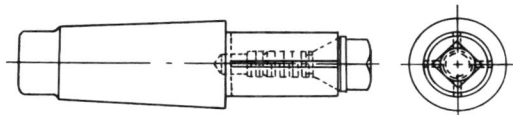

A. EXPANSION-STUD MANDREL

B. THREADED-STUD MANDREL

FIG. 24-5 TYPES OF TAPER-SHANK MANDRELS WHICH ARE MOUNTED IN THE LATHE HEADSTOCK SPINDLE

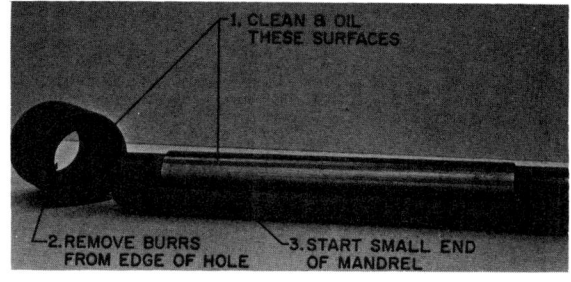

FIG. 24-6 PREPARING THE MANDREL AND WORKPIECE BEFORE MOUNTING

Section 2 Machining Between Centers

6. If an arbor press is available, place the work on the arbor press table, preferably with a machined surface down.

   The hole is then at right angles to table.

7. Press the mandrel firmly, but not too tightly, into the workpiece, figure 24-7. If the mandrel is forced into the hole too tightly, it may distort or break the workpiece.

8. If an arbor press is not available, use a lead or babbit hammer to drive the mandrel into the hole.

MACHINING WORK ON A MANDREL

The workpiece is pressed on a solid mandrel and is held securely only by friction. Therefore, whenever possible, take all cuts towards the large end of the mandrel. This tends to keep the workpiece tight on the mandrel. If it is necessary to cut towards the small end, only light cuts should be taken. The following precautions should be observed when mounting a mandrel on a lathe, and also during the machining operation.

1. Use an indicator to check that the lathe *headstock center is running true.*

   If the center is not true, the mandrel runs out of true, and the work is not turned or faced true with the hole.

2. Thoroughly clean the lathe and mandrel centers.

3. Check the alignment of the lathe centers with a test bar and dial indicator to insure that parallel work is turned.

4. Mount a lathe dog on the *large end* of the mandrel (the size is stamped on this end).

5. Mount the mandrel and workpiece on the lathe and carefully adjust the lathe center tension.

   Use a rotating tailstock center whenever possible to avoid overheating and ruining the lathe center and the mandrel.

6. When machining the outside diameter of a part, always cut towards the large end of the mandrel so that the part is forced on and not off the mandrel.

7. When machining small parts, use cutting fluid wherever possible to keep the heat of the workpiece to a minimum. If the part is overheated, it may expand and slip on the mandrel.

8. Light cuts should be taken on large diameter work mounted on a small mandrel to eliminate slippage, overcome chattering and springing, or bending the mandrel.

If possible, drive the workpiece directly from the driveplate by means of a suitable stud, figure 24-8. This prevents the workpiece from slipping or turning on the mandrel.

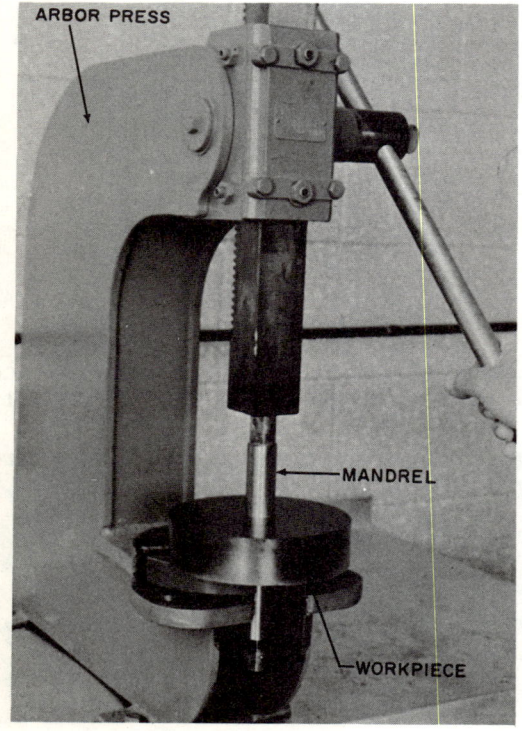

FIG. 24-7 PRESSING A SOLID MANDREL INTO A WORKPIECE ON AN ARBOR PRESS

9. When facing the sides of a workpiece, be sure not to cut into and damage the mandrel with the toolbit.

   a. Before the facing operation, set the toolbit point close to the mandrel diameter with a .002 feeler gage.

   b. Note the reading on the crossfeed graduated collar and be sure this setting is not exceeded when taking subsequent facing cuts.

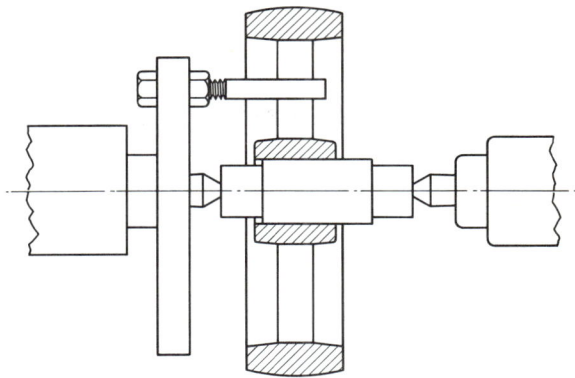

FIG. 24-8 LARGE WORK TURNED ON A SMALL MANDREL SHOULD BE DRIVEN DIRECTLY FROM THE DRIVEPLATE BY MEANS OF A SUITABLE STUD.

REVIEW QUESTIONS

1. What purpose does a mandrel serve?
2. What is the difference between a mandrel and an arbor?

Types of Mandrels

3. Make a neat diagram of a 1 1/4" standard solid mandrel 8" long. Label the nominal size, small-end diameter, and large-end diameter of the mandrel.
4. What is the purpose of:
   a. Turning the ends smaller than the mandrel diameter?
   b. Recessing the ends?
5. Why are expansion mandrels used?
6. Name and explain the principle of two types of expansion mandrels.
7. What is a gang mandrel and for what purpose is it used?
8. How can an internal threaded part be held squarely on a threaded mandrel?
9. Name two types of taper-shank mandrels. What is the purpose of each?

Mounting Work on a Mandrel

10. What is the procedure for preparing the workpiece and mandrel before they are mounted?
11. How may the mandrel be forced into the hole in the part to be machined? What care should be exercised?

Section 2   Machining Between Centers

Machining Work on a Mandrel

12. Why should the headstock center be running true?

13. In which direction should cuts be taken on work mounted on a mandrel? Explain why.

14. What precautions should be observed when machining large-diameter work on a small-diameter mandrel?

15. When facing the sides of a part, what precaution can be taken to insure that the toolbit does not damage the mandrel?

Unit 25 ECCENTRICS

An eccentric shaft has two or more turned diameters, the axes of which are parallel but not coincident with the normal axis of the shaft. An eccentric, figure 25-1, is used to convert rotary motion into reciprocating motion or vice versa. A crankshaft, figure 25-2, can be used for the same purpose, but generally it has a larger throw. Eccentrics are used in feed mechanisms and locking devices, while crankshafts are found in automobile and steam engines.

The *throw* of an eccentric is the amount that the center holes have been offset from the normal axis of the work. In figure 25-1 the throw is 1/4", while in figure 25-2 the throw of the crankshaft is 1 1/4". *The total travel of the reciprocating part is always twice the amount of throw of the eccentric or crankshaft.*

Before an eccentric shaft can be turned, it is necessary to lay out and drill the centers of the axes. There are three different types of eccentrics, each of which requires special procedures in laying out the centers and machining. Examples of these are:

- When the throw is large enough that all centers can be located on the ends of the workpiece at one time.

- When the throw is so small that the centers are too close and cannot be located on the workpiece at the same time

- When the throw is too large for all centers to be located on the ends of the workpiece

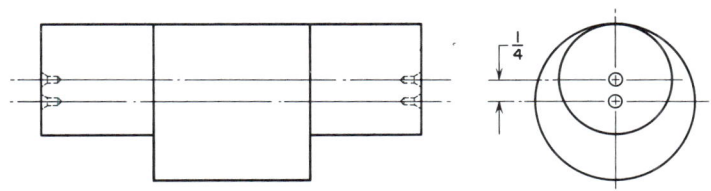

FIG. 25-1 AN ECCENTRIC GENERALLY HAS A SMALLER THROW AND IS USED TO CONVERT ROTARY MOTION INTO RECIPROCATING MOTION.

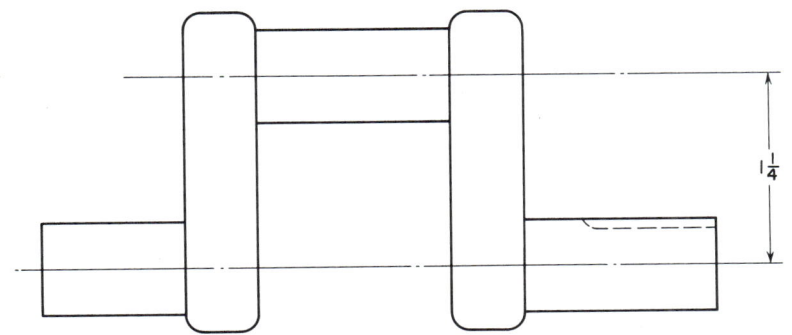

FIG. 25-2 THE LARGER THROW OF A CRANKSHAFT GENERALLY CONVERTS RECIPROCATING MOTION INTO ROTARY MOTION.

Section 2 Machining Between Centers

EXAMPLE 1

Procedure: Turning an Eccentric Where all Centers Can be Located on the Workpiece

1. Determine rough diameter of stock.

   Calculations for Shafts With One Eccentric Diameter. To the throw, add one-half of each eccentric diameter plus 1/8" for machining. For example, the size of material required for the eccentric in figure 25-3 would be $T + R_1 + R_2 + 1/8''$ or $5/16'' + 1/2'' + 1/2'' + 1/8'' = 1\ 7/16''$.

2. Face the work to length (6") in a three-jaw chuck.

3. Apply layout dye to both ends of the work.

4. Mount the workpiece in a V-block and set it on a surface plate.

5. Set the vernier height gage scriber to the top of the work and note the vernier reading.

6. Subtract one-half the diameter of the work from this reading and set the height gage to this dimension.

7. Scribe a centerline on each end of work.

8. Rotate the workpiece 90° and scribe another centerline at each end with the same height gage setting.

9. *Lower or raise* the height gage setting by .312 (5/16) and scribe the lines for the offset centers on both ends, figure 25-4.

10. Carefully center punch the intersections of all lines. Check the accuracy of the punch marks with a magnifying glass and correct if necessary.

11. Center drill both sets of center holes.

12. Mount the work on the lathe and turn the largest diameter first. *Note:* If all diameters are the same, it is good practice to turn the middle diameter first.

13. Mount the work on the offset centers and turn the eccentric to the required diameter.

EXAMPLE 2

Procedure: Cutting an Eccentric Having a 1/8" Throw

When the centers are too close to permit the drilling of all the centers on the workpiece at the same time, the following procedure should be used.

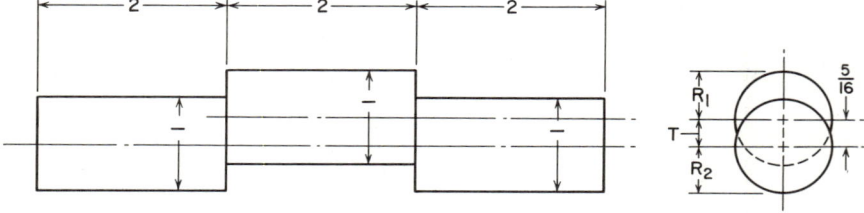

FIG. 25-3 AN ECCENTRIC SHAFT HAVING A 5/16" THROW

1. Determine the diameter of the rough stock required.

2. Cut off the rough stock 3/4" longer than the required length.

FIG. 25-4 LAYING OUT THE CENTERS OF AN ECCENTRIC WITH A VERNIER HEIGHT GAGE

Unit 25 Eccentrics

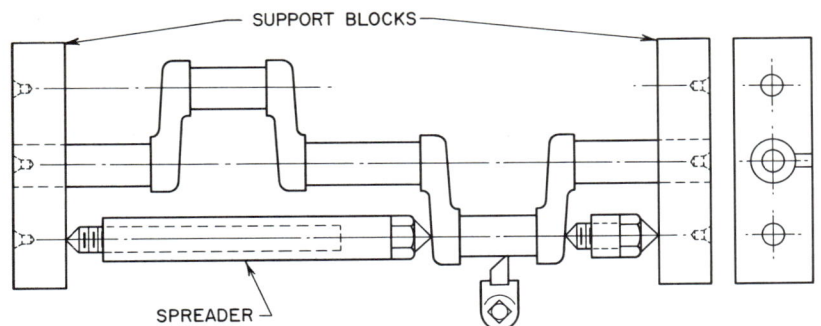

FIG. 25-5 TURNING CRANKSHAFT USING SUPPORT BLOCKS AND SPREADERS

3. Face both ends of the workpiece only enough to provide a smooth surface for the layout.
4. Drill one set of center holes, with the workpiece in a three-jaw chuck.
5. Turn the journals to size.
6. Remove both center holes by facing.
7. Relocate the centers of the finished journals with a vernier height gage.
8. Lay out the centers for the eccentric (1/8" throw).
9. Drill the offset center holes on the drill press.
10. Turn the eccentric diameters to size.

EXAMPLE 3

Procedure: Turning an Eccentric (Crank) Having a Large Throw

A crankshaft having a large throw is generally forged, and not turned, from a solid piece of round stock.

1. Lay out and drill the center holes for the journals.
2. Mount the work on the lathe and turn the journals.
3. Lay out and drill a set of centers having the required throw in a set of support blocks, figure 25-5.
4. Bore the center hole in the support blocks to be a snug fit on the journals. Use a setscrew in the block to secure the block to the shaft.

5. Align the offset centers of both support blocks with the offset section to be turned and lock the blocks in position with the setscrews.
6. Mount the work and the blocks between centers of the lathe.
7. Apply a counterbalance to the faceplate to prevent vibration when machining.
8. Turn the journals to size.
9. Adjust spreaders between cranks and support blocks, figure 25-5.
10. Machine the offset diameters and readjust the spreaders as required.

Procedure: Turning Eccentrics in a Four-Jaw Chuck

When a specified amount of throw is desired on a relatively short, single-ended eccentric as shown in figure 25-6, it can be held and machined in a four-jaw chuck as follows:

FIG. 25-6 USING A FOUR-JAW CHUCK TO MACHINE AN ECCENTRIC

Section 2   Machining Between Centers

1. Turn the large diameter to size for the full length of the workpiece.
2. Mount the workpiece in a four-jaw chuck, using soft metal between the chuck jaws and the finished diameter.
3. Using the concentric rings on the chuck face, offset two opposite jaws the approximate distance of the throw.
4. Mount a long-range dial indicator on the carriage or toolpost. *Note:* The indicator should have a range of at least twice the throw of the eccentric.
5. Rotate the work by hand until the indicator spindle bears against the low spot.
6. Turn the crossfeed handle until the indicator registers approximately .020 on the low spot and set the bezel to zero.
7. Revolve the work exactly one-half turn until the indicator registers on the high spot.
8. Adjust the two opposite chuck jaws until the dial indicator registers *double the amount of throw* between the low and high spots on the work.
9. Run the indicator along the workpiece to see that it is parallel to the centerline of the lathe.
10. Tighten all jaws uniformly and recheck for eccentricity and parallelism.
11. Proceed to machine the eccentric diameter.

REVIEW QUESTIONS

1. Define an eccentric. What is its purpose?
2. How does a crankshaft differ from an eccentric?
3. Define the throw of a crankshaft.
4. How is the diameter of the stock required to turn a shaft having one eccentric calculated?
5. What is the procedure for laying out the ends of an eccentric having a 1/2" throw?
6. What procedure should be followed for laying out and turning an eccentric with a 3/32" throw?
7. What accessories are required to machine a crankshaft with a 4" throw?
8. What is the procedure for setting up a workpiece for a 1/8" throw in a four-jaw chuck?

SECTION 3

Machining in a Chuck

Unit 26 MOUNTING AND REMOVING CHUCKS

Lathe chucks are devices, which when mounted on a lathe spindle, hold workpieces that are difficult or impossible to mount between centers due to their size or shape. The jaws of the chuck are adjustable so that various sizes of workpieces can be held. The most common chucks used on a lathe are the three-jaw, the four-jaw, and the spring-collet chuck. The construction and use of each chuck is described in unit 2 on lathe accessories.

LATHE SPINDLE TYPES

There are three types of lathe spindle noses upon which accessories such as chucks are mounted. They are the threaded spindle nose, the tapered spindle nose, and the cam-lock spindle nose.

The Threaded Spindle Nose

Many older lathes are provided with spindle noses which have an external thread. The threaded spindle receives a threaded, flanged disc, figure 26-1, fitted to a chuck, which serves as an adaptor and a drive. The flanged disc, commonly known as an *adaptor plate,* is accurately fitted to the lathe spindle and chuck. The lathe chuck is then held true to the axis of the spindle by the fit on the threads, and it is squared with the spindle when the adaptor plate bears against the square shoulder of the spindle.

The American Standard Tapered Spindle Nose

The American Standard, or type L, tapered spindle nose, figure 26-2, has a 3 1/2" taper per foot. This steep taper permits spindle accessories, such as chucks and driveplates, to be readily cleaned and easily mounted

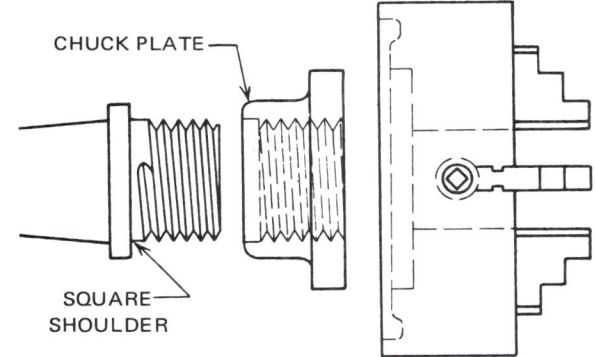

FIG. 26-1 THREADED SPINDLE NOSE

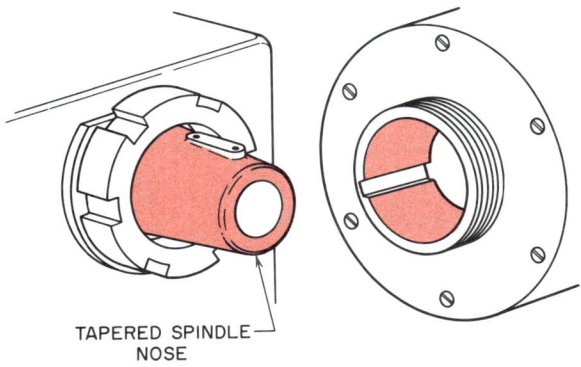

FIG. 26-2 AMERICAN STANDARD (TYPE L) TAPER SPINDLE NOSE

Section 3 Machining in a Chuck

and removed. A key drive locates the accessory on the spindle nose while the lock ring is used to fasten the chuck or driveplate securely on the spindle. When the lock ring is tightened, the chuck is drawn tightly onto the taper and is aligned to an accuracy of one or two thousandths. Accessories from other lathes, if they have the same taper, can be interchanged readily and will run true.

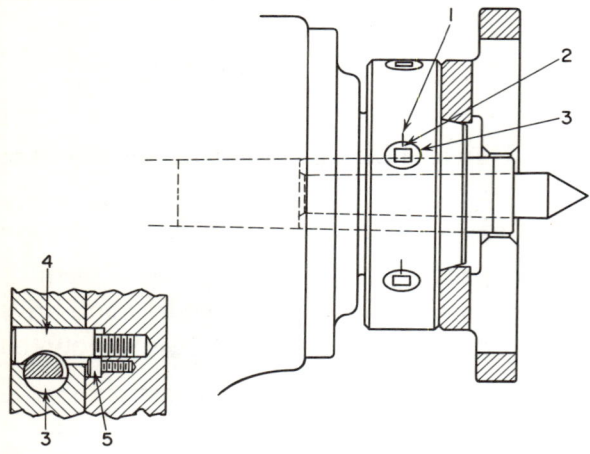

1. SPINDLE NOSE REGISTRATION LINES
2. CAM LOCK REGISTRATION LINES
3. CAM LOCKS
4. CHUCK OR FACEPLATE STUDS
5. RETAINING SCREW

FIG. 26-3 CAM-LOCK SPINDLE WITH TYPE D-1 SPINDLE NOSE (Courtesy Pratt & Whitney, Inc.)

The Cam-Lock Spindle Nose

The cam-lock, or type D, spindle nose, figure 26-3, has a short tapered section of 3" taper per foot. The face of the spindle may have three or six clearance holes to receive the cam-lock studs which protrude from spindle accessories, such as chucks and faceplates. When the cam locks in the spindle are tightened, the accessory is drawn firmly onto the taper and against the flange face. A partial turn of the cam locks releases the studs and allows the accessory to be removed quickly and easily.

MOUNTING AND REMOVING CHUCKS

Care should be exercised when mounting or removing chucks and spindle accessories to prevent damage which might destroy the accuracy of either the lathe or accessory. The following points should be observed when mounting and removing chucks and accessories.

1. Never use power when mounting or removing accessories from a threaded spindle nose.

   This practice is not only dangerous, but may force the accessory on the spindle, causing it to jam and possibly damage the threads.

2. Thoroughly clean the spindle and the mating parts of the spindle accessory before mounting to insure that the accessory is mounted correctly and not damaged, figure 26-4.

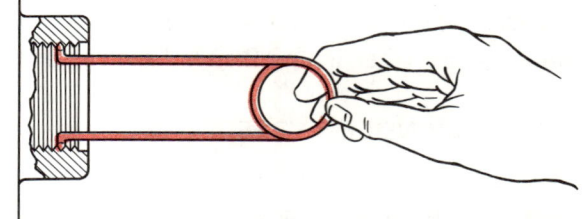

FIG. 26-4 USING A SPRING THREAD CLEANER TO REMOVE DIRT AND CHIPS FROM A THREADED ADAPTOR

Dirt or chips left on either of the mating parts can be jammed between the spindle and accessory. This causes the accessory to run out of true and can permanently damage the threads or tapers on the spindle or accessory.

3. Always place a wooden block or cradle between the chuck and lathe bed to prevent damage to the lathe ways, figure 26-5.

   If a chuck is dropped on the ways of the lathe, it nicks or burrs the ways. The lathe cuts slight tapers on the workpiece whenever the carriage is deflected by the nicks or burrs on the ways.

Unit 26   Mounting and Removing Chucks

4. When spindle accessories are not in use, they should be cleaned and stored carefully to prevent them from being damaged by other tools or equipment.

Procedure: Removing a Chuck

The following procedure for removing a chuck applies to the removal of any spindle-mounted accessory such as drive and faceplates.

1. Set the lathe to the slowest spindle speed.
2. *Disengage the electrical switch at the lathe.* This prevents the lathe from being started accidentally while the chuck is being removed.
3. Place a wooden block or cradle between the chuck and lathe bed, figure 26-5.
4. Depending on the type of lathe spindle nose, remove the chuck (or accessory) by following the procedure outlined in the steps listed under a, b, or c.

   a. *Threaded Spindle Nose*

   1) Revolve the lathe spindle until a chuck-wrench socket is in an upper position.

FIG. 26-5   A WOODEN BLOCK OR CRADLE PREVENTS DAMAGE TO THE LATHE WAYS IN CASE THE ACCESSORY IS DROPPED

FIG. 26-6   A HARDWOOD BLOCK BEING USED TO REMOVE A CHUCK FROM A THREADED SPINDLE NOSE (Courtesy South Bend Lathe, Inc.)

Insert the proper chuck wrench into the hole and *sharply* pull it towards the front of the lathe *(counterclockwise)* to loosen the chuck on the spindle.

OR

Place a hardwood block or a soft metal step block under a chuck jaw as illustrated in figure 26-6.

By hand, revolve the lathe spindle *clockwise* until the chuck is loosened on the lathe spindle.

2) Carefully remove the chuck from the lathe spindle.

3) Place cloth or paper into the chuck threads to keep dirt and chips out of the internal threads.

4) Store the chuck carefully where it can not be damaged by other tools or equipment.

Section 3 Machining in a Chuck

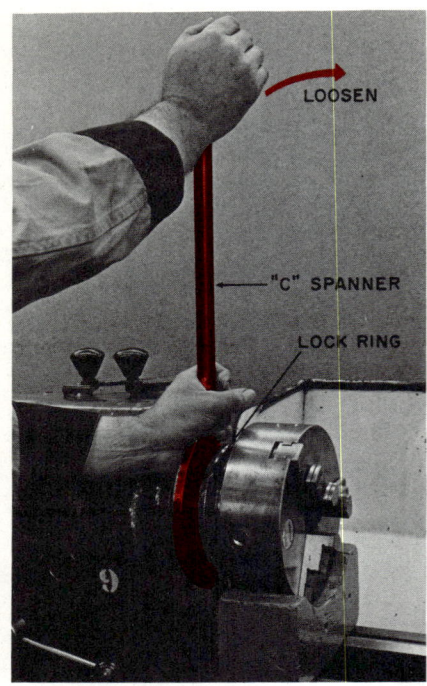

FIG. 26-7 A C-SPANNER WRENCH IS USED TO LOOSEN THE LOCK RING HOLDING A CHUCK ON A TAPER SPINDLE NOSE.

b. *Taper Spindle Nose*

1) Secure the proper size C-spanner wrench to fit the lock ring.

2) Place the spanner wrench around the *front* of the lock ring with the handle in an upright position, figure 26-7. *Note:* The lip of the wrench should be in one of the slots of the lock ring.

3) Place one hand on the curve of the spanner wrench to keep it from slipping off the lock ring.

4) *Sharply* strike the handle of the wrench in a *clockwise* direction with the other hand to loosen the lock ring.

5) Remove the wrench and hold the chuck with one hand while turning the lock ring clockwise with the other hand.

6) If the lock ring becomes tight, use the spanner wrench to loosen the taper contact between the spindle nose and chuck.

7) Place cloth or paper in the taper of chuck adaptor to keep dirt or chips out of the internal taper.

8) Store the chuck carefully, with the jaws up, where it can't be damaged.

c. *Cam-Lock Spindle Nose*

1) Secure the proper size chuck wrench to fit the cam-lock sockets.

2) Turn each cam lock *counterclockwise* until its registration line is in a vertical position or coincides with the spindle nose registration line.

3) Hold the chuck with one hand, and *sharply* strike the top of the chuck with the palm of the other hand to break the taper contact between the spindle nose and the chuck. *Note:* A soft-faced hammer may be used to strike the top of the chuck.

4) Remove the chuck, clean, and store it in a suitable storage compartment.

Procedure: Mounting a Chuck

The following procedure for mounting a chuck applies to the mounting of any spindle-mounted accessory such as drive and faceplates.

1. Set the lathe to the slowest spindle speed.

2. *Disengage the electrical switch at the lathe.* This prevents the lathe from being accidentally started while the chuck is being mounted.

3. Thoroughly *clean all the mating surfaces* of the lathe spindle and chuck.

4. Place the chuck on a wooden cradle block and slide it close to the headstock spindle.

5. Depending on the type of lathe spindle nose, mount the chuck (or accessory) by following the procedure outlined in the steps listed under a, b, or c.

   a. *Threaded Spindle Nose*

      1) Revolve the lathe spindle *counterclockwise by hand* and bring the chuck into contact with the spindle threads. *Never start a chuck on the lathe spindle by power.*

      2) The chuck should easily thread onto the lathe spindle if the chuck and spindle are correctly aligned and have been thoroughly cleaned. If the chuck is difficult to thread on the spindle, remove, and thoroughly clean the threads, and restart it on the spindle.

      3) When the back of the adaptor plate is within 1/16" of the spindle shoulder, quickly turn the chuck so that it seats snugly against the spindle shoulder. *Note:* Do not jam or force a chuck too tightly against the shoulder, otherwise the internal threads may be damaged and the chuck is then difficult to remove.

   b. *Taper Spindle Nose*

      1) Revolve the lathe spindle or chuck *by hand* until the key on the tapered spindle nose matches the keyway in the chuck adaptor plate.

      2) Slide the chuck onto the spindle nose and turn the lock ring counterclockwise.

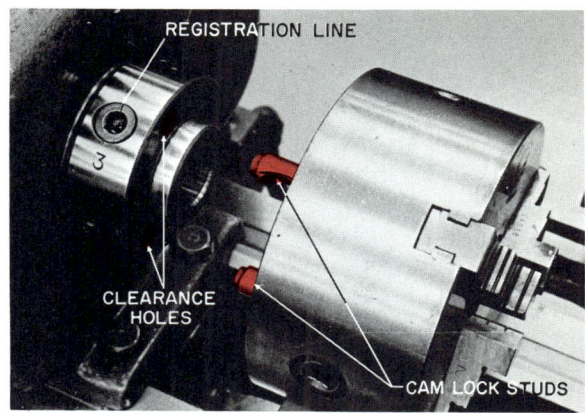

FIG. 26-8 MOUNTING A CHUCK ON A CAM-LOCK SPINDLE NOSE

   3) If the lock ring does not thread easily, remove the chuck, clean the threads of the chuck adaptor and lock ring, and remount the chuck.

   4) Standing in front of the lathe, place the spanner wrench around the lock ring and strike it sharply downward with the palm of one hand to tighten it securely.

   c. *Cam Lock Spindle Nose*

      1) Align the registration line of each cam lock vertically or with the corresponding registration line on the spindle nose, figure 26-8.

      2) Turn the lathe or chuck by hand until the clearance holes of the spindle align with the cam-lock studs of the chuck, figure 26-8.

      3) Slide the chuck on the lathe spindle.

      4) Securely tighten each cam lock in a clockwise direction.

REVIEW QUESTIONS

Lathe Spindle Types

1. What is the purpose of an adaptor plate?

2. How does the American Standard taper spindle compare with the cam-lock spindle nose?

169

Section 3 Machining in a Chuck

Mounting and Removing Chucks

3. Why should care be exercised when mounting or removing chucks and spindle accessories?

4. What is the result of mounting spindle accessories which have not been thoroughly cleaned?

5. Why should a wooden block or cradle be used when mounting or removing chucks?

Removing a Chuck

6. What precaution should be taken to insure that a lathe cannot be started accidentally while a chuck is being removed?

7. What is one method of removing a chuck from a threaded spindle nose? Explain fully.

8. What precaution should be taken to keep dirt and chips out of the internal threads of a chuck?

9. How is a chuck removed from a taper spindle nose?

10. What is the procedure for removing a chuck from a cam-lock spindle nose?

Mounting a Chuck

11. Why should all the mating surfaces of a chuck and lathe spindle nose be thoroughly cleaned before mounting?

12. What should be done if a chuck is difficult to turn on a threaded spindle nose?

13. Explain why a chuck or spindle accessory should not be jammed on a threaded spindle.

14. What is the purpose of the key and lock ring on a taper spindle nose?

15. What is the procedure for mounting a chuck on a cam-lock spindle nose?

Unit 27   MOUNTING WORK IN A CHUCK

Work that cannot be held between centers can be held in a chuck for machining on a lathe. Although there are several types of lathe chucks and the method of mounting work in each may vary, it is important that *the workpiece and chuck jaws be cleaned before work is mounted.* This not only insures greater accuracy in machining, but also helps to preserve the accuracy of the chuck.

THREE-JAW CHUCK WORK

Round and hexagonal work can be quickly and conveniently mounted in a three-jaw universal chuck. Since all jaws move simultaneously, the work is automatically centered to within about .003. As the scroll plate becomes worn, the chuck may lose this accuracy. When this occurs, work should be mounted in a four-jaw independent chuck to maintain the desired accuracy. Shims may be used under one or two jaws of the universal chuck to position the work to the accuracy required.

To further preserve the accuracy of a universal chuck, work which has an outer scale should not be gripped without the protection of soft shims under the chuck jaws.

Procedure: Mounting Work in a Three-Jaw Chuck

1. Clean the surface of the workpiece and the jaws of the chuck.

2. Using the *proper chuck wrench,* open the jaws slightly more than the size of the work. *Note:* Do not use a chuck wrench that is too small since the sockets in the chuck can be damaged as well as the chuck wrench.

3. Insert the work into the chuck, leaving the proper amount protruding. This amount depends on the operation to be performed. If the surface of the workpiece is finished, a piece of thin, soft sheet metal should be wrapped around the surface to protect the finish.

4. With the left hand, tighten the chuck wrench while the right hand slowly rotates the workpiece. This causes the work to center properly in the chuck.

5. Tighten the chuck jaws securely by using the chuck wrench only.

Do not use an extension on the chuck wrench handle. This exerts too much pressure on the jaws and causes the scroll plate to wear which results in premature inaccuracy in the chuck. Excessive pressure can also spring some types of work.

There are some types of universal chucks that are equipped with soft jaws which are mounted on the base jaw of the chuck with cap screws. A plug of the same diameter as the workpiece is gripped by the base jaws, and then the soft jaws are machined to the size and shape of the workpiece to be turned, figure 27-1. This procedure insures that the work is held securely and accurately.

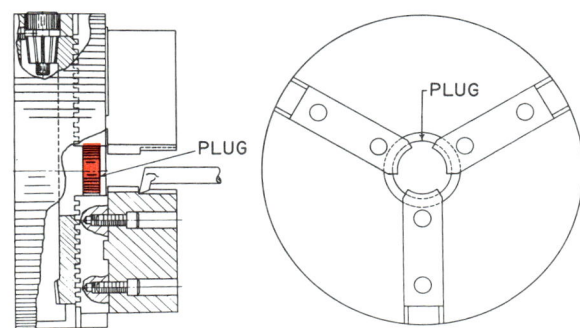

FIG. 27-1   A PLUG GRIPPED BY THE BASE JAWS WILL POSITION THE SOFT JAWS FOR BORING.

Section 3   Machining in a Chuck

Hints on Using a Three-Jaw Universal Chuck

1. Finish all diameters before removing the workpiece from a three-jaw chuck. If the work is removed and then replaced for further machining, it probably will not run true and the turned diameters will not be concentric.
2. Remove the live center and the spindle sleeve before mounting a chuck.
3. Clean the lathe spindle and the chuck adaptor before mounting a chuck.
4. Tighten the chuck jaws around the most rigid part of the work to prevent distortion of the workpiece.
5. If the work projects more than three times the diameter of the stock, it should be supported by a center or steady rest.
6. Never grip the work on a diameter smaller than the diameter to be machined unless absolutely necessary, otherwise the work may be bent.
7. Never use an air hose to clean a chuck.
8. Oil the chuck sparingly.
9. Store chucks in suitable compartments when not in use.
10. Never leave a chuck wrench in a chuck.

Assembling the Jaws in a Three-Jaw Universal Chuck

When chuck jaws are changed in a universal chuck, care must be taken to assemble them in the proper order, otherwise the chuck jaws do not run true. All chucks and the jaws supplied with each chuck are marked with the same serial number. It is very important to match the serial numbers of the chuck and the jaws being inserted in it.

1. Thoroughly clean the jaws and the jaw slides in the chuck.
2. Turn the chuck wrench clockwise until the start of the scroll thread is almost showing at the back edge of slide 1. *Note:* Be sure that the start of the scroll thread does not protrude into the jaw slide.
3. Insert jaw 1 (in slot 1) and press down with one hand while turning the chuck wrench clockwise with the other.
4. After the scroll thread has engaged in the jaw, continue turning the chuck wrench clockwise until the start of the scroll thread is near the back edge of groove 2.
5. Insert jaw 2 and repeat steps 3 and 4.
6. Insert the third jaw in the same manner.

FOUR-JAW INDEPENDENT CHUCKS

When it is necessary for a workpiece to run absolutely true, it is generally mounted in a four-jaw independent chuck. Since each jaw can be adjusted independently, work can be trued to within .001 accuracy or less. The jaws of this chuck are reversible and permit a wide range of work to be gripped either externally or internally.

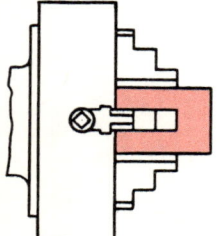
A. JAWS IN NORMAL POSITION GRIPPING THE OUTSIDE OF THE WORKPIECE

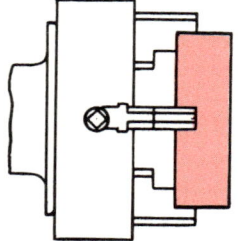
B. CHUCK JAWS REVERSED TO GRIP LARGE DIAMETER WORK

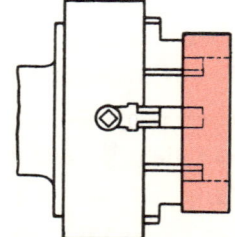
C. JAWS IN NORMAL POSITION BUT GRIPPING ON THE INSIDE DIAMETER OF WORKPIECE

FIG. 27-2   METHODS OF GRIPPING WORK IN A FOUR-JAW INDEPENDENT CHUCK

Unit 27   Mounting Work in a Chuck

Round, square, octagonal, hexagonal, and irregularly-shaped workpieces can be held for machining in a four-jaw independent chuck. Work can be adjusted to run concentric or off center as required. The face of the chuck has a number of evenly spaced concentric grooves which permit quick and approximate positioning of the chuck jaws.

Procedure: Truing Work in a Four-Jaw Chuck (Approximate Methods)

1. Measure the diameter of the workpiece.
2. Adjust the chuck jaws to the approximate size of the workpiece by using the rings on the face of the chuck.
3. Place the work in the chuck and tighten the jaws lightly on the work.
4. Adjust all jaws so that they are in the same relation to one ring in the chuck face.
5. True the workpiece, using any of the following methods.

   a. *Chalk Method*
      1) Start the lathe and slowly bring a piece of chalk towards the revolving workpiece until it lightly marks the surface, figure 27-3A.
      2) Stop the lathe and see if the chalk mark is uniform around the work circumference. A uniform mark indicates that the work is true.
      3) If the chalk mark appears only on a portion of the circumference, adjust the workpiece by loosening the jaw opposite the chalk mark and tightening the jaw where the chalk mark appears.
      4) Repeat this procedure until the chalk mark appears lightly around the entire circumference.

   b. *Surface Gage Method*
      1) Place a surface gage on the lathe carriage and adjust the scriber point until it is close to the work circumference, figure 27-3B.
      2) Revolve the lathe by hand to find the low spot on the work.
      3) Loosen the jaw *nearest* the low spot and tighten the jaw *opposite* the low spot to adjust the work closer to center.
      4) Repeat steps 2 and 3 until the work is running true.

A. CHALK METHOD   B. SURFACE GAGE METHOD   C. TOOLHOLDER METHOD

FIG. 27-3   APPROXIMATE METHODS OF ALIGNING WORK IN A FOUR-JAW CHUCK

Section 3 Machining in a Chuck

c. *Toolholder Method*

The back of a lathe toolholder is another method of approximately truing a workpiece. Once the toolholder is set as in figure 27-3C, follow steps 2 to 4 of the surface gage method to true the work in the chuck.

6. Tighten all chuck jaws uniformly so that the work is held securely.

7. When long pieces of work are mounted in the chuck, it is necessary to check both ends of the work for concentricity. When the work runs reasonably true near the chuck, check and true the other end by tapping the high spot with a soft-faced hammer. Repeat this procedure until both ends of the work run true.

Procedure: Truing Work in a Four-Jaw Independent Chuck Using a Dial Indicator

When work must be set up accurately in a chuck, a dial indicator should be used to check the concentricity of the work, figure 27-4.

1. Mount the work and true it approximately by means of the chalk, surface gage, or toolholder method.

2. Mount a dial indicator in the toolpost of the lathe, figure 27-4, with its plunger in a horizontal position.

FIG. 27-4 TRUING A WORKPIECE IN A FOUR-JAW INDEPENDENT CHUCK USING A DIAL INDICATOR

3. Adjust the crossfeed handle until the indicator needle registers about .020

4. Take the lathe "out-of-gear" and revolve the spindle slowly by hand.

5. Note the highest and lowest reading on the indicator.

6. Loosen the jaw at the low reading no more than one-eighth of a turn and tighten the opposite jaw (at the high reading).

7. Continue to adjust *only* these two jaws until the indicator reading is the same for both of them. *Note:* When setting up work in a four-jaw chuck, *only adjust two opposite jaws at any one time,* otherwise difficulty is experienced in truing the workpiece.

8. Adjust the other two jaws in the same manner until the indicator needle shows no movement at any point around the circumference.

9. Tighten all jaws evenly and recheck the indicator reading on the workpiece.

COLLET CHUCKS

Collet chucks are used on lathes for holding and machining small parts, since they allow the parts to be mounted quickly and accurately. Because it is easy to impair the accuracy of the collets, the following precautions should be observed when using a draw-in or spring collet chuck.

1. Be sure the surface of the work is true and free of scale and burrs.

2. Since spring collets are precision ground, the size of the stock for which the collet is designed should not be more than .002-.003 from this size.

3. Never tighten a collet unless there is a workpiece in it. If pressure is left on the collet, it may spring out of shape.

Unit 27 Mounting Work in a Chuck

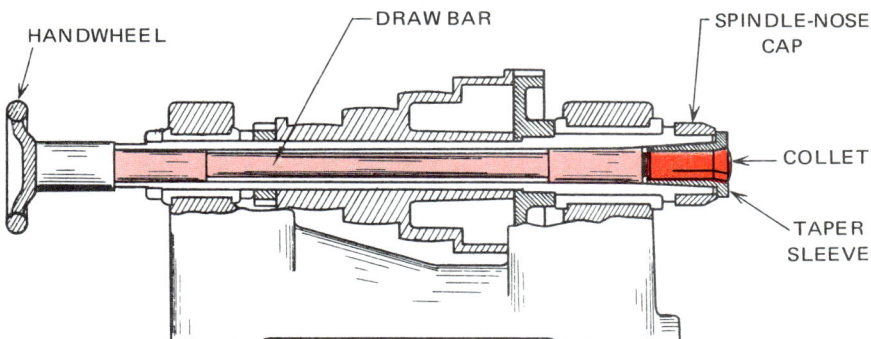

FIG. 27-5 A DRAW-IN COLLET MOUNTED IN A LATHE SPINDLE

4. Indicate the first workpiece in each lot to check for runout and taper before machining. Should either of these occur, remove the collet chuck and thoroughly clean and remove any burrs from the collet, chuck, and the lathe spindle.

Use of Collet Chucks

Procedure: Draw-in Collet Chucks

1. Clean the spindle nose and the bore of the lathe spindle.
2. Clean the bearing surface of the collet chuck spindle adaptor.
3. Mount the collet chuck on the spindle nose and insert the draw bar in the back of the lathe spindle, figure 27-5.
4. Insert the proper collet into the chuck.
5. Start the thread of the draw bar onto the thread of the collet.
6. Insert the work into the collet, leaving a maximum of three times the diameter protruding, unless the end of the work is to be supported by the dead center.
7. Turn the handwheel to tighten the collet onto the work.
8. Proceed with the machining operations.

Spring Collet Chucks

Spring collet chucks, figure 27-6, differ from draw-in collets in that they are mounted on the spindle in the same manner as a chuck or driveplate. Since they are not held in place by a draw bar, spring collet chucks can accommodate larger size stock than the draw-in collet chuck. The principle and application of spring collet chucks are the same as the draw-in collet chuck.

The Jacobs Spindle Nose Collet Chuck

The Jacobs collet chuck, figure 27-7, has a wider range than the spring collet chucks. The rubberflex collet consists of a series of metal inserts molded into rubber. Since these collets are very flexible, each collet has a range of approximately 1/8'' (1/16 over and under the nominal size). This allows a much wider range of workpieces to be accommodated with fewer collets.

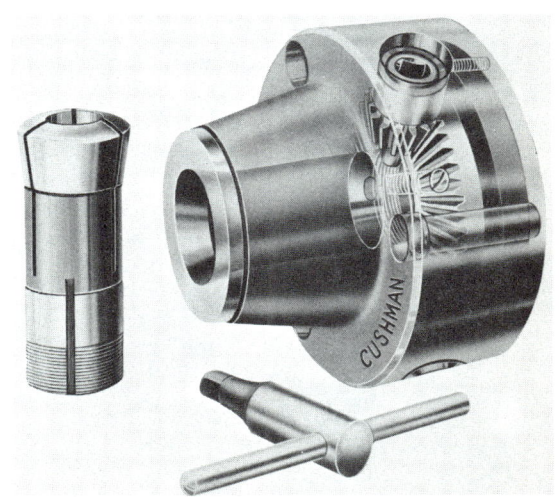

FIG. 27-6 A SPRING COLLET CHUCK
(Courtesy Cushman Industries, Inc.)

175

Section 3   Machining in a Chuck

FIG. 27-7   THE JACOBS SPINDLE NOSE CHUCK WITH RUBBER-FLEX COLLET (Courtesy Jacobs Mfg. Co.)

Procedure: Mounting Work in a Spindle Nose Collet Chuck

1. Clean the lathe spindle nose and the taper in the chuck adaptor plate.
2. Set the speed-change levers at a low spindle speed to lock the spindle.
3. Mount the chuck on the spindle nose.
4. Clean the body cone and the inside of the nose.
5. Clean the internal and external surfaces of the collet.
6. Place the proper collet in the body cone.
7. Place the nose collar over the collet and thread the nose into the chuck by rotating the handwheel clockwise.
8. Insert the work into the collet and rotate the handwheel clockwise to tighten the collet jaws onto the workpiece. *Note:* If the workpiece does not extend at least 3/4" into the rubberflex collet, it is advisable to place a plug of the same diameter as the workpiece in the back of the rubberflex collet. This insures that the work is held securely since the collet grips the work along its entire length.
9. Press the lock ring in toward the spindle nose and turn the handwheel slightly until the lock ring snaps into place. This locks the chuck in position.
10. Set the proper spindle speed and machine the workpiece.

REVIEW QUESTIONS

1. Why should the workpiece and chuck jaws be cleaned before work is mounted?

Three-Jaw Chuck Work

2. What happens if a three-jaw chuck is tightened excessively over a period of time?
3. What procedure should be followed when holding a workpiece having a scaly surface?

Mounting Work in a Three-Jaw Chuck

4. Why is it unwise to use an undersized chuck wrench when mounting work in a chuck?
5. What procedure insures that the work seats properly as the chuck jaws are tightened?
6. Why is it advised not to use an extension on a chuck wrench?
7. Describe soft jaws. What is the purpose of these jaws?

8. What procedure should be followed when boring out soft jaws to a specific size?

Hints on Using a Three-Jaw Chuck

9. Why is it important that all diameters on a workpiece be finished in one setup before work is removed from a three-jaw chuck?

10. What are four of the most important hints on the use of a three-jaw universal chuck?

Assembling the Jaws in a Three-Jaw Chuck

11. What is the procedure for inserting or reversing the jaws in a three-jaw universal chuck?

Four-Jaw Chuck

12. Compare the advantages and disadvantages of a four-jaw chuck with those of a three-jaw chuck.

13. What purpose do the concentric rings on the face of the chuck serve?

14. Describe one method of approximately aligning work in a four-jaw chuck.

15. How is truing accomplished on long workpieces held in a four-jaw chuck?

16. What is the procedure for truing a piece of work to within .001 in a four-jaw chuck?

Collet Chucks

17. What are three precautions that should be observed when using a draw-in or spring collet chuck?

18. How much stock should protrude beyond the collet when machining?

19. What is the advantage of a rubberflex collet over a spring collet?

20. What precaution should be taken when setting up a workpiece where only 1/2" of its length is held in the rubberflex collet?

Unit 28 CENTER HOLE DRILLING AND TRUING

Center holes can be drilled in round work by holding the workpiece in a three-jaw chuck and having the center drill in a drill chuck mounted in the tailstock. This method is quick, accurate, and it eliminates the necessity of laying out the center hole location. Center drill sizes, and the procedure for drilling center holes in a three-jaw chuck are fully explained in Unit 8, Centering Work.

TRUING A CENTER HOLE

When machines or components are disassembled for repairs, the center holes in shafts and rods often are damaged. If this part is to be mounted between centers for machining, it is then necessary to true the damaged center hole, otherwise the part does not run true.

Procedure: Truing a Center Hole

1. Set up the work in a four-jaw chuck and true the outside diameter with a dial indicator. *Note:* If the workpiece is long, a steady rest should be used to support and true the right-hand end of the work.

FIG. 28-1A   A SPOTTING TOOL

2. Grind a 60° spotting tool, figure 28-1A. Be sure to grind sufficient side clearance to prevent it from rubbing in the center hole.
3. Mount the spotting tool in a carbide toolholder, set the toolbit on center and in line with the axis of the work, figure 28-1B.
4. Carefully align the cutting edge to 30° (one-half the included angle of the center hole).
5. Set the lathe to approximately 200 r.p.m.
6. Start the lathe and slowly feed the toolbit into the center hole with apron handwheel.
7. With the crossfeed handle, gradually feed the toolbit out until it trues the damaged center.
8. Mount a countersink in the tailstock and lightly drill the center hole to bring it to the correct angle (60°). Use cutting fluid for this operation.

FIG. 28-1B   A SPOTTING TOOL SET UP TO TRUE A DAMAGED CENTER

Procedure: Truing a Center Hole With a Special Countersink

Center holes can be trued quickly and fairly accurately with a special 60° countersinking tool held in a drill chuck mounted in the tailstock, figure 28-2. This centering tool can be made from a center drill with a broken point, but whose 60° angular portion is undamaged. A flat, parallel with one cutting face, is ground across the end of the center drill so that only one lip or cutting edge remains.

1. Mount the workpiece in a four-jaw chuck and true the diameter with a dial indicator.

2. Mount a drill chuck into the tailstock spindle which has been aligned with the headstock.
3. Hold the centering tool short in the drill chuck. It should not extend more than 1/2" beyond the chuck jaws.
4. Set the tailstock spindle so that it extends approximately 1" beyond the tailstock.
5. Slide the tailstock towards the workpiece until the centering tool is about 1/2" from the work and lock it in position.
6. Start the lathe revolving at approximately 600 r.p.m.
7. Tighten the tailstock spindle clamp until a slight drag is felt when turning the tailstock handwheel.

This prevents the tailstock spindle from deflecting when the center hole is being corrected.

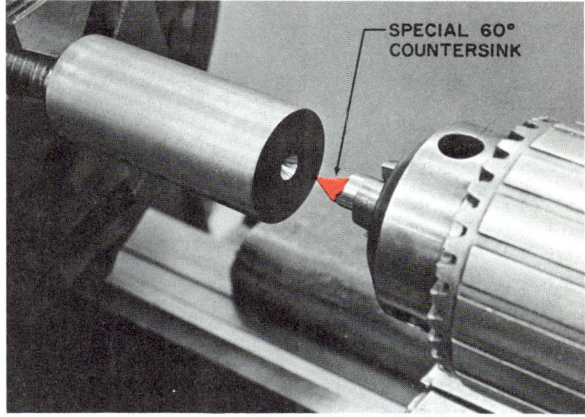

FIG. 28-2 A CENTER HOLE MAY BE TRUED QUICKLY AND FAIRLY ACCURATELY BY USING A SPECIAL 60° COUNTERSINK.

8. Apply cutting fluid and *slowly* bring the centering tool into the damaged center by turning the tailstock handwheel.
9. Continue feeding the centering tool only until the center hole runs true.

REVIEW QUESTIONS

1. Why is it important that center holes in a piece of work be drilled to the correct size?
2. What size center drill should be used for a piece of work 1 1/4" diameter and how deep should the hole be drilled?
3. How far should work extend beyond the chuck jaws while center drilling?

Truing a Center Hole

4. How should a long shaft be set up prior to truing a damaged center hole?
5. Make a neat sketch of a spotting tool.
6. Why is it desirable to mount a spotting tool in a carbide toolholder when truing a center hole?
7. Why should a countersink be used to finish the center hole?

Truing a Center Hole With a Special Countersink

8. How can a special 60° countersinking (centering) tool be made?
9. Why should the tailstock be aligned with the headstock when truing a center hole?
10. How should the centering tool and tailstock be set up for truing a center hole?

Unit 29 FACING IN A CHUCK

Facing in a chuck is the process of machining an end or face to produce a smooth, flat surface which is square with the axis of the work. This operation is similar to facing between centers and is performed for the following reasons:

- To smooth and square an end surface
- To have an accurate surface from which to take measurements
- To cut the work to the desired length

Work faced in a chuck generally consists of the sides of flanges, shoulders, and larger surfaces than those faced between centers. Chuck work can be faced more conveniently since the lathe centers do not interfere. Therefore, the entire surface can be machined. The workpiece can be held in either a three- or four-jaw chuck which allows both external and internal surfaces to be faced, figure 29-1. Another advantage of facing work in a chuck is that a larger variety of cutting tools can be used.

Procedure: Facing in a Chuck

1. Mount and true the workpiece in the chuck.

   The workpiece should not extend more than three times the diameter beyond the chuck jaws. This prevents chattering and also reduces the possibility of bending the work.

2. Set the compound rest 30° to the right if only one surface is being faced.

   When the compound rest is set at 30° and the compound rest feed screw is advanced .020, the cutting tool moves sideways .010 or *always one-half the amount fed.*

   OR

   Set the compound rest at 90° to the cross-slide if a series of short steps or shoulders must be faced to accurate lengths.

   When the compound rest is set at 90° (parallel to the lathe ways) accurate spacing of shoulders is possible by moving the

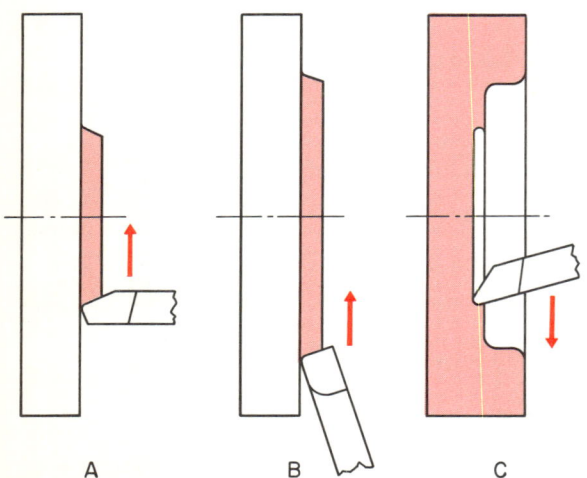

FIG. 29-1 THE TOOLBIT IS FED RADIALLY TO FACE A SURFACE FLAT AND SQUARE WITH THE AXIS.

FIG. 29-2 IN SOME CASES, ONE TOOLBIT CAN BE USED FOR BOTH ROUGHING AND FINISHING CUTS WHEN FACING IN A CHUCK.

Unit 29 Facing in a Chuck

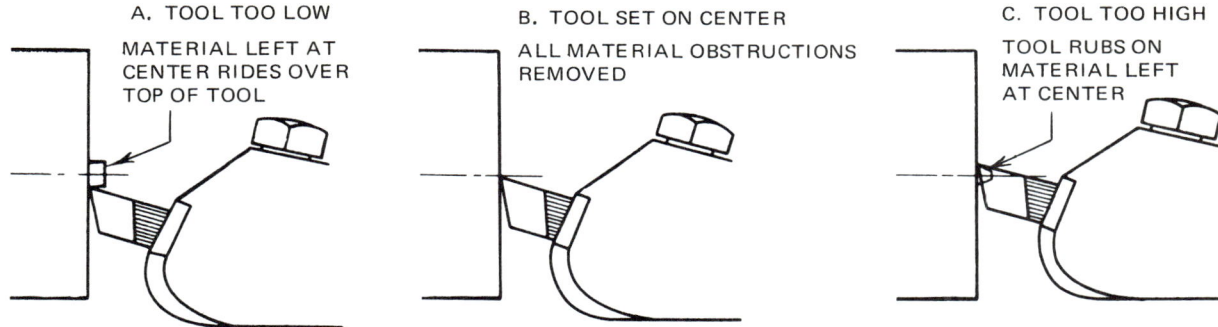

FIG. 29-3 THE IMPORTANCE OF SETTING THE POINT OF THE TOOLBIT EXACTLY ON CENTER FOR FACING WORK

amount required on the compound rest graduated collar. The toolbit advances lengthwise *the same amount which is fed.*

3. Fasten a suitable facing tool in the lathe toolholder, figure 29-2.

   The toolbit shown in figure 29-2 may be used for both roughing and finishing cuts when one surface is being cut. For the roughing cut, the toolbit is fed from the outside to the center. It is fed from the center outward for the finish cut.

4. Adjust the toolholder so that its side clears the surface to be faced.

5. Set the point of the facing tool exactly on center, figure 29-3B.

6. Set the lathe speed to suit the diameter and type of material being cut.

7. Start the lathe and move the carriage so that the point of the facing tool is just touching the surface to be faced.

8. Turn the crossfeed handle and bring the toolbit out to clear the work diameter.

9. Lock the carriage in position and set the depth of the roughing cut to within .015 of the finished size with the compound rest handle. The roughing cut should be as large as possible; if necessary, take more than one roughing cut. *Note:* The roughing cut should be deep enough to get under the scale of the surface being faced. This is especially important when facing castings.

10. Take the roughing cut feeding the toolbit in towards the center with the crossfeed handle or automatic crossfeed.

11. Measure the work to determine the amount yet to be removed.

12. Set the depth of the finish cut with the compound rest handle.

13. Take the finish cut feeding from the center to the outside diameter using the crossfeed handle.

14. With a file, remove the burrs from the edge of the workpiece.

15. If the other end must be faced parallel, reverse the work in the jaws and tap the piece snugly against the chuck face.

    If the work is shorter than the length of the chuck jaws, place suitable parallels or a plug between the chuck face and the work. Tighten the chuck jaws and tap the work snugly against the parallels. *Be sure to remove the parallels before starting the lathe.*

Procedure: Facing Several Workpieces to the Same Length

1. Face one surface on each workpiece only enough to clean up the face.

2. Set the toolbit to the finished length with a rod, standard, or gage of the suitable length against the chuck face, parallels, or

Section 3   Machining in a Chuck

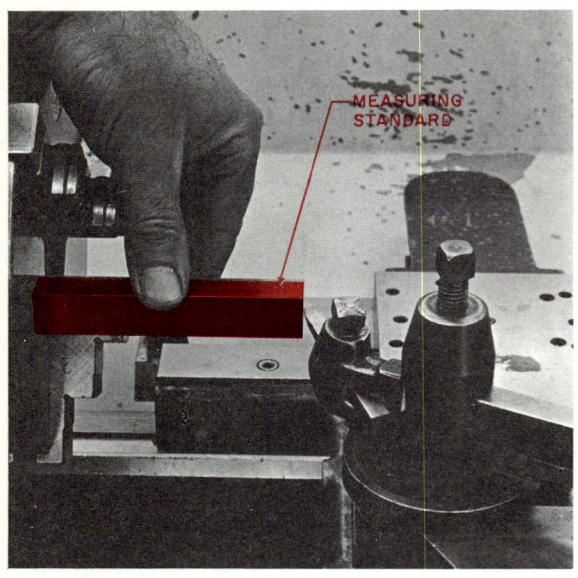

FIG. 29-4   USING A STANDARD TO SET THE FACING TOOL A DEFINITE LENGTH FROM THE CHUCK FACE

plug, figure 29-4. *Note:* A micrometer carriage stop may also be used for setting the toolbit to the required length.

3. Lock the carriage in position, recheck the accuracy of the toolbit setting, and then set the compound rest collar to zero.

4. Turn the compound rest screw counterclockwise one-half turn to remove the backlash, and then clockwise to within .030 of zero.

5. Start the machine and take the roughing cut, machining from the outside to the center of the workpiece.

6. Turn the crossfeed screw clockwise to zero while the lathe is running.

7. Take the finish cut, machining from the center towards the outside diameter.

REVIEW QUESTIONS

1. What are three reasons for facing work?
2. What type of work is generally faced in a chuck?

Facing Work in a Chuck

3. Why should the compound rest be set at 30° when facing?
4. What is the relationship between the amount of compound rest feed and the amount of metal removed when the compound rest is set at 30°?
5. How many cuts are desirable when facing a workpiece?
6. In which direction is the cutting tool fed when facing a workpiece?
7. What precaution should be taken to insure that the carriage does not move when facing the work?
8. What precaution should be taken after the workpiece has been faced?
9. How should the workpiece be set up if both ends must be machined parallel?

Facing Several Workpieces to the Same Length

10. How may the toolbit be set when machining several workpieces to the same length?
11. How is the toolbit set for the roughing cut?
12. How is the finish cut to be taken?

Unit 30   TURNING IN A CHUCK

Much lathe work is turned between centers for the convenience of setup for various operations. A great deal of the work machined on a lathe cannot be held between centers and must be held in some other manner for machining purposes. This type of work is generally held in a three- or four-jaw chuck.

The actual machining operation is the same as for parallel turning between centers, unit 13, and wherever possible, the work should be cut to size in two cuts (one rough and one finish cut). Generally, the roughing or first cut is taken to within 1/32" of the finish size.

PRECAUTIONS

By observing the following precautions, many problems which occur when machining work in a chuck can be eliminated.

1. Slowly rotate the work as the chuck jaws are tightened. This helps to seat the work properly in the chuck.

2. Tighten the chuck sufficiently so that the workpiece is not moved into the chuck by the pressure of the cut.

    This is especially important if the end of the work is being supported by the dead center since it could result in bent work.

3. Support the end of the work with a dead center if more than three times the diameter of the workpiece protrudes beyond the chuck jaws.

4. Since a dead center cannot be adjusted too accurately when the work is held in a chuck, a revolving dead center should be used.

    Unlike the standard dead center, the revolving dead center is adjusted snugly into the center hole, and the tailstock spindle clamp is then tightened.

5. Always set the toolbit on center. If the toolbit is set too low, the work may be bent unless the end of the work is supported by a dead center.

6. Always position the toolpost on the left side of the compound rest. This prevents the jaws from striking the compound rest when cutting close to the chuck.

7. Set the toolholder straight in (90° to the work) or slightly to the right to prevent the tool from "digging in" on a heavy cut.

    Note: The toolholder should point to the left only when finishing cuts are taken.

8. Move the carriage until the toolbit is at the extreme left end of travel and the **toolbit is within 1/8" of the work surface.** Rotate the chuck one turn by hand to see that the jaws do not strike the compound rest.

9. The r.p.m. at which work is revolved is calculated in the same manner as when the work is held between centers.

    The only exception is when the workpiece is irregular in shape and would tend to vibrate if turned at a high speed. In this case the speed should be reduced.

10. All machining operations should be performed before the work is removed from the chuck.

    This is especially important when machining work held in a three-jaw chuck, since it is difficult and time-consuming to retrue the work in a chuck.

Section 3 Machining in a Chuck

11. When rough turning, the depth of cut and feed should be as heavy as possible.

    Care should be taken to see that the force of the cut does not move the work into the chuck jaws.

12. When finish turning, the depth of cut should be about .030 – .050 and the feed fine enough (generally .005) to produce the desired finish.

13. When the work is too long to be held between centers, it should be held in a chuck and the right-hand end supported with a steady rest (see unit 23).

14. A follower rest should be used when machining long, slender work over the entire length (see unit 23).

REVIEW QUESTIONS

1. When should work which is held in a chuck be supported with the dead center?

2. What are three advantages of a revolving dead center?

3. How is the revolving dead center adjusted in the workpiece center?

4. What precautions should be taken when setting up the toolbit for machining a workpiece held in a chuck?

5. When does the rule for calculating the correct spindle speed not apply?

6. Why should all machining operations be performed before work is removed from a chuck?

7. What precautions should be taken regarding depth of cut and feed?

8. When should a steady rest be used?

9. When should a follower rest be used?

Unit 31 CUTTING-OFF OR PARTING

When several identical parts have to be made on a lathe, they are often made from bar stock inserted through the lathe spindle and held by a chuck or collet. When the part is completely machined, it is separated from the bar stock with a cutting-off or parting tool mounted in the toolpost.

Although most work is cut off using an inserted-blade type cutting-off tool, figure 31-1A, a parting tool ground from a square toolbit, figure 31-1B, is often more satisfactory for small work.

HINTS ON CUTTING-OFF (PARTING)

Because of the shape of the blade, cutting-off tools are more fragile than most lathe tools and certain precautions should be observed in their use.

1. Work must always be held in a chuck or collet and *never between centers.*

2. Work must be gripped securely in the chuck or collet.

3. Work should protrude only enough to permit the cut to be made as close to the chuck jaws as possible. (Use a right-hand offset toolholder.)

4. Work having more than one diameter should be gripped on the large diameter when grooving or parting.

   Never grip work on the small diameter and attempt to cut off or groove the large diameter. This results in bent work.

5. Do not extend the blade beyond the holder any more than is necessary to cut off the workpiece (generally one-half the diameter of the work plus 1/8" for clearance is sufficient).

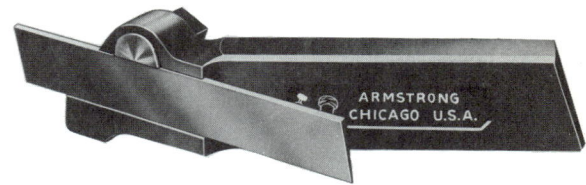

FIG. 31-1A INSERTED-BLADE TYPE CUTTING-OFF TOOL WITH RIGHT-HAND OFFSET (Courtesy Armstrong Brothers Tool Co.)

6. Grip the cutting toolholder as short as possible with the blade set on center. It is preferable that there be no back rake on the top of the cutting blade.

7. Grooving or parting operations require a *steady feed* of the cutting tool.

   Too light or an intermittent feed tends to dull the cutting edge. Too heavy a feed causes jamming and tool breakage.

8. Always wear goggles when using a cutting-off tool.

9. Use cutting fluid on steel. Brass and cast iron should be cut off dry.

FIG. 31-1B A TOOLBIT GROUND AS A PARTING TOOL FOR USE WITH STRAIGHT AND OFFSET TOOLHOLDERS

Section 3   Machining in a Chuck

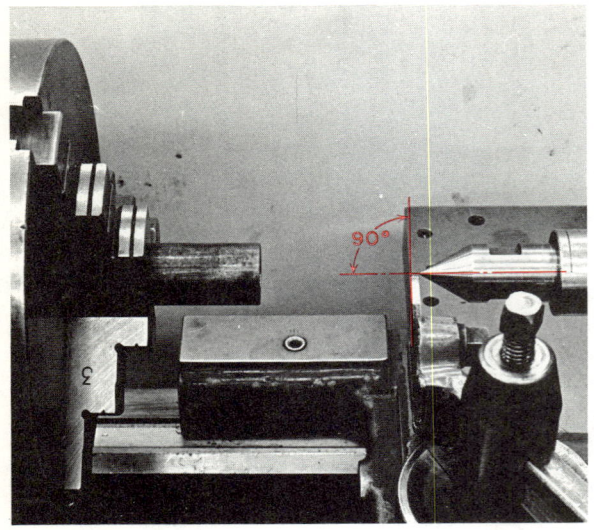

FIG. 31-2   A CUTTING-OFF TOOL IS MOUNTED TO THE LEFT OF THE COMPOUND REST, ON CENTER, AND SQUARE WITH THE WORK.

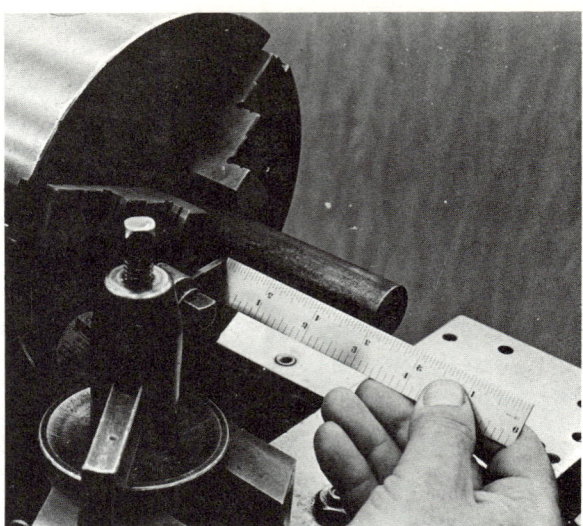

FIG. 31-3   POSITIONING THE CUTTING-OFF TOOL FOR A PIECE OF WORK 3" LONG

Cutting-Off (Parting) Procedure

1. Mount the work in a chuck or collet chuck with as little material as possible protruding beyond the chuck.

   The amount of material protruding should be the length of the material to be parted off plus 1/2".

2. Mount a right-hand offset toolholder on the left side of the compound rest, figure 31-2. Grip the toolholder as short as possible to prevent chatter.

3. Adjust the parting tool so that it extends out of the toolholder approximately one-half the diameter of the work plus 1/8" for clearance.

4. Set the cutting-off tool *exactly on center* and with as little back rake as possible.

   If the cutting tool is too high, it does not cut through the workpiece. If too low, the work may be bent and the cutting tool damaged.

5. Set the blade of the parting tool square (90°) with the work so that it does not rub against the sides of the groove as it is fed into the work.

6. Set the lathe spindle speed to one-half the speed for turning.

7. Move the carriage so that the *right-hand side of the blade is at the point where the work is to be cut off,* figure 31-3.

8. Start the lathe, and with the crossfeed handle, feed the toolbit *steadily* into the work. For cutting-off operations in steel, cutting fluid should be used. Cast iron and brass are parted dry.

9. As the tool is fed into the work, it is advisable to move the cutting tool back and forth a few thousandths with the carriage handwheel to prevent the tool from binding in the groove. It should be done on all parting operations and when grooves deeper than 1/4" are required.

10. Remove the burrs from both sides of the groove with a file just before the parting operation is complete.

11. Continue to feed the tool into the work until the part is severed.

    When the work is almost cut off, the tool should be fed slowly.

REVIEW QUESTIONS

1. Name two types of parting tools. Where is each type used?

Hints on Cutting-Off (Parting)

2. What are six precautions that should be observed when using a parting tool?

3. How far should the cutting-off blade extend beyond the toolholder?

Cutting-Off (Parting) Procedure

4. How far should the work protrude beyond the chuck jaws?

5. Why is a right-hand offset toolholder better than a straight toolholder for most parting operations?

6. What will happen if the cutting tool is set
    a. Too low?
    b. Too high?

7. Why should the cutting-off tool be set square with the work?

8. What procedure should be followed to set the cutting-off tool for the length of work to be cut?

9. What precaution should be taken to prevent the cutting-off tool from binding when cutting-off or grooving?

Unit 32 DRILLING

Work held in a chuck or mounted on a faceplate can be drilled quickly and fairly accurately in a lathe. The drill, held in a drill chuck or fitted directly into the taper of the tailstock spindle, is brought against the revolving work by turning the tailstock handwheel. This method should always be used when it is necessary to drill a hole in the center of a round piece of work. Drilling generally precedes other lathe operations such as boring, reaming, counterboring, and tapping.

Various methods can be employed to hold drills when drilling on a lathe, depending on the type and size of drill being used. The most common methods are:

- In a drill chuck for straight-shank drills
- Directly into the taper of the tailstock spindle for taper shank drills
- In a drill holder for large taper shank drills

The drilling procedure for each method differs slightly and is discussed below.

FIG. 32-1 A CENTER HOLE PROVIDES A GUIDE AND TENDS TO KEEP A DRILL FROM WANDERING OFF CENTER.

Procedure: Drilling With the Drill Held in a Drill Chuck

1. Align the lathe centers.
2. Mount the workpiece in the proper chuck or on a faceplate.

   If the outside diameter has been finished and a hole must be machined concentric with the periphery, the work should be set up in a collet chuck or a four-jaw chuck and trued with an indicator.

3. Face the workpiece.
4. Drill a starting hole with a center drill mounted in the tailstock, figure 32-1.

   If the drill hole location is to be extremely accurate, the hole should first be spotted with a spotting tool, unit 28.

5. Mount the proper size drill in the drill chuck.
6. Extend the tailstock spindle no more than 1" beyond the tailstock to permit maximum drilling depth.
7. Move the tailstock to the left until the point of the drill is about 1/4" from the end of the workpiece.
8. Clamp the tailstock to the bed of the lathe.
9. Set the spindle speed to the correct r.p.m. for the size of drill and the material being drilled.
10. Start the lathe and turn the tailstock handwheel only until the drill cuts to its full diameter (the drill point is completely in the metal).
11. Note the reading of the graduation on the tailstock spindle, or place a pencil mark on the spindle adjacent to the tailstock casting. The depth of the hole can be

Unit 32   Drilling

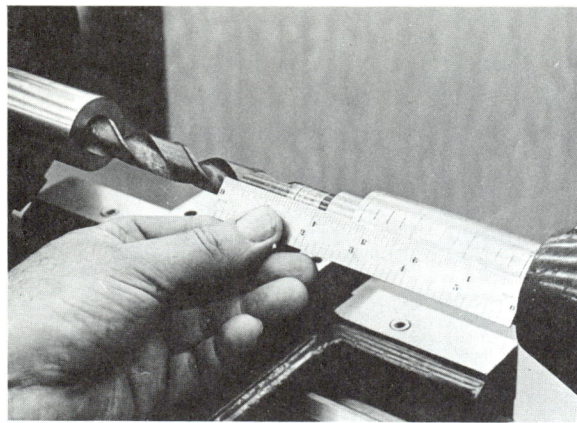

FIG. 32-2 MEASURING THE DEPTH OF A HOLE BEING DRILLED

FIG. 32-3 SUPPORTING THE END OF THE DRILL WILL PREVENT ITS WOBBLING WHEN STARTING A HOLE.

measured by the tailstock graduation or from the pencil line with a rule, figure 32-2.

12. Apply cutting fluid and drill the hole to the required depth.

It is good practice to back out the drill occasionally to clear the chips and apply cutting fluid.

In some cases where the drill is sufficiently rigid, it is possible to start the drill without the use of a center hole. When this method is used, the drill should be supported by the end of the toolholder, figure 32-3, to prevent the drill from wobbling as it starts into the workpiece.

Procedure: Drilling With the Drill Mounted in the Tailstock Spindle

If a hole larger than the capacity of the drill chuck is required, it is necessary to mount a tapered shank drill in the tailstock spindle, figure 32-4A, or in a drill holder, figure 32-4B.

Because of the wide web on a large drill, there is a tendency for the drill to wander, particularly at the start of the hole. To overcome this, a pilot hole, slightly larger than the web of the drill, is first drilled in the metal. The use of a pilot hole reduces the tendency for the drill to wander since the point of the drill does not contact the metal and cannot

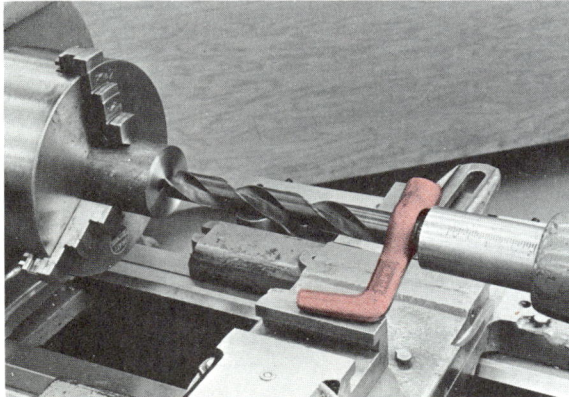

FIG. 32-4A A TAPERED SHANK DRILL MOUNTED IN THE TAILSTOCK SPINDLE. NOTE THE USE OF A LATHE DOG.

FIG. 32-4B DRILLING A HOLE WITH A DRILL MOUNTED IN A DRILL HOLDER

Section 3 Machining in a Chuck

pivot off center. By using a pilot hole, a drill cuts fairly close to size, providing it is correctly ground.

1. Face and drill a center hole in the workpiece.
2. Drill the proper size pilot hole to the required depth.
3. Remove the drill chuck and mount the proper size drill in the tailstock spindle. A lathe dog should be attached near the shank of the drill with the tail supported on the back of a toolholder or the compound rest, figure 32-4A. This prevents the drill shank from turning in the tailstock spindle and scoring it.

OR

Mount the drill in a drill holder which is supported by the dead center and the toolholder, figure 32-4B. The point of the drill is supported by the center hole of the work.

4. Set the proper spindle speed.
5. Apply cutting fluid and drill the hole to the required depth.

Note: **Feed the drill slowly when it is breaking through the workpiece. This is especially important when using a drill holder, figure 32-4B.**

REVIEW QUESTIONS

1. How does drilling in a lathe and in a drill press differ?
2. What are three methods of holding a drill in a lathe?

Drilling With a Drill Held in a Drill Chuck

3. How should the work be mounted if the hole must be concentric with the diameter?
4. What are two methods by which a starting hole can be machined prior to drilling?
5. How is the drill positioned near the work prior to drilling?
6. What are two methods by which the depth of the drilled hole can be measured?
7. If a center drill is not used, how is the point of the drill supported to prevent it from wobbling as it starts into the work?

Drilling With the Drill Mounted in the Tailstock Spindle

8. What are two methods by which large drills can be supported for drilling holes?
9. What is the purpose of a pilot hole and how is its size determined?
10. What precautions should be observed when the taper-shank drill is mounted in the tailstock spindle?
11. What precautions should be observed when drilling with a drill mounted in a drill holder, figure 32-4B?

Unit 33 BORING

Boring is the process of enlarging and truing a drilled or cored hole with a single-point cutting tool. When a hole is drilled in a workpiece which has been accurately set up in a lathe, the hole may not be concentric with the outside. This occurs if the drill becomes dull or wanders due to hard spots or sand holes in the metal. If the hole location and size must be accurate, it is necessary to bore the hole before reaming so that the hole is concentric with the diameter of the workpiece. Boring is also used to finish an off-size hole for which there is no drill or reamer available.

BORING TOOLS AND HOLDERS

Since boring is a turning operation performed on the inside of a hole, the boring tool is similar to a turning tool. It is generally shaped like a turning tool which is ground to cut from left to right.

Boring tools may be of two types, the solid forged type and the boring bar type with a suitable toolbit mounted in the end of the bar.

The solid forged boring tool, figure 33-1, is generally made from high-speed steel. The end is forged and then ground to resemble a left-hand turning tool. Boring tools of this type are generally used for light duty and are held in a special toolholder which is mounted in the toolpost, figure 33-1.

A boring bar toolholder, figure 33-2, is mounted in the toolpost and is used for heavier cuts than the forged boring tool. Various size bars, having broached holes to accommodate square toolbits at an angle of 30°, 45° or 90°, can be mounted in the tool

FIG. 33-1 SOLID FORGED BAR IN A TOOLHOLDER USED FOR LIGHT BORING (Courtesy Armstrong Bros. Tool Co.)

FIG. 33-2 A BORING BAR TOOLHOLDER USED FOR GENERAL-PURPOSE BORING (Courtesy J.H. Williams Co.)

A. PLAIN BORING BAR (Courtesy J. H. Williams Co.)

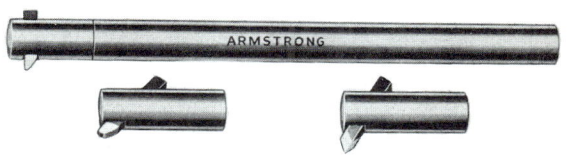

B. END-CAP BORING BAR (Courtesy Armstrong Bros. Tool Co.)

FIG. 33-3 COMMON TYPES OF BARS USED FOR THE BORING OPERATION

holder. The boring bars may be of the plain type, figure 33-3A, or the end-cap type, figure 33-3B. In the plain type, the cutting tool is held in position by a setscrew, while in the end-cap type, it is held in position by the wedging action of a hardened plug.

Section 3 Machining in a Chuck

A **heavy-duty boring bar set**, figure 33-4, is mounted in the T-slot of the compound rest and can accommodate three different sizes of boring bars. The construction of this unit permits easy changing of the boring bars to suit the size of hole being bored. This allows the operator to use the largest bar possible for each job which permits the use of greater speeds and feeds, thereby increasing production.

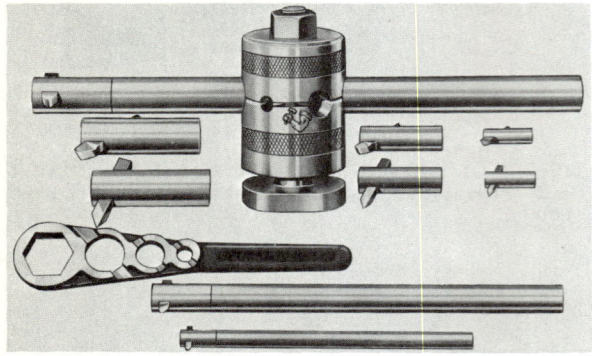

FIG. 33-4 A HEAVY-DUTY BORING BAR SET (Courtesy Armstrong Bros. Tool Co.)

Procedure: Boring

1. Mount the workpiece in a chuck.
2. Face and center drill the workpiece.
3. Set the lathe to the proper speed for drilling.
4. Drill the hole to the required depth and to within 1/16" of the finished diameter.
5. Mount the boring bar holder on the left side of the compound rest.
6. Mount the largest boring bar which can be accommodated in the drilled hole. *Grip the boring bar as short as possible to minimize chatter.* The boring bar should be level and parallel to the centerline of the lathe, figure 33-5.
7. Set the point of the toolbit just slightly above center since there is a tendency for the tool to spring down when cutting.
8. Set the lathe to the proper speed for the material to be bored and select a medium feed (.008-.010) for rough boring.

The speed for boring is the same as that for turning. The r.p.m. are calculated on the diameter of the hole.

9. Start the lathe and turn the crossfeed handle *counterclockwise* until the cutting tool touches the inside diameter of the hole.
10. Take a light trial cut about .005 deep and about 1/4" long at the right-hand end of the work.
11. Stop the lathe and measure the diameter of the bored section, using a telescopic gage or an inside micrometer.
12. Calculate the amount of material to be removed from the hole for the roughing cut.

    Leave the hole .010-.020 undersize for a finish cut.
13. Feed the cutting tool *out* (counterclockwise), using the crossfeed handle for a distance equal to one-half the amount of material to be removed.
14. Take a trial cut 1/4" long and recheck the diameter.

FIG. 33-5 A BORING BAR SET UP TO BORE A HOLE (Courtesy Colchester Lathe Co. Ltd.)

15. Take the roughing cut for the required length.

    Should vibration or chatter occur during a boring operation, it will be necessary to reduce the spindle speed and increase the feed slightly.

16. Stop the lathe and move the carriage to the right until the boring tool clears the hole. *Do not move the crossfeed handle.*

17. Set a fine feed (.005) for the finish cut.

18. Adjust the cutting tool to the finished diameter, using the crossfeed graduated collar.

19. Start the lathe, take a cut about 1/4" long, and check the diameter for size.

20. If the size is correct, take the finishing cut.

Due to the spring in the boring bar or boring tool, the hole is often slightly enlarged at the start of the hole. This condition is known as *bell-mouthing.* Should this occur, several cuts should be taken along the hole without adjusting the depth of cut. This procedure should eliminate the bell-mouthing.

SQUARING THE SHOULDER OF A BORED HOLE

If a bored hole is blind or has a step which must be at right angles to the bore, the surface must be faced. This is generally accomplished with a toolbit mounted in the 45° hole of the boring bar. See unit 34 for a detailed explanation of how to perform this operation.

REVIEW QUESTIONS

1. Define boring.
2. Why is a single-point cutting tool used in boring?
3. What may cause a drill to wander during a drilling operation?
4. What sequence of operations should be followed when machining a hole that must be accurate to size, shape, and location?

Boring Tools and Holders

5. What are two types of boring tools and what is the purpose of each?
6. Name and describe two types of boring bars.

Boring Procedure

7. What size bar should be used for any boring operation? Explain why.
8. What four points should be observed when setting up the boring bar and tool?
9. At what speeds and feeds should the work be rough bored?
10. How much material should be left in a hole for the finish cut?
11. How can boring bar chatter be minimized when boring a hole?
12. What is bell-mouthing and how can it be eliminated?
13. What is the procedure for squaring the shoulder of a bored hole?

Unit 34 COUNTERBORING

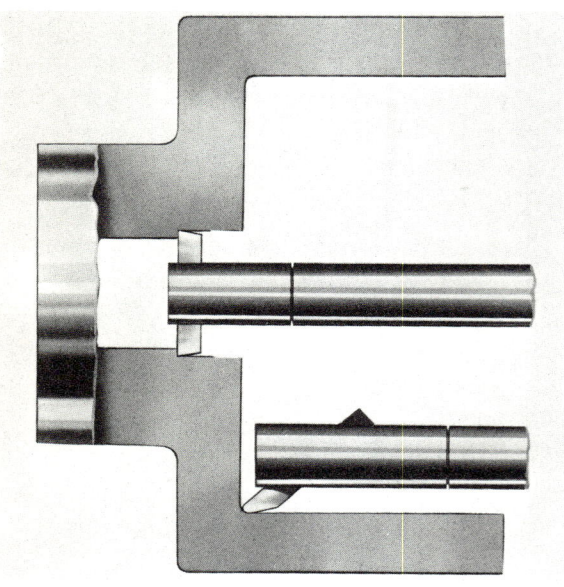

FIG. 34-1 A BORING TOOL FOR FINISH BORING AND SQUARING THE SHOULDER OF A COUNTERBORE (Courtesy Armstrong Bros. Tool Co.)

FIG. 34-2 SET THE COMPOUND REST TO 90° AND LOCK THE CARRIAGE IN ORDER TO MACHINE ACCURATELY A COUNTERBORE TO DEPTH.

Counterboring is the process of enlarging a hole for a definite depth. Holes are counterbored to provide recesses for bolts, nuts, differences in shaft diameters, and to act as a depth guide when internal threading. A counterbored hole has parallel sides and the shoulder between the bored and the counterbored hole is square.

Procedure: Counterboring

1. Set the spindle speed for the diameter of the counterbored hole.

2. Using a regular boring tool mounted in the 90° hole in the boring bar, rough out the hole within 1/32" of the finished diameter and depth.

3. Mount a suitably ground finishing tool in the 45° hole in the boring bar, figure 34-1.

4. Start the machine, feed the boring tool until it touches the internal diameter, and note the graduated collar setting.

5. Face (square) the internal shoulder, taking successive cuts from the center outwards. The crossfeed handle should be turned counterclockwise to the original setting plus .020 to provide a slight undercut at the shoulder.

If the depth of the shoulder must be accurate, set the compound rest 90° to the cross-slide and use the compound rest collar to gage the depth, figure 34-2. *The carriage must be locked if accuracy is to be maintained.*

6. Take a light trial cut about 1/8" long at the start of the hole.

7. Stop the machine and check the diameter of the counterbore.

8. Set the boring tool to the finished diameter, using the crossfeed graduated collar.

9. Start the machine, take a trial cut, and recheck the diameter.

10. Machine the internal diameter to the shoulder.

Unit 34 Counterboring

REVIEW QUESTIONS

1. Define counterboring. What is the purpose of counterboring?

2. How should a counterbore be roughed out?

3. Describe the tool used for finish counterboring.

4. What is the procedure for squaring the shoulder of a counterbore?

5. What is the procedure for accurately controlling the depth of the shoulder?

Unit 35 REAMING

Reaming is the process of enlarging a drilled or bored hole to produce an accurately sized and shaped hole with a good surface finish. This operation is performed with a finishing tool called a reamer, figure 35-1. Reamers are divided into two classes: machine reamers and hand reamers.

MACHINE REAMERS

Machine reamers can be divided into two types:
- Roughing reamer which cuts on the ends only
- Finishing reamer which cuts on the ends and along the length of the tooth.

FIG. 35-1 THE ROSE REAMER CUTS ON THE END ANGLE ONLY. (Courtesy Cleveland Twist Drill Co.)

FIG. 35-2 THE FLUTED CHUCKING REAMER CUTS ON THE ENDS AS WELL AS ALONG THE FULL LENGTH OF THE LAND. (Courtesy Cleveland Twist Drill Co.)

FIG. 35-3 THE JOBBER'S REAMER HAS A TAPERED SHANK AND LONGER FLUTES THAN THE CHUCKING REAMER.

FIG. 35-4A THE CHUCKING EXPANSION REAMER HAS PROVISION FOR EXPANDING THE DIAMETER. (Courtesy Cleveland Twist Drill Co.)

FIG. 35-4B AN ADJUSTABLE REAMER HAS INSERTED BLADES. (Courtesy Cleveland Twist Drill Co.)

Roughing Reamers

Roughing reamers are designed to produce a hole quickly and to within a few thousandths under the nominal size of the hole.

The rose reamer, figure 35-1, is the most commonly used machine reamer. The ends of the teeth are chamfered to about 45° and are provided with clearance so that all the cutting is done at the end of the reamer. The lands have no clearance and are used to guide the reamer in the hole. Because of their construction, rose reamers remove material quickly, and are particularly suited to the machining of cored holes. They are often made about .003-.005 under the nominal size so that the hole can be finished to size with a finishing reamer.

Finishing Reamers

Finishing reamers are used to finish a hole which has been bored or rough-reamed to within about .005 of the finish size.

Fluted chucking reamers, figure 35-2, have more teeth than a rose reamer for a given diameter and may have straight or tapered shanks. They are designed to cut along the full length of the land to produce a smooth and accurately sized hole. The ends of the reamer teeth are slightly chamfered for end cutting.

The jobber's reamer, figure 35-3, is available with straight or helical flutes. This reamer cuts along the full length of the lands, which are longer than those of a chucking reamer. It is used as a precision reamer and is designed to remove only a few thousandths from the hole.

Expansion reamers are sometimes used as finishing reamers and can be adjusted slightly to compensate for wear. The most common expansion reamers are the chucking expansion, figure 35-4A, and the adjustable blade reamer, figure 35-4B.

Taper Reamers

Taper machine reamers, figure 35-5, are manufactured in all standard tapers and with tapered shanks, so that they can be mounted directly in the spindle of the machine. At one time, both roughing and finishing taper machine reamers were used to produce tapered holes. The operation performed by a roughing taper machine reamer has been replaced by step drilling, boring, a special spiral-fluted reamer containing less flutes, or by using a worn finishing reamer.

REAMER CARE

Proper care of a reamer is imperative if it is to produce accurate and smoothly finished holes. The following points should be observed if the reamer is to give top performance.

1. Store reamers in separate containers or compartments when not in use. Plastic or cardboard tubes provide good reamer storage

2. Clean the reamer and coat it with a light film of oil before storing.

3. Never place reamers on metal surfaces such as machines and bench tops. Lay them on a cloth or a piece of masonite.

4. *Never* turn a reamer backwards; this dulls the cutting edges.

5. Always use cutting fluid when reaming steel. This prolongs the life of the reamer and produces a smoother hole.

6. Use helical fluted reamers for deep holes or holes which have keyways or grooves.

7. For best results and to preserve the life of a finishing reamer, always use a roughing reamer before using a finishing reamer.

ROUGHING REAMER

FINISHING REAMER

FIG. 35-5 TAPER MACHINE REAMERS

REAMING ALLOWANCES

The method by which a hole has been produced in the workpiece determines the amount of material to be left for reaming. Rough-drilled or punched holes should be rough-reamed before a finishing reamer is used. The amount of material left in a hole for reaming purposes depends on factors such as the condition of the machine, the type of material, the finish desired, the speed and feed of the reamer, and the use of cutting fluid. Table 35-1 shows suggested stock allowances for machine reaming.

MACHINE REAMING ALLOWANCES	
Hole Diameter	Allowance
1/4"	.010
1/2"	.015
1"	.020
1 1/2"	.025
2"	.030
3"	.045

TABLE 35-1

REAMING SPEEDS AND FEEDS

Although the recommended reaming speed is about one-half the drilling speed for similar materials, the speed is affected by several factors:

- The rigidity of the setup
- The type of material
- The finish and tolerance required

Higher reaming speeds are used on rigid setups, while less rigid setups require slower

Section 3 Machining in a Chuck

speeds. Holes requiring a fine finish and close tolerance require slower speeds. The use of cutting fluid permits higher speeds and produces better surface finishes.

Table 35-2 gives suggested reaming speeds for high-speed steel reamers. Carbide-tipped reamers can be operated at about twice the speed shown. If the reamer chatters while cutting at any of these speeds, reduce the speed until the chatter is eliminated.

RECOMMENDED REAMING SPEEDS (High-Speed Steel)	
Material	Speed (f.p.m.)
Aluminum	140-200
Brass	130-180
Bronze	50-100
Cast Iron	50-70
Machine Steel	50-70
Stainless Steel	40-50
Magnesium	170-250

TABLE 35-2

The feeds used for reaming are generally about twice the feeds used for drilling. Feeds which are too slow produce a glazed surface, excessive reamer wear, and often chatter. Too high a feed reduces hole accuracy and produces a poor surface finish. When tapered holes are being reamed, it is necessary to feed the reamer slowly, since the reamer is cutting along its entire length.

REAMING

Although basically the reaming of various holes is similar, different procedures should be used when reaming straight or tapered holes with machine or hand reamers.

Procedure: Reaming a Straight Hole

1. Align the lathe centers
2. Mount the work in the lathe.
3. Face and center-drill the workpiece.
4. Select the proper size drill.
   a. If the hole is to be positioned accurately, it should be bored after drilling. In this case leave from 1/32" to 1/16" material on the inside of the hole so that it can be trued by boring.
   b. Holes which must be accurate should be rough-reamed and then finish-reamed.
   c. If extreme accuracy of the hole location is not necessary, select a drill about 1/64" less than the finished hole size and ream directly after drilling.
5. Mount the drill, set the proper spindle speed, and machine the hole to the required depth. Use cutting fluid for this operation.
6. Remove the drill and mount the proper reamer in the lathe.
7. Reduce the spindle speed to one-half that of drilling.
8. Slide the tailstock until the reamer is close to the work and lock the tailstock.

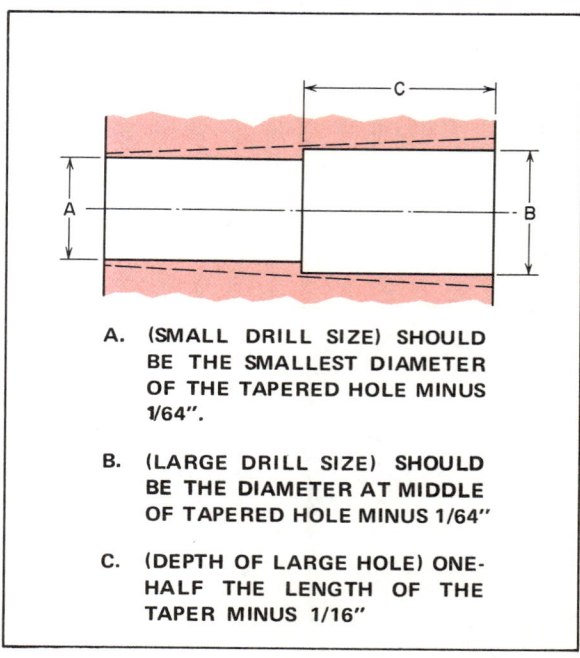

A. (SMALL DRILL SIZE) SHOULD BE THE SMALLEST DIAMETER OF THE TAPERED HOLE MINUS 1/64".

B. (LARGE DRILL SIZE) SHOULD BE THE DIAMETER AT MIDDLE OF TAPERED HOLE MINUS 1/64"

C. (DEPTH OF LARGE HOLE) ONE-HALF THE LENGTH OF THE TAPER MINUS 1/16"

FIG. 35-6 DRILL SIZES FOR TAPERED HOLES

The tailstock spindle clamp should be "snugged up" to prevent any side movement of the tailstock spindle.

9. Apply cutting fluid and feed the reamer steadily into the work.
10. Ream to the required depth and back out the reamer

    Never turn a reamer backwards or run a machine in reverse when removing a reamer since this dulls the cutting edges.
11. Shut off the machine and check the hole size.
12. Remove the burr from the edge of the hole with a scraper.

Procedure: Reaming a Tapered Hole

When a tapered hole is to be produced in a workpiece, it is necessary to step-drill the hole prior to reaming. The proper sizes of drills should be calculated as shown in figure 35-6.

1. Align the lathe centers.
2. Mount the work in the lathe.
3. Drill a hole for the full depth, 1/64" smaller than the smallest diameter of the tapered hole. (See "A" in figure 35-6.)
4. Select a drill about 1/64" smaller than the size of the finished hole measured at a point one-half way along the length of the tapered section. (See "B" in figure 35-6.)
5. Drill this hole to a depth of about 1/16" less than half the length of the tapered hole. (See "C" in figure 35-6.)
6. Remove the drill and mount a roughing taper reamer in the tailstock. Fasten a dog to the neck of the reamer with the tail resting on the compound rest to prevent the reamer from turning.
7. Set the spindle speed to about one-half the speed for straight reaming.
8. Apply cutting fluid and rough-ream the tapered hole to about .005 undersize.
9. Remove the roughing reamer and mount a finish taper reamer in the tailstock.
10. Apply cutting fluid and finish-ream the hole to size.
11. Remove the burr from the edges of the hole with a scraper.

HAND REAMING

If a hole is produced slightly undersize by a machine reamer, it can be brought to size by a hand reamer, figure 35-7.

Hand reamers may have straight or helical teeth and are used to remove *no more than .005* from a hole. A square on the end of the shank provides a means of turning the reamer into the hole with a wrench. Hand reamers are slightly tapered at the end of the reamer to permit easy starting. Since this taper has a tendency to dig into the work, the hand reamer should *never be used with power.*

Procedure: Reaming a Hole With a Hand Reamer

1. Drill or bore the hole to within .005 of the finished size.
2. Lock the headstock spindle or set it to the slowest speed.
3. Mount a center in the tailstock spindle.
4. Place the cutting end of the reamer into the hole and slide the tailstock up to engage the lathe center in the center hole of the reamer shank.
5. Lock the tailstock in position.
6. Apply cutting fluid to the reamer.

FIG. 35-7 STRAIGHT AND HELICAL FLUTED HAND REAMERS (Courtesy Cleveland Twist Drill Co.)

7. Using a wrench, turn the reamer *clockwise with the left hand.* At the same time, slowly turn the tailstock handwheel to provide support and advance the reamer into the hole.

    Never turn a reamer counterclockwise; the cutting edges can be dulled.

8. Ream the hole to the required depth.
9. Remove the reamer by turning it clockwise and pulling it from the hole at the same time.
10. Remove any burrs from the edge of the hole with a scraper.

REVIEW QUESTIONS

1. What is the process of reaming?
2. What are the two classes of reamers?

Machine Reamers

3. What are the two types of machine reamers?

Roughing Reamers

4. What is the difference in the cutting action of a roughing reamer and a finishing reamer?
5. Describe a rose reamer. What is the purpose of this reamer?

Finishing Reamers

6. How much material should be removed with a finishing reamer?
7. What are four types of finishing reamers?
8. Describe a fluted chucking reamer.
9. How does a jobber's reamer differ from a fluted chucking reamer?
10. What is the advantage of a chucking expansion reamer?

Taper Reamers

11. Name and describe two types of taper machine reamers.

Reamer Care

12. What are six points to observe in the care of reamers?

Reaming Allowances

13. List four factors which affect reaming allowances.
14. How much allowance should be left when reaming the following holes:
    a. 5/16" diameter?
    b. 3/4" diameter?
    c. 2 1/2" diameter?

Reaming Speeds and Feeds

15. What are three factors that affect reaming speed?

16. What procedure should be followed if the reamer chatters at the recommended speed?

17. At what speed should a 1 1/4" reamer be operated when reaming a hole in machine steel?

18. What is the effect of
    a. Too low a feed?   b. Too high a feed?

Reaming Procedure

19. What is the procedure to ream a hole that must be positioned accurately?

20. What precaution must be observed in regard to the rotation of the reamer?

Reaming Tapered Holes

21. How are the step drill sizes determined for a tapered hole?

22. At what spindle speed should taper reaming be performed? Why does this differ from straight reaming?

Hand Reamers

23. What is the procedure for setting up a hand reamer to ream in a lathe?

24. In which direction should a hand reamer always be turned?

Unit 36 TAPPING

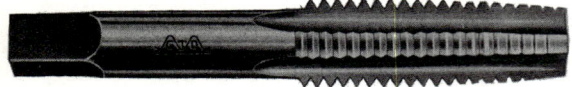
A. TAPER

B. PLUG

C. BOTTOMING

FIG. 36-1 A SET OF HAND TAPS (Courtesy Greenfield Tap and Die Co.)

Small internal threads required on work mounted in a lathe are generally cut with a tap supported by the tailstock center. The advantages of using a tap to cut internal threads are:

- It is the only method of threading a small hole.
- The thread form is accurately reproduced.
- The thread is cut square with the axis of the workpiece.
- It is a fast method of cutting an internal thread.

The main disadvantage of cutting internal threads with a tap is that the thread may not be exactly concentric with the outside diameter of the work. Large internal threads are often cut close to finish size on the lathe and then sized by means of a tap.

TYPES OF TAPS

Internal threads can be produced by means of hand taps and special-purpose taps.

Hand taps, figure 36-1, are available in sets containing three taps: taper, plug, and bottoming. When using hand taps to cut internal threads on a lathe, it is important that the tap be rotated into the work by hand and not by power.

Special-purpose taps, figure 36-2, are available in a variety of types and sizes to suit various metals or operations. Most of these taps can be used by hand or in a machine for power tapping. The shape of the initial cutting edge on spiral point taps causes the chips to advance ahead of the tap.

A. TWO-FLUTED SPIRAL POINT

B. THREE-FLUTED SPIRAL POINT

C. SPIRAL FLUTED

FIG. 36-2 SPECIAL PURPOSE TAPS (Courtesy Butterfield Division, Litton Industries)

Unit 36   Tapping

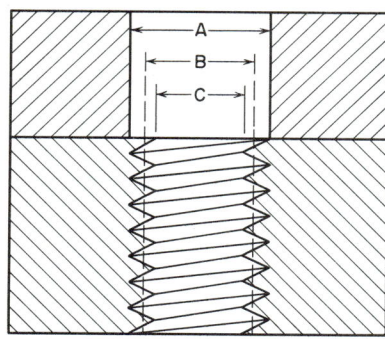

A = BODY SIZE
B = TAP DRILL SIZE
C = MINOR DIAMETER

FIG. 36-3 A CROSS SECTION OF A TAPPED HOLE

TAP DRILL SIZE

Before a tap is used to cut an internal thread, a hole must be drilled in the workpiece to the correct *tap drill size*. This drill size leaves the proper amount of material in the hole for a tap to cut the thread. The tap drill, *which is always smaller than the tap diameter,* leaves enough material in the hole to produce approximately 75% of a full thread.

The tap drill size for a National Form thread (75% of a full thread) can be calculated as follows:

$$\text{Tap drill size} = \text{Major Dia.} - \frac{1}{\text{No. thds/inch}}$$

$$= D - \frac{1}{N}$$

Example: Calculate the tap drill size required for a 1-8 – NC thread

$$\text{T.D.S.} = D - \frac{1}{N}$$

$$= 1'' - \frac{1}{8}$$

$$= \frac{7''}{8}$$

Should a full thread be required (100% bearing), the tap drill size should be equal to the major diameter minus the double depth of thread.

Procedure: Tapping a Hole by Hand

1. Align the lathe centers
2. Mount and true the work in a chuck.
3. Face and center drill the workpiece.

   The hole should be drilled deep enough so that the top of the center hole is slightly larger than the major diameter of the tap. This protects the start of the thread.

4. Mount the correct tap drill size in a drill chuck or tailstock spindle.
5. Drill the tap drill size hole to the correct depth.

FIG. 36-4 CUTTING AN INTERNAL THREAD WITH A TAP

6. Mount a dead center in the tailstock spindle or a stub center in the drill chuck.
7. Set the lathe to its slowest speed or lock the spindle.
8. Place the proper size tap into the hole of the workpiece.
9. Slide the tailstock up until the dead or stub center bears in the center hole of the tap or tap wrench, figure 36-4, and lock the tailstock in place.

Section 3 Machining in a Chuck

On small taps having no center hole, a T-handle tap wrench with a center hole should be used.

10. Apply cutting fluid and turn the tap into the work, keeping the dead center snug in the center hole. The dead center must be kept in engagement with the center hole to insure that the tap remains parallel with the axis of the hole.

11. Back off the tap frequently to break the chips. Continue tapping until the hole is tapped.

It is good practice to remove the tap from the hole occasionally to clear the chips.

12. If a blind hole is being tapped, it is necessary to use the taper, plug, and bottoming taps in that sequence.

Procedure: Tapping a Hole by Power

Holes can be tapped by power on lathes where the spindle rotation can be reversed by holding a spiral-pointed (gun) tap in a drill chuck mounted in the tailstock. Extreme caution should be used for this operation, and it is not recommended for blind holes.

1. Align the lathe centers.
2. Face and center drill the workpiece.
3. Drill the hole to the correct tap drill size.
4. Set the lathe to the lowest speed.
5. Mount a spiral-pointed tap in a drill chuck mounted in the tailstock spindle.
6. Start the lathe and apply cutting fluid to the tap.
7. Slide the tailstock to the left and bring the tap into the hole. Maintain hand pressure on the tailstock until the tap has started the thread. *The tailstock should be free and not clamped to the lathe bed for this operation.*
8. Allow the tap to advance approximately 1/2" into the workpiece and then reverse the spindle rotation to bring the tap out of the hole. As the tap is coming out of the hole, apply hand pressure to force the tailstock to the right until the tap clears the hole.
9. Clear the chips from the flutes and apply cutting fluid.
10. Continue the tapping procedure until the hole is threaded to the correct depth.

CAUTION: Whenever a tap turns in the drill chuck, quickly reverse the spindle to bring the tap out of the hole.

REVIEW QUESTIONS

1. What are the advantages of using a tap to cut internal threads?

2. What procedure should be followed to cut large internal threads?

Types of Taps

3. Describe a spiral-point tap. What is the purpose of this tap?

Tap Drill Size

4. Define a tap drill.

5. Calculate the tap drill size for the following threads:

    a. 7/8-9 NC   b. 7/16-14 NC   c. 1/4-28 NF

Tapping a Hole by Hand

6. How deep should a center hole be drilled prior to tapping? Explain why.

7. What procedure should be followed to insure that the tap remains in alignment during the tapping operation?

Tapping by Power

8. What type of tap should be used when tapping by power?

9. How is the tap fed into the work when using power?

10. What three precautions should be observed when tapping by power?

Unit 37 TAPER TURNING

Some external tapers (depending upon the nature of the workpiece), and all internal tapers can be machined readily when the work is held in a chuck. Tapers can be machined on work held in a chuck by using either the taper attachment or the compound rest. The compound rest is used to machine short, steep tapers or angles, while the taper attachment is used for long, slow tapers.

TAPER CALCULATIONS

To set the taper attachment or compound rest for taper turning, the taper per foot or the angle of the workpiece must be calculated. The taper attachment can be set for either taper per foot or degrees of taper. The compound rest can only be set for degrees of taper and, therefore, it is necessary to calculate the angle of a taper.

Taper per Foot Calculations

In many cases, the taper per foot of an object is specified on the blueprint. However, if it is not specified, the taper per foot of a workpiece can be calculated by applying the following formula:

$$\text{Taper per foot} = \frac{(\text{large dia.} - \text{small dia.}) \times 12}{\text{length of the taper}}$$

See unit 18 for a more detailed explanation of taper per foot calculations.

Angular (degree) Calculations

When a taper is to be machined by the compound rest, the compound rest must be set to the proper angle. This angle (the angle with the centerline, or one-half the included angle) can be calculated by using the approximate method, or by trigonometry, depending upon the accuracy required.

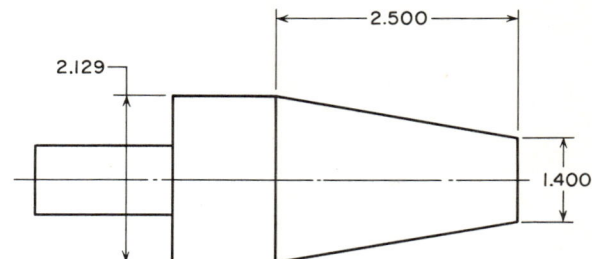

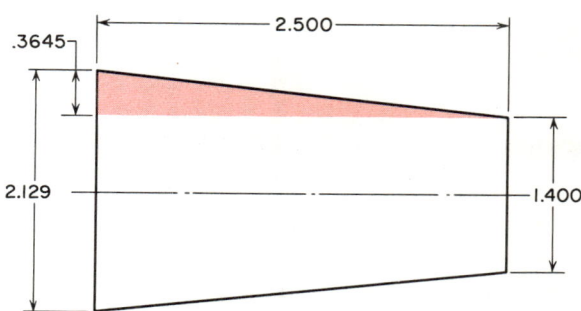

FIG. 37-1 A SHORT TAPER WHICH CAN BE CUT USING THE COMPOUND REST OR TAPER ATTACHMENT

Approximate Method

TPF x 2°23' = angle with centerline (in deg.)

This method uses a constant of 2°23' which, when multiplied by the taper per foot, produces an answer in units of degrees and minutes. It is quite accurate when the taper per foot is in inches or half inches (e.g. 1'', 1½'' TPF, etc.). It is reasonably accurate when calculating other fractional tapers such as 5/8'', 1 ¼'', etc.

Trigonometry

$$\frac{\text{TPF}}{24} = \text{tangent of angle (with centerline)}$$

OR

$$\frac{\text{Lgth. Side Opposite}}{\text{Lgth. Side Adjacent}} = \text{Tangent of Angle}$$

Unit 37 Taper Turning

For the workpiece illustrated in figure 37-1A, calculate the taper per foot and the angular setting required for the compound rest or taper attachment.

$$\text{Taper per Foot} = \frac{(2.129 - 1.400) \times 12}{2.5}$$

$$= \frac{.729 \times 12}{2.5} = 3.499 \text{ or } 3\frac{1}{2}''$$

Compound Rest or Taper Attachment Angular Setting

Approximate method

$$= 3.5 \times 2°23' = 7°80' = 8°20'$$

Trigonometry

$$= \frac{3.5}{24} = \tan .1458 = 8°19'$$

OR

$$= \frac{.3645}{2.5} = \tan .1458 = 8°19'$$

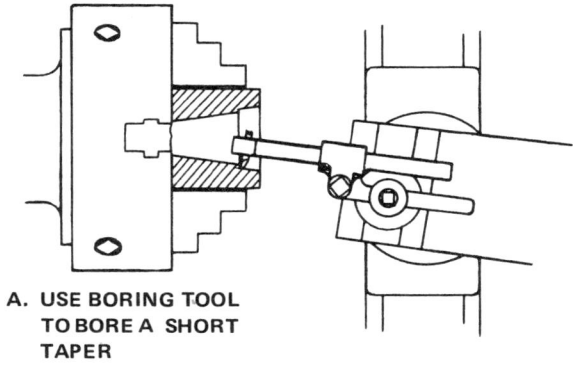

A. USE BORING TOOL TO BORE A SHORT TAPER

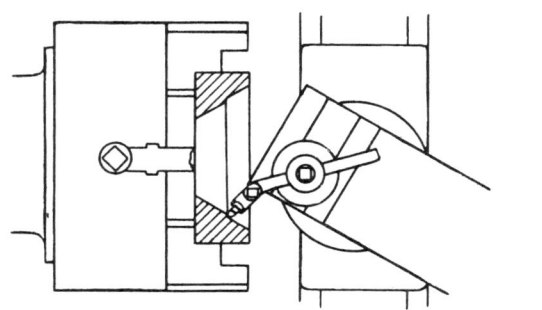

B. USING A RIGHT-HAND OFFSET TURNING TOOL TO BORE LARGE ANGULAR OPENINGS

FIG. 37-2 THE COMPOUND REST CAN BE USED TO MACHINE SHORT, STEEP TAPERS.

TAPER TURNING (COMPOUND REST)

A convenient way to cut short, steep internal or external tapers is by swivelling the compound rest to the desired angle and feeding the cutting tool with the compound rest feed screw, figure 37-2. Steeper tapers can be machined with the compound rest than is possible with the taper attachment. The length of the taper which can be cut is limited to the amount of travel of the compound rest slide.

The cutting tool can be mounted in a boring bar, figure 37-2A, for most internal tapers or in a toolholder, figure 37-2B, for large internal tapers and most external tapers. The tools are mounted and set up in the lathe toolpost as for boring or turning angular surfaces.

Accurate Angular Settings

A compound rest or taper attachment can be accurately set to a specific angle by using a test bar and dial indicator.

Assume that a 1° angle must be machined on a piece of work 1" long, figure 37-3A. By simple calculation, the rise of a 1° angle (x) in 1" can be found:

$$x = \text{side adjacent} \times \tan \text{of angle}$$
$$= 1.000 \times \tan 1°$$
$$= 1.000 \times .01745$$
$$= .01745$$

Therefore, for every degree of taper in a 1" length, the rise is .01745. For example, in a 5° taper the rise equals 5 × .01745 or .0872.

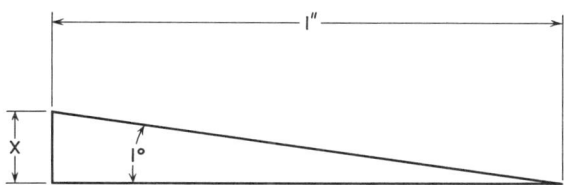

FIG. 37-3A A 1° ANGLE WILL PRODUCE A RISE OF .01745 FOR EVERY INCH OF LENGTH.

Section 3 Machining in a Chuck

FIG. 37-3B A COMPOUND REST OR TAPER ATTACHMENT CAN BE ACCURATELY SET TO AN ANGLE WITH A TEST BAR AND DIAL INDICATOR.

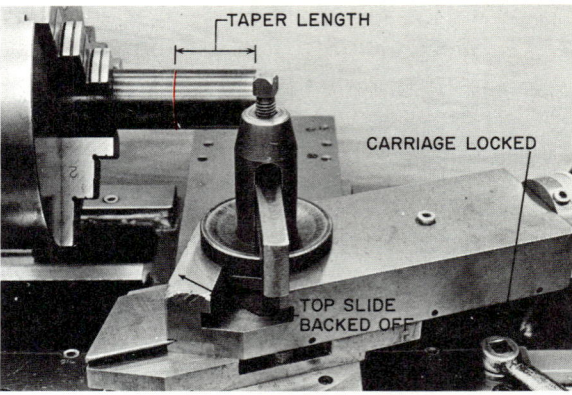

FIG. 37-4 THE TOP SLIDE OF THE COMPOUND REST SET TO INSURE SUFFICIENT MOVEMENT TO MACHINE THE LENGTH OF THE TAPER

Procedure: Setting a Compound Rest or Taper Attachment to an Accurate Angle

1. Mount a test bar and dial indicator in the lathe.
2. Align the lathe centers.
3. With a pencil, mark off a 1" length on the test bar, figure 37-3B.

   A 2" or 3" length is preferable since the compound rest or taper attachment setting can be made more accurately over a longer length.
4. Calculate the angle and set the compound rest or taper attachment to this angle.
5. Move the carriage until the indicator point is exactly on the right-hand 1" line, figure 37-3B.
6. Bring the indicator into contact with the test bar until it registers approximately one-quarter of a turn and set the bezel to zero.
7. Use the compound rest feed screw to move the dial indicator exactly one inch. If the taper attachment is used, move the carriage exactly one inch.
8. If the angular setting is correct, the indicator should register .01745 for every degree in 1" of length (e.g. in 10° the indicator should register 10 x .01745 or .1745).
9. Adjust the compound rest or taper attachment until the angle is correct.

Procedure: Machining A Taper Using the Compound Rest

1. Check the blueprint to determine the angle of the taper. Calculate this if it is not given.
2. Loosen the compound rest screws.
3. Swing the compound rest to the required angle.

   If the included angle is given on the part, swivel the compound rest to one-half the included angle.
4. Tighten the compound rest screws, using only two-finger pressure to avoid stripping the threads.
5. Set the proper cutting tool (boring or turning) on center and position it so that *the point and not the side does the cutting.*
6. Back off the top slide of the compound rest to insure that there is enough movement to machine the length of the taper, figure 37-4.

Unit 37 Taper Turning

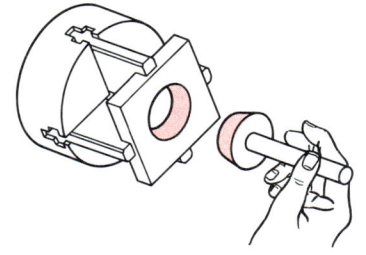

A. USING A PLUG GAGE

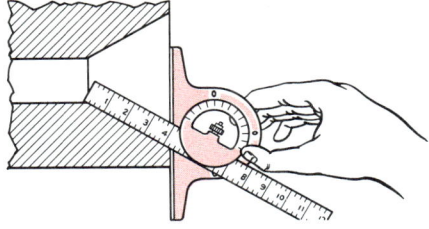

B. A PROTRACTOR USED TO CHECK THE ANGLE OF A TAPERED HOLE

FIG. 37-5 METHODS OF CHECKING THE ACCURACY OF INTERNAL TAPERS

7. Move the carriage to position the cutting tool at the start of the taper and then lock the carriage.
8. Rough-turn the taper section (internal or external) by feeding the cutting tool using the *compound rest feed screw.*
9. Check the accuracy of the taper and readjust the compound rest setting if necessary.
10. Finish-turn the taper section.
11. Check the accuracy of the taper with a ring or plug gage, figure 37-5A, protractor, figure 37-5B, sine bar, or taper micrometer, depending upon the accuracy required.

See unit 18 for a detailed explanation of the various methods and procedures for measuring tapers.

Machining a Taper Using the Taper Attachment

The taper attachment can be used to machine internal or external tapers quickly and accurately. Set the taper attachment accurately to the degree or taper per foot required and machine the taper as outlined in unit 18.

REVIEW QUESTIONS

1. What type of tapers are generally cut with
   a. the compound rest?   b. the taper attachment?

Taper Calculations

2. A tapered workpiece having the following dimensions is required: Large diameter = 2 1/4'', Small diameter = 1 7/8'', Length of taper = 3''. Calculate
   a. The taper per foot
   b. The angle required to set the compound rest or taper attachment by:
      1) The approximate method   2) Trigonometry

Taper Turning (Compound Rest)

3. What is the advantage of using the compound rest to machine tapers?
4. What are two methods of setting up tools for machining internal tapers?
5. How can a taper attachment or compound rest be set *accurately* to a 7° angle?
6. How should the compound rest be set to insure that the entire length of taper may be machined?
7. What are two methods of checking the accuracy of a taper?
8. How can the taper attachment be set up accurately to machine a taper?

Unit 38 INTERNAL THREADING

Many internal threads are finished with a tap; however, if a certain size of tap is not available, or when it is essential that the thread is concentric with the diameter, it must be cut on the lathe. The workpiece to be threaded may be held in a collet, chuck, on a faceplate, or supported in a steady rest. The internal threading operation is very similar to the operation of cutting external threads on a lathe. However, extra care should be taken due to the spring of the boring bar and the inability of the operator to see the operation as easily as when cutting an external thread. Before proceeding with the threading operation, it is necessary to calculate the tap drill size of the hole to be threaded. See unit 36 on how the tap drill size is calculated.

Procedure: Cutting a 1 1/4" 7-NC Internal Thread

1. Calculate the tap drill size of the thread.

$$TDS = D - \frac{1}{N}$$
$$= 1.250 - \frac{1}{7}$$
$$= 1.250 - .143$$
$$= 1.107$$

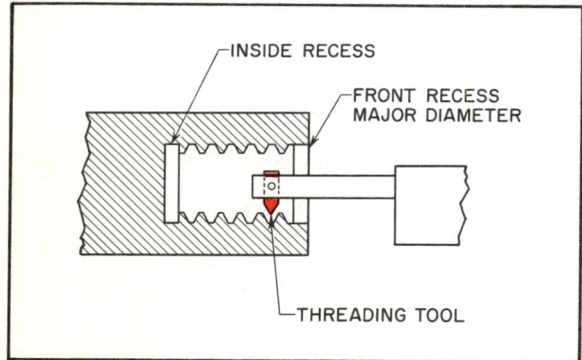

FIG. 38-1 A RECESS AT THE FRONT AND BACK OF HOLE IS DESIRABLE WHEN INTERNAL THREADING.

2. Mount the workpiece in a collet, chuck, or on a faceplate.

3. Drill a hole about 1/16" smaller than the tap drill size, that is, 1.107 – .062.
    $$= 1.045$$
    $$= 1 \ 3/64" \text{ approximately}$$

4. Set up a boring tool and bore the hole to 1.107.

   The boring bar should be as large as possible and should be gripped short to minimize vibration or chatter.

5. Counterbore the end of the hole to the major diameter of the thread (1.250) for a distance of about 1/16". This serves as a guide to the depth of thread during the threading operation.

6. If the hole to be threaded does not go through the metal (a blind hole), it is necessary to cut a recess at the end of the thread to provide clearance for the threading tool at the end of the cut, figure 38-1. This recess should be cut slightly deeper than the major diameter and should be wide enough to permit the toolbit to clear the thread when the split nut is disengaged.

7. Set the compound rest to *29° to the left*, figure 38-2.

8. Set the quick-change gearbox to 7 T.P.I.

9. Engage the lead screw.

10. Mount a threading tool in the boring bar and set the point of the tool on center. The boring bar should be parallel to the centerline of the machine.

11. Square the threading tool with a center gage, figure 38-3.

210

12. The depth of hole to be threaded should be marked on the boring bar, measuring from the threading tool. When this mark is even with the edge of the workpiece, the split-nut lever must be disengaged.

13. Start the lathe and turn the crossfeed handle until the point of the toolbit touches the diameter of the hole.

14. Set the crossfeed and compound rest graduated collars to zero and clear the cutting tool from the hole with the carriage handwheel.

15. Turn the compound rest handle *counterclockwise* to move the toolbit out .003 - .005, and take a trial cut.

16. At the end of each cut (when the mark on the boring bar is even with the edge of the workpiece), disengage the split-nut lever and turn the crossfeed handle *clockwise* to clear the toolbit from the thread.

17. Move the carriage to the right until the toolbit is clear of the work.

18. Check the pitch of the thread with a screw pitch gage.

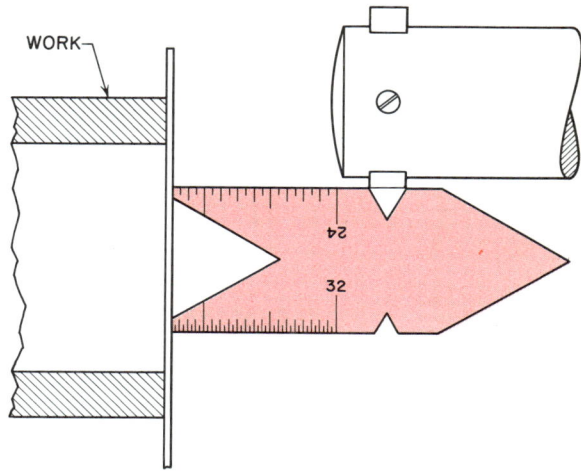

FIG. 38-3 SQUARING THE THREADING TOOL WITH A CENTER GAGE

19. Return the crossfeed handle to zero and set the depth of cut by feeding the compound rest counterclockwise about .015.

20. Take successive cuts, decreasing depths until the thread is to the proper depth. If the point of the threading tool is the correct width, the amount of compound rest feed can be calculated by applying the following formula:

$$\text{Compound rest feed} = \frac{.750}{N}$$

$$= \frac{.750}{7}$$

$$= .107$$

(Table 21-1 in unit 21 lists the amount of compound rest feed required for various National Form Threads).

21. As the thread becomes deeper, it is necessary to decrease the amount of compound rest feed to decrease the spring of the boring bar.

The last few cuts should only be .001 deep in order to eliminate any spring remaining in the bar.

22. Check the thread for fit with a thread plug gage or a bolt.

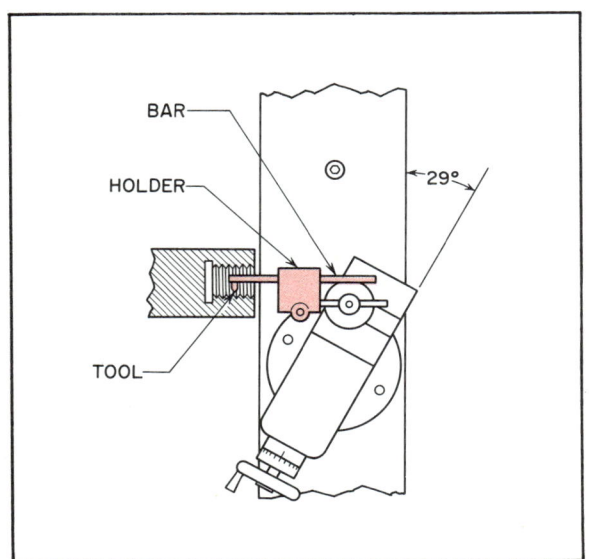

FIG. 38-2 THE COMPOUND REST IS SWUNG TO 29° TO THE LEFT FOR CUTTING RIGHT-HAND INTERNAL THREADS.

Section 3 *Machining in a Chuck*

REVIEW QUESTIONS

1. What are two reasons why care should be taken when cutting an internal thread on a lathe?

2. Calculate the tap drill size for a 1½" − 6 NC thread.

3. Why is the end of the hole counterbored to the size of the major diameter?

4. Why is a recess necessary at the end of a blind hole to be threaded?

5. Compare the setting of the compound rest for cutting internal and external threads. Why do these settings differ?

6. Why should the boring bar be marked when internal threading? How is this done?

7. Why is the depth of cut decreased with each successive cut?

8. Calculate the amount of compound rest feed required to cut the following National Form threads:

    a. 3/4" − 10   c. 1 1/4" − 6
    b. 1" − 8   d. 2" − 4 1/2"

Unit 39　FACEPLATE WORK

Many types of lathe work cannot be gripped in a chuck or mounted between centers because of their unusual shapes. These workpieces are usually clamped to a faceplate for machining operations. Faceplates are especially useful for holding large, flat, irregularly-shaped workpieces, castings, jigs, and fixtures.

FACEPLATES AND ACCESSORIES

A *faceplate,* figure 39-1, is similar to a lathe driveplate except that it is larger in size. It contains a number of T-slots and/or elongated holes to accommodate bolts and clamps. When the faceplate is mounted on the lathe spindle, its face or surface is at right angles to the centerline of the lathe.

FIG. 39-1 WORK MAY BE HELD ON A FACEPLATE FOR MACHINING OPERATIONS.

Faceplate Accessories

Since the type and shape of workpiece which can be mounted on a faceplate varies greatly, a variety of faceplate accessories are used when mounting work.

An angle plate, figure 39-2, can be fastened to the faceplate to hold flat workpieces for machining. The machined surface of the workpiece is clamped to the angle plate and the subsequent machining operations are all parallel to the machined surface.

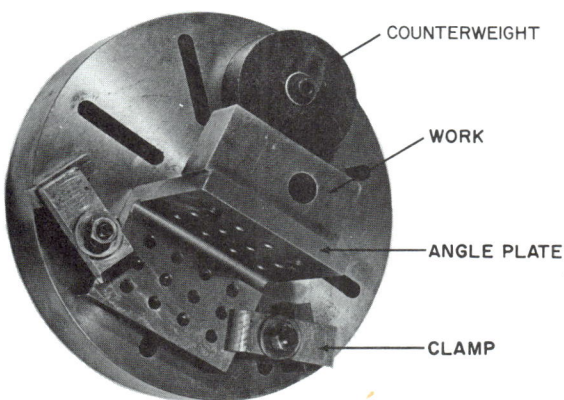

FIG. 39-2 AN ANGLE PLATE BEING USED TO HOLD A SQUARE WORKPIECE FOR MACHINING

A V-block, figure 39-3, can be fastened to the surface of the angle plate to hold round workpieces. The machining operations performed on workpieces held in this manner are all at 90° (right angles) to the centerline of the workpiece.

A variety of bolts, clamps, parallels, stop blocks, and counterweights are used to set up and fasten work to a faceplate, figures 39-2 and 39-3.

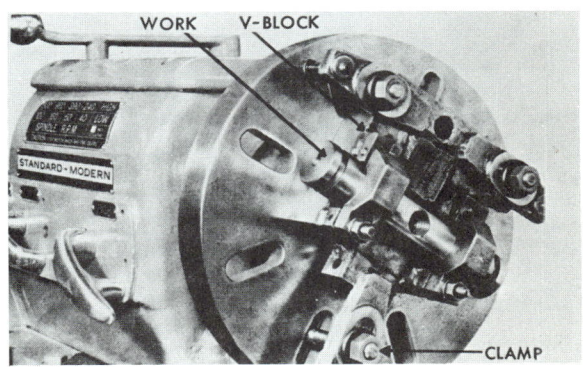

FIG. 39-3 A V-BLOCK FASTENED TO THE ANGLE PLATE IS USED TO HOLD ROUND WORKPIECES.
(Courtesy Standard-Modern Tool Co.)

Section 3 Machining in a Chuck

MOUNTING THE FACEPLATE

A faceplate should be mounted carefully, using the same procedure as for mounting chucks, unit 26. It is important that any nicks or burrs be removed from the surface of the faceplate with a file before a workpiece is mounted. When extreme accuracy is required, it is good practice to take a light facing cut from the surface of the faceplate before mounting the workpiece. This removes any nicks and burrs, and insures that the faceplate surface is true and at 90° to the centerline of the lathe.

MOUNTING THE WORKPIECE

Work can be mounted on a faceplate while it is on the lathe spindle or while it is on a bench, figure 39-4. If the workpiece is heavy or awkward to hold, it is recommended that the workpiece be mounted when the faceplate is on the bench, and then the entire unit mounted on the lathe spindle.

HINTS FOR MOUNTING A WORKPIECE

1. Remove all nicks and burrs from the surfaces of the faceplate and the workpiece with a file.

2. Place a piece of paper, slightly larger than the workpiece, on the faceplate. This reduces the danger of the work slipping during the machining operation.

3. Center the workpiece (center punch mark or hole) approximately on the faceplate.

4. Parts which do not lie flat should be shimmed to prevent rocking and distortion when the work is clamped.

5. Lightly fasten the workpiece to the faceplate with suitable bolts and clamps.

   The position of the bolts and clamps is very important to clamp a workpiece effectively. Figure 39-5A illustrates the proper clamping procedure. Note that the bolt is close to the workpiece and a step or packing block is a little higher than the workpiece. In figure 39-5B the bolt is closer to the block and, therefore, most of the clamping pressure is on the block and not the workpiece.

6. Mount the faceplate on the lathe spindle.

7. True the center punch mark or the hole with the centerline of the lathe.

8. Tighten all clamping nuts securely.

   If a number of duplicate pieces are to be machined, after the first part has been

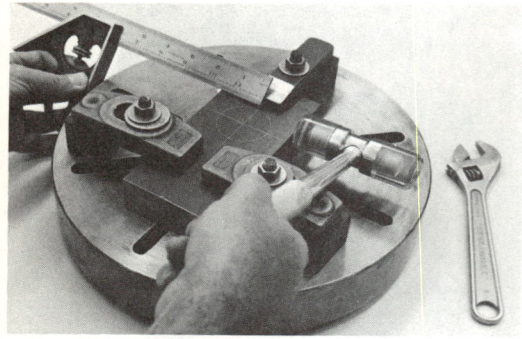

FIG. 39-4 LOCATING AND CLAMPING WORK TO A FACEPLATE BEFORE MOUNTING IT ON THE LATHE SPINDLE

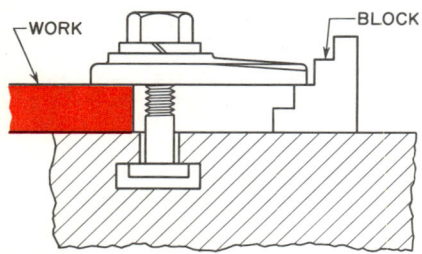

FIG. 39-5A CORRECT CLAMPING SETUP INSURES THAT THE WORKPIECE IS HELD SECURELY.

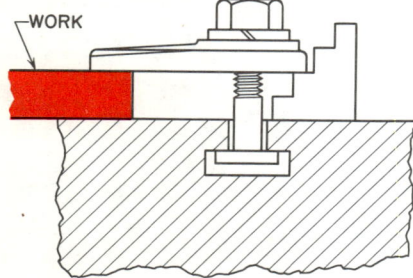

FIG. 39-5B WORK INCORRECTLY CLAMPED WITH MOST PRESSURE BEING APPLIED TO THE BLOCK AND NOT THE WORKPIECE

Unit 39 Faceplate Work

located, the faceplate can be set up as a fixture, using parallel strips and stop blocks against the sides of the workpiece. Succeeding parts need only to be brought against the strips and stop blocks to be accurately located.

9. Mount suitable counterweights to balance the location of the workpiece on the faceplate and to overcome vibration when the lathe is in operation, figure 39-2.

ALIGNING THE WORKPIECE

After the workpiece has been fastened to a faceplate, it is necessary to locate the center punch mark, hole, or diameter with the axis of the lathe. Punch marks can be aligned with a center tester, or with a wiggler and indicator if accuracy is required. An indicator should be used to true holes or diameters.

Procedure: Truing a Center Punch Mark with a Center Tester

1. Mount a center tester in the toolpost as shown in figure 39-6.
2. Bring the left-hand point of the center tester into the center punch mark.
3. Adjust the cross-slide and center tester until the right-hand point of the center tester is close to the point of the tailstock center.
4. Revolve the lathe spindle by hand and observe how much the point of the center tester revolves around the tailstock center.
5. With a hardwood block or brass rod, tap the workpiece until the point of the center tester does not move around the tailstock center when the spindle is revolved.
6. Tighten all bolts securely and recheck the accuracy of the center punch mark.

Procedure: Truing a Center Punch Mark with a Wiggler and Dial Indicator

1. Mount a drill chuck in the tailstock spindle.
2. Grip the shank of the center finder in the chuck and adjust the tailstock spindle until the point of the wiggler fits into the center punch mark, figure 39-7.
3. Lock the tailstock and tailstock spindle in position.
4. Mount an indicator on the toolpost and bring its point into contact with the wiggler body as close to the point as possible.
5. Revolve the lathe spindle by hand and stop when the indicator is at the high spot.
6. With a hardwood block or brass rod, tap the workpiece until the indicator registers one-half the run-out.
7. Continue to tap the workpiece until there is no movement of the indicator needle when the work is revolved.
8. Tighten all bolts securely and recheck the accuracy of the center punch mark.

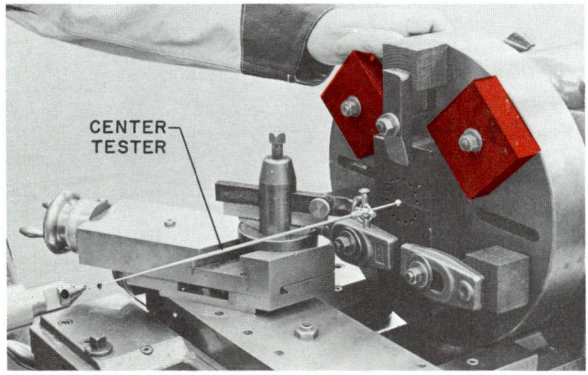

FIG. 39-6 USING A CENTER TESTER TO ALIGN A CENTER PUNCH MARK

FIG. 39-7 USING A WIGGLER AND DIAL INDICATOR TO TRUE A CENTER PUNCH MARK

Section 3 Machining in a Chuck

FIG. 39-8 TRUING A DIAMETER ON A FACEPLATE WITH A DIAL INDICATOR

Procedure: Truing a Hole or Diameter with a Dial Indicator

1. Mount a dial indicator with an internal or external attachment, as required, on the toolpost, figure 39-8.

2. Bring the indicator into contact with the diameter to be trued until the needle registers approximately one-half turn.

3. Revolve the lathe spindle by hand and note the high reading on the dial indicator.

4. With a hardwood block or brass rod, tap the workpiece one-half the difference between the high and low indicator reading.

To prevent damage to the indicator, *always tap the workpiece away from and not towards the indicator.*

5. Continue tapping the workpiece one-half the difference between the high and low reading until there is no movement of the indicator needle when the spindle is revolved.

6. Tighten all bolts securely and recheck the accuracy of the setup.

HINTS FOR TURNING WORK ON A FACEPLATE

1. Be sure that the faceplate is thoroughly cleaned and mounted securely on the spindle.

2. Use the shortest bolts and clamps possible to avoid hitting the lathe carriage during the machining operation.

3. Use sufficient counterweights to balance the work accurately on the faceplate and to reduce vibration.

4. Always turn the lathe spindle one full turn by hand to insure that the work on the faceplate clears the carriage.

5. The lathe speed should be set so there is no excessive vibration while the work is revolving.

REVIEW QUESTIONS

1. For what purpose are faceplates used?

Faceplates and Accessories

2. Describe a standard faceplate.

3. Name three faceplate accessories. What is the purpose of each?

Mounting the Faceplate

4. How should a faceplate be prepared prior to mounting on a lathe spindle?

Mounting the Workpiece

5. Why is it advisable to use paper between the work and faceplate?

6. How should parts which are not flat be prepared prior to clamping to a faceplate?

Unit 39 Faceplate Work

7. How should the bolt, clamp, and block be placed for effective clamping? Illustrate with a diagram.

8. How can the faceplate be set up to locate quickly a number of duplicate parts?

Aligning the Workpiece

9. Explain an accurate method of aligning a center punch mark.

10. What is the procedure for aligning a hole or diameter?

Hints for Turning Work on a Faceplate

11. What are three important points which should be observed when turning work on a faceplate?

SECTION 4

Special Operations

Unit 40 SPECIAL FORM TURNING

When special curved or irregularly-shaped surfaces are required on lathe work, they can be produced by several means.

- By manipulating the lathe controls *freehand,* see unit 19.
- By using a *form tool* ground to the desired shape to produce the form, see unit 19.
- By rotating the *compound rest* on its swivel base.
- By placing a *rod* between the headstock and cross-slide, or between the tailstock and cross-slide.
- Manually, with the use of a *suitable template and follower*

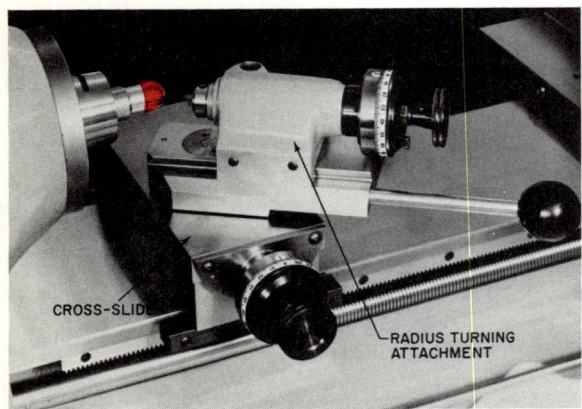

FIG. 40-1 A RADIUS TURNING ATTACHMENT SHOWN MOUNTED ON THE CROSS-SLIDE FOR TURNING CONCAVE OR CONVEX FORMS (Courtesy Hardinge Bros., Inc.)

COMPOUND REST

Concave and convex forms can be accurately machined by loosening the compound rest studs *only* enough to allow the compound rest to be swivelled in an arc. A certain amount of tension or drag is required on the studs to prevent the compound rest from swivelling too freely. The size of the concave or convex radius is dependent upon the distance that the cutting tool is from the center of the compound rest swivel. This distance can be adjusted by using the compound rest feed screw. If the cutting tool is ahead of the center, a concave radius is produced. A convex radius is cut when the cutting tool is behind the center of the compound rest swivel.

A special radius turning attachment, figure 40-1, can be mounted on the cross-slide of the lathe to produce precision concave or convex forms. The top slide, which holds the cutting tool, can be adjusted the required distance from the center to produce the size and type of radius required. The handle is used to rotate the swivel slide, upon which the cutting tool is mounted, through 360°.

FACE CONCAVE AND CONVEX TURNING

Many times it is necessary to cut a concave or convex form on the *face* of only a few workpieces. Since the quantity required does not justify making a template and/or the purchase of a tracer attachment, a more economical method must be used to produce the desired form. A simple, yet very effective method of cutting large concave or convex radii on the *face* of a workpiece is illustrated in figures 40-2A and B.

A standard lathe can be easily adapted to the cutting of concave or convex radii on the face of a workpiece by drilling four center holes, one in the headstock, one in the tailstock, and one on each side of the cross-slide, figures 40-2A and B. A rod, preferably one that can be adjusted to various lengths, is necessary to produce the size of the radius required.

FIG. 40-2A A CONVEX RADIUS IS PRODUCED ON THE FACE OF A WORKPIECE WHEN A ROD IS PLACED BETWEEN THE HEADSTOCK AND CROSS-SLIDE.

FIG. 40-2B A CONCAVE RADIUS IS PRODUCED ON THE FACE OF A WORKPIECE WHEN A ROD IS PLACED BETWEEN THE TAILSTOCK AND CROSS–SLIDE.

Procedure: Cutting a Convex Face Radius

1. Mount the workpiece in a chuck or on a faceplate.

2. Place the proper length of rod for the radius required in the center holes of the headstock and cross-slide.

   If an accurate radius is required, it is necessary to clamp a block having a center hole on the lathe bed with its edge even with the face of the work to be machined. A rod of the proper length, corresponding to the radius required, must be placed between the block and the cross-slide.

3. Set the compound rest at 90° to the cross-slide.

4. Set the cutting tool on center and adjust the cross-slide until the tool point is exactly aligned with the center of the rod and the workpiece, figure 40-2A.

   For an accurate radius, the point of the cutting tool must be directly over the *end of the rod.*

5. Bring the cutting tool close to the workpiece and set the depth of cut with the compound rest screw.

6. Apply sufficient pressure on the carriage handwheel to keep the cross-slide against the rod constantly.

7. Feed the cutting tool across the face of the workpiece with the crossfeed screw handle.

8. Check the accuracy of the radius and adjust the rod or cutting tool as required.

Cutting a Concave Face Radius

The setup and operation for cutting a concave radius is very similar to that required for a convex radius. The basic difference is that the tailstock is locked in position and a rod of the required length is placed between the tailstock and the cross-slide.

Section 4 Special Operations

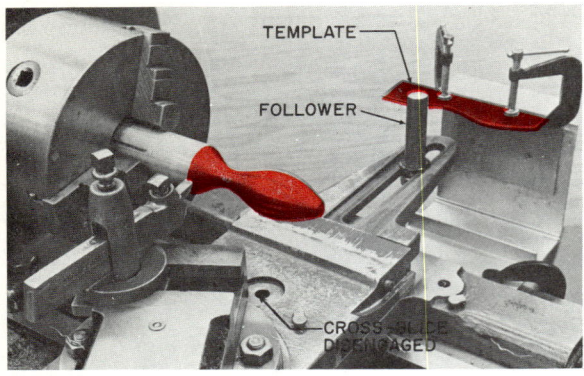

FIG. 40-3 A FOLLOWER ON THE CROSS-SLIDE FOLLOWS A TEMPLATE IN ORDER TO PRODUCE A SPECIAL FORM.

USE OF A TEMPLATE AND FOLLOWER

When a few pieces of a special form are required, they can be accurately produced by using a template of the desired form and a follower fastened to the cross-slide of the lathe. The accuracy of the template which must be made determines the accuracy of the form produced.

Procedure: Cutting a Special Form Using a Template and Follower

1. Make an accurate template to the form desired.
2. Mount the template on a bracket fastened to the back of the lathe, figure 40-3.
3. Position the template lengthwise in relation to the workpiece.
4. Mount a suitable cutting tool in the toolpost.
5. Fasten a follower, the face of which must have the same form as the point of the cutting tool, on the cross-slide of the lathe.
6. Rough-out the form on the workpiece, keeping the follower close to the template by manually operating the carriage and cross-slide controls.

   While taking roughing cuts, keep the distance between the follower and the template fairly constant. For the final roughing cut, the follower should be kept within approximately 1/32" of the template at all times.
7. Disconnect the cross-slide from the cross-feed screw.
8. Apply hand pressure on the cross-slide to keep the follower in contact with the template.
9. Engage the automatic carriage feed and take a finish cut from the workpiece while the follower is being kept in contact with the template.
10. Check the part for size and, if further cuts are required, set the depth of cut using the compound rest feed screw.

An indicator mounted on the cross-slide can be used instead of the follower to trace the contour of the template. Although this method is employed for certain applications, it takes considerable practice to maintain a constant indicator reading on the template contour.

REVIEW QUESTIONS

1. What are the four methods used to produce curved or irregular forms on a workpiece?

Compound Rest

2. How can a compound rest be used to cut concave or convex radii?
3. How is the cutting tool set in order to produce

   a. A concave radius?   b. A convex radius?

4. Describe a radius turning attachment.

Face Concave and Convex Turning

5. How can a convex radius be produced on the face of a workpiece?

6. How does the setup for cutting a concave radius on a face differ from that required for a convex radius?

7. What precautions should be observed when cutting a concave or convex radius?

Use of a Template and Follower

8. What is the procedure for roughing out a form using a template and follower?

9. Explain the lathe setup and procedure for taking a finish cut.

Unit 41 TURNING A COMMUTATOR

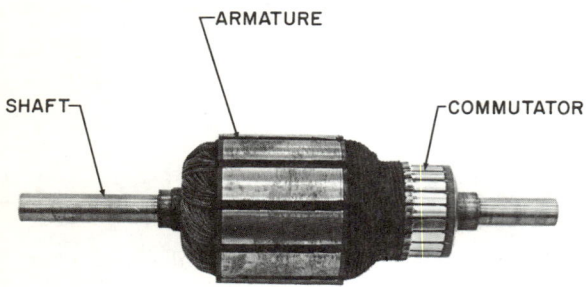

FIG. 41-1 PARTS OF A ROTOR

Most electric motors and generators have an inside rotating member called a rotor, which is composed of an *armature* and a *commutator* mounted on a *shaft,* figure 41-1.

The commutator, which is divided into segments by means of mica insulation, provides the bearing surface for the motor brushes. With prolonged service, the commutator becomes worn and the insulation between the segments projects above the surface of the commutator. This prevents a good contact between the brushes and the commutator, and arcing results. Continued arcing results in rapid deterioration of the commutator which must be machined to restore it to a proper operating condition. When machining is necessary, it is important that the commutator be machined smooth, true, and concentric with the shaft to insure good brush contact, and to minimize the vibration as the rotor revolves at high speed.

METHODS OF MOUNTING ROTORS FOR MACHINING

Rotors can be mounted by any of the three following methods, when it is necessary to turn the commutator.

Between Centers. Often the rotor shaft has center holes in each end. If this is so, the rotor can be held between the lathe centers for the turning operation. Before the commutator is turned by this method, the condition of the center holes should be checked. The rotor should also be rotated on centers and checked with an indicator to see that it is not bent.

In a Four-Jaw Chuck and a Steady Rest. If the shaft does not have center holes, the end opposite the commutator can be gripped and trued in a four-jaw chuck. The commutator end must then be supported with a steady rest. (See unit 27 for mounting and truing work in a four-jaw chuck, and unit 23 for the use of a steady rest.)

If the shaft is small enough, it can be gripped in a collet chuck in lieu of a four-jaw chuck. This reduces the amount of setup time. The work can also be held in a three-jaw chuck which is in good condition. If a three-jaw or collet chuck is used, it is good practice to test the end of the shaft held in the chuck for concentricity with an indicator.

A Jacob's Chuck Commutator Turning Kit. A set of special chucks capable of holding shafts up to 3/4" diameter, are available for turning commutators, figure 41-2.

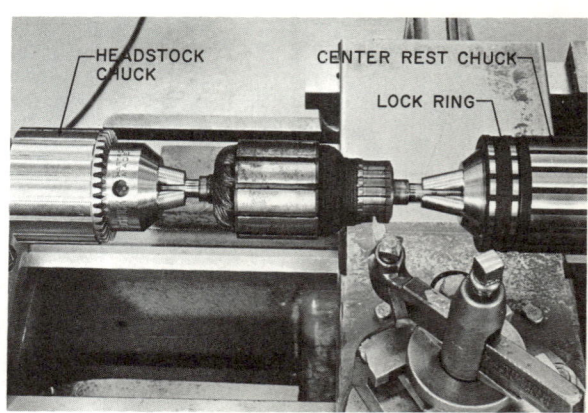

FIG. 41-2 A ROTOR HELD FOR MACHINING BY MEANS OF SPECIAL COMMUTATOR TURNING CHUCKS

Unit 41   Turning a Commutator

The headstock chuck is designed to fit into the lathe spindle by means of a tapered shank. Some headstock chucks may thread directly onto a threaded spindle nose.

The center rest chuck is mounted in the tailstock by means of a tapered arbor. Three stationary bronze jaws provide bearing surface and an accurate support for the end of the shaft. The jaws can be adjusted to the diameter of the shaft and then locked to maintain this size.

Procedure: Setting Up a Rotor Using a Commutator Turning Kit

1. Clean the headstock and tailstock spindle holes. Check for burrs and remove.
2. Align the centers of the lathe using a test bar and dial indicator.
3. Mount the headstock chuck in the headstock spindle.
4. Mount the center rest chuck in the tailstock spindle.
5. Mount the end of the shaft furthest from the commutator in the headstock chuck and tighten the chuck onto the shaft *by hand.*
6. Slide the tailstock up to the commutator end of the shaft and lock the tailstock in position.
7. Adjust the tailstock spindle until approximately 1" of the shaft is inside the center rest chuck. Tighten the tailstock spindle clamp.

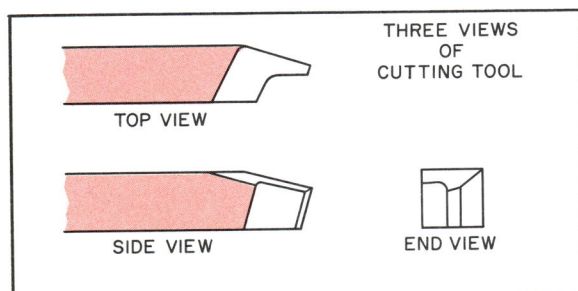

FIG. 41-3   A SUITABLE CUTTING TOOL FOR TURNING COMMUTATORS

8. Apply oil to the shaft where it is supported by the center rest chuck.
9. Rotate the lathe spindle by hand and adjust the center rest chuck sleeve until the chuck jaws lightly contact the shaft. Proper adjustment is indicated when light lines appear on the oiled surface.
10. Lock the chuck sleeve in this position by tightening the small sleeve or lock ring and check for play between the shaft and chuck jaws.

   Be sure that the chucks grip the rotor shaft at the bearing mounting points.

11. Tighten the driving chuck with the key.

Procedure: Turning a Commutator

1. Align the lathe centers.
2. Mount the cutting tool, figure 41-3, on center.
3. Clean the bearing surfaces on the shaft.
4. Mount the rotor by one of the previously mentioned methods.
5. Set the lathe spindle to the speed used for machining copper.
6. Set the lathe for the finest feed.
7. Start the machine and carefully adjust the tool until it touches the diameter of the commutator.
8. Carefully feed the toolbit in until it just cleans up the commutator at the part showing the greatest wear.

   It is important that as little material as possible be removed from commutator.

9. Note the reading on the graduated collar and back the toolbit away from the diameter.
10. Move the carriage to the right until the tool clears the end of the work; reset the graduated collar to the original setting plus .005.

223

Section 4 Special Operations

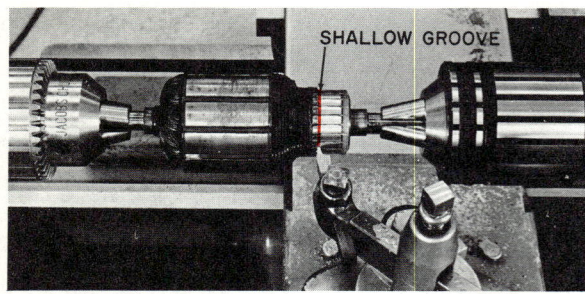

FIG. 41-4 A SMALL GROOVE AT THE END OF THE MACHINED SURFACE FACILITATES UNDERCUTTING THE MICA INSULATION.

FIG. 41-5 A HACKSAW BLADE GROUND TO UNDERCUT MICA BETWEEN THE SEGMENTS OF A COMMUTATOR

11. Engage the automatic feed and take a cut for the distance that the commutator had been previously machined. *Do not cut into the grooves into which the copper wires are soldered.*

12. If the commutator has not cleaned up, take another *light cut.*

13. At the end of the cut, machine a small, very shallow groove to facilitate undercutting the mica insulation, figure 41-4.

14. Using a specially ground hacksaw blade, figure 41-5, carefully undercut the mica.

15. Polish the turned surface with fine sandpaper.

Do not use emery cloth since the emery dust may stick between the segments and cause a short.

REVIEW QUESTIONS

1. When should a commutator be reconditioned?

2. What precautions should be observed when machining a commutator to insure good brush contact?

Methods of Mounting Rotors for Machining

3. What are two common methods of holding rotors for machining? Briefly describe each method.

4. How can the rotor be tested to see that it is not bent?

5. Describe the two chucks that comprise the commutator turning kit.

Setting Up a Rotor Using a Commutator Turning Kit

6. How is a rotor set up using a commutator turning kit?

7. Why should the chucks grip the rotor shaft at the bearing-mounting points?

Turning a Commutator

8. How is it possible to determine when the absolute minimum amount of material has been removed?

9. What two precautions must be observed when taking a cut across the commutator?

10. Why is a small, shallow groove machined near the end of the commutator?

11. Why is sandpaper rather than emery cloth used to polish the commutator?

Unit 42 GRINDING

FIG. 42-1 A TOOLPOST GRINDER SET UP FOR EXTERNAL GRINDING

FIG. 42-2 A TOOLPOST GRINDER WITH AN INTERNAL GRINDING ATTACHMENT (Courtesy The Dumore Co.)

FIG. 42-3 DRESSING A GRINDING WHEEL FOR CYLINDRICAL GRINDING (Courtesy The Dumore Co.)

FIG. 42-4 DRESSING A WHEEL FOR INTERNAL GRINDING (Courtesy The Dumore Co.)

Internal and external grinding can be performed on a lathe if a proper grinding machine is not available. A *toolpost grinder*, figure 42-1, mounted on the compound rest, can be used for cylindrical and taper grinding, as well as for angular grinding, such as lathe centers. An *internal grinding attachment,* figure 42-2, permits the production of straight and tapered holes. Grinding should be performed on the lathe only if a cylindrical grinder is not available, or when a small grinding operation does not warrant the setting up of a cylindrical grinder.

Procedure: Dressing the Wheel

To perform properly and efficiently, any grinding wheel should be trued and dressed. Grinding wheels on toolpost grinders are quickly and easily dressed as follows:

1. Set the lathe in the lowest gear, or lock the spindle.

2. Mount the diamond dresser on the workpiece or a piece of stock held in a chuck, making sure that the diamond point is set to center height, figures 42-3 and 42-4.

Section 4 *Special Operations*

Note the position of the diamond for dressing the wheel for cylindrical (external) and internal grinding.

3. Mount the toolpost grinder with its spindle on center and parallel to the centerline of the lathe. This is important if the periphery of the wheel is to be ground square with the face.

4. Wear safety glasses and then start the grinder.

5. Bring the grinding wheel into *light contact* with the stationary diamond dresser and move the carriage back and forth to dress the grinding wheel.

6. Move the wheel in towards the diamond about .001 using the crossfeed handle and take another pass across the diamond.

7. Continue steps 5 and 6 until the full surface of the wheel has been dressed.

Procedure: External Grinding on a Lathe

1. Clean the ways of the lathe and cover with canvas or cloth to protect them from grinding dust.

2. Adjust the tailstock center as required for parallel or tapered grinding.

   If a taper is required, it can be produced by one of the following means: offsetting the tailstock, engaging the taper attachment, or by the compound rest.

3. Mount the toolpost grinder in the compound rest and adjust the spindle to center height.

4. Mount the proper grinding wheel on the grinder.

5. Select and mount the proper pulleys on the grinder and motor spindle to produce the *proper* speed for the size of the grinding wheel to be used.

   Be sure the wheel-guard has been mounted over the wheel prior to grinding.

6. Place a small pan of water under the grinding wheel to catch as much abrasive dust as possible.

7. True and dress the grinding wheel. *Be sure to wear safety glasses.*

8. Set a fairly slow lathe spindle speed. (About 50 r.p.m. is satisfactory for most work; however, the speed depends on the diameter of the work.)

9. Start the lathe spindle rotating in reverse (clockwise).

10. Set the carriage feed .020 to .040 depending on the width of the grinding wheel.

11. Mount the work in the lathe.

    If the work has been hardened and is to be ground between centers, it is first necessary to remove all the scale from the center holes by cleaning with a special 60° center hone.

12. Start the grinder and carefully bring the grinding wheel up to the revolving workpiece until it sparks slightly.

13. Slowly feed the carriage by hand along the length of the workpiece to remove any high spots.

14. Move the carriage until the grinding wheel is at the right-hand end of the surface to be ground.

15. Set the grinding wheel for .001-.002 cut, engage the automatic feed, and take a pass across the work for the required length.

16. At the end of the cut, reverse the feed and take a .001 to .002 cut in the opposite direction.

17. Using a micrometer, check the workpiece for size and taper or parallelism.

    Do not measure the work while it is hot as it may be undersize when it cools.

18. Rough-grind the work to within .002 of finish size and *then be sure to cool the workpiece.*
19. Redress and true the grinding wheel.
20. Finish-grind the work to within .0002 of finish size.

    Without setting any further depth of cut, feed the grinding wheel back and forth across the work until the wheel "sparks out."

Procedure: Grinding a Lathe Center

Lathe centers can be ground conveniently on a lathe using a toolpost grinder.

1. Remove the chuck or driveplate from the lathe spindle.
2. Clean and remove all burrs from the taper on the center, the spindle bushing, and the headstock spindle.
3. Mount the center and the spindle bushing in the headstock spindle.
4. Set the lathe spindle speed to about 50 r.p.m.
5. Swing the compound rest to 30° with the centerline of the lathe, figure 42-5.
6. Clean the ways of the lathe and protect them with a cloth or a piece of canvas. Place a pan of water beneath the lathe center.
7. Mount the toolpost grinder so that its spindle is parallel to the angle of the center. The center of the spindle should be set to center height.
8. Mount the proper grinding wheel and true and dress it, using the compound rest feed.
9. Start the lathe with the spindle rotating in reverse.
10. Start the grinder and adjust the grinding wheel until it just touches the revolving center.

FIG. 42-5 A LATHE SET UP TO REGRIND A LATHE CENTER

11. Lock the carriage to the lathe bed.
12. Feed the grinding wheel in .001, using the crossfeed handle.
13. Using the compound rest feed, move the grinding wheel along the angle of the center.
14. Stop the lathe and check the angle of the center with a center gage. Adjust the compound rest if necessary.
15. Continue to grind the center until all wear marks are removed.
16. If a high finish is required on the center, polish it with fine abrasive cloth at a high spindle speed.

Procedure: Internal Grinding on a Lathe

Straight and tapered holes can be finished on a lathe by using a toolpost grinder equipped with an internal grinding attachment, figure 42-2. Tapered holes can be ground using the

Section 4 Special Operations

taper attachment or by means of the compound rest feed handle after the compound rest has been set to one-half the included angle of the taper.

1. Mount the work in a chuck or on a faceplate and dial indicate if necessary.
2. Set up the toolpost grinder to center height.

   The wheel size should be from 2/3 to 3/4 the size of the hole to be ground.
3. If a tapered hole is required, set the taper attachment or the compound rest, as required.
4. Cover the ways of the lathe with a cloth or a piece of canvas.
5. Start the grinder and dress the wheel, figure 42-4.
6. Set the lathe to turn at about 50 r.p.m.
7. Start the lathe, being sure that the spindle is rotating in a forward direction (as for turning).
8. Move the grinding wheel into the hole and, using the crossfeed handle, carefully bring it into contact with the surface of the hole.
9. Feed the grinding wheel about .001, using the crossfeed handle, and take a pass along the length of the hole.

   To avoid bell-mouthing, do not let more than half the width of the grinding wheel overlap either end of the hole.
10. Take light cuts and continue grinding until the hole is within about .002 of size. Let the wheel "spark out."
11. Finish-grind to size. Be sure to let the wheel "spark out."

Hints on Toolpost Grinder Use

1. Select a wheel in which the bond wears away as fast as the grains become dull. If the grains dull faster than the bond breaks down, the wheel is too hard and glazes. If the bond wears away before the grain dulls, the wheel is too soft and loses its size too rapidly, causing a taper in the work.
2. If the wheel appears too hard, increase the speed of the work or decrease the speed of the wheel. If the wheel is too soft and is wearing rapidly, increase the speed of the wheel, or decrease the speed of the work.

GRINDING PROBLEMS AND REMEDIES		
Problem	Cause	Remedy
Tapered Work	Inaccurate machine setup Wheel too soft	Align centers Use harder wheel Decrease work speed Increase wheel speed
Burning of work, wheel glazing	Wheel too hard	Use softer wheel Decrease wheel speed Increase work speed
Chatter	Defective grinder belt Work not properly mounted Worn bearings on machine or grinder	Change belt Recheck setup Check and correct

TABLE 42-1

Unit 42 Grinding

ACCURATE DEPTH SETTINGS USING THE COMPOUND REST		
Compound Rest Setting	Compound Rest Feed	Depth of Cut
30°	.001	.00086
45°	.001	.0007
60°	.001	.0005
70°	.001	.00034
75°	.001	.00026
80°	.001	.00017
84°16′	.001	.0001

TABLE 42-2

CAUTION: Do not increase the wheel speed above that recommended by the manufacturer.

3. Always cool the work to room temperature before checking the size.

4. When grinding blind holes, a micrometer carriage stop should be used to prevent the grinding wheel from striking the bottom of the hole.

5. Table 42-1 lists some of the common grinding problems and suggested remedies.

6. The compound rest may be used to set accurately depths of cuts less than .001. Consult table 42-2 for the settings and depths required.

REVIEW QUESTIONS

1. When should a lathe be used for grinding purposes?

Dressing the Wheel

2. What two precautions must be observed when setting up the toolpost grinder for dressing the wheel?

3. What depth of cut should be used when dressing a wheel?

External Grinding on a Lathe

4. What precautions should be taken to protect the lathe from abrasive dust when grinding?

5. How should the centers of the hardened work be prepared if the work is to be ground between centers?

6. What precaution should be observed when measuring work?

7. What is the procedure for finish-grinding the work to size?

Section 4 Special Operations

Grinding a Lathe Center

8. What precautions should be observed when mounting the lathe center prior to grinding?

9. How is the grinder set up for grinding a lathe center?

10. In what direction should the lathe spindle rotate for external grinding?

Internal Grinding on a Lathe

11. To what angle should the compound rest be set when grinding an internal taper or angle?

12. How can bell-mouthing be prevented when internal grinding?

Hints on Toolpost Grinder Use

13. What are the results if a wheel is too hard? Too soft?

14. What precaution should be taken when grinding blind holes?

15. What are the possible causes and remedies for the following problems:
    a. Tapered work?
    b. Chatter?
    c. Wheel glazing?

Unit 43 SPRING WINDING

One of the jobs often required of the machinist is the making of coil springs. If a spring-winding machine is not available, certain sizes of coil springs can be made on a lathe. In this operation, spring or music wire is wound around a mandrel of the *proper* diameter in an operation similar to left-hand thread cutting. The selection of the proper size mandrel depends on several factors:

- The type of spring wire used
- The size spring wire used
- The inside diameter required in the coil spring
- The outside diameter required on the coil spring
- The tension applied to the wire when winding the spring

Much information on spring winding is available in the *Machinery's Handbook,* and the lathe operator is advised to refer to this for the selection of wire size, coils per inch, mandrel size, etc.

METHODS OF HOLDING MANDRELS

After the proper size of mandrel has been selected for the job, it can be held by one of three methods.

1. Between centers. This type of mandrel, driven by a lathe dog, should be accurately centered and should have a hole drilled through the diameter about 1" to 1 1/2" (depending on the size of the mandrel) from the left end of the mandrel.

2. Mandrels of under 3/8" can be held in a drill chuck mounted in the headstock, with the right-hand end supported by the tailstock center.

3. Mandrels larger than 3/8" diameter can be held in a universal three-jaw chuck, with the right-hand end being supported by the tailstock center. The end of the wire can be fitted into the mandrel hole, or it can be held between the mandrel and the chuck jaw. The latter method tends to damage the chuck jaws and, in time, makes the chuck inaccurate.

Procedure: Winding a Coil Spring

1. Determine the correct wire size, mandrel size, and the number of coils per inch from *Machinery's Handbook*.

2. Mount the mandrel by one of the methods above.

3. Set the lathe spindle speed to about 50 r.p.m.

4. Set the feed directional lever for left-hand threads.

5. Set the quick-change gearbox on the lathe to the required number of threads per inch.

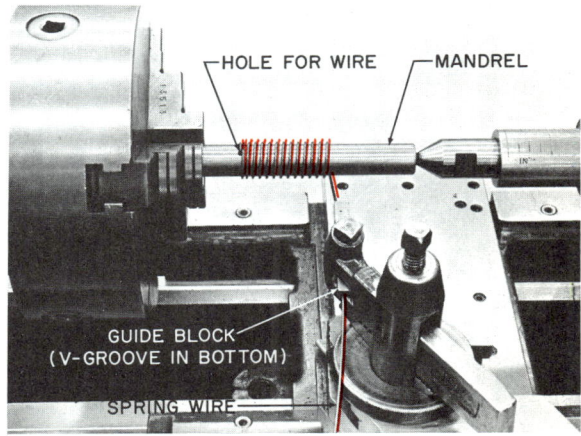

FIG. 43-1 THE LATHE SET UP FOR ONE METHOD OF WINDING COIL SPRINGS

Section 4 Special Operations

After the spring has been wound and the tension is removed, the spring expands and elongates, so that it is larger with fewer coils per inch. It is therefore necessary to allow for this action when setting the machine for spring winding. For example, if six coils per inch are required on the finished spring, it is necessary to set the quick-change gearbox to more than six, possibly seven, threads per inch.

6. Mount a wire guide block with the V-slot on the bottom in a left-hand offset lathe toolholder.

The guide block can be made from a piece of 5/16" or 3/8" square cold rolled steel with a shallow V-slot cut lengthwise to guide the wire. Tension can be applied to the wire by the toolholder screw. Another method of supporting the wire is between two wooden blocks mounted in the lathe toolpost. Tension can be applied to the wire by tightening the toolpost screw.

7. Move the carriage until the end of the toolholder is opposite the hole in the mandrel.

8. Uncoil sufficient wire and feed it through the V-slot in the guide block and into the hole of the mandrel.

9. Apply tension on the spring wire with the toolholder or toolpost screw and then engage the split-nut lever.

10. Start the machine and carefully let the wire feed through the block until the desired length of spring is wound, figure 43-1.

Be sure the wire is clear, and there is no danger of it catching on anything, including clothing.

11. Shut off the lathe and then disengage the split-nut lever.

If the spring must have close coils at either end, it is necessary to start the lathe and then engage the split-nut lever at the left end of the spring. When the spring is of the desired length, the split-nut lever is disengaged, and the machine is shut off after sufficient close-coils have been formed at the end.

12. Release the tension on the spring by loosening the toolpost or toolholder screw.

13. Using a pair of end nippers, carefully cut the spring wire between the mandrel and the wire guide block.

14. Cut the left end of the spring with end nippers or remove the mandrel from the lathe and grind through the spring wire.

REVIEW QUESTIONS

1. What five factors affect the size of the mandrel when spring winding?

2. Consult *Machinery's Handbook*. What is the number of coils per inch for compression springs made from the following Brown and Sharpe wire sizes:
   a. #8?
   b. #12?
   c. #15?
   d. #20?

Methods of Holding Mandrels

3. What are three methods by which a mandrel can be held for spring winding?

4. What are two methods by which the end of the spring wire can be secured for spring winding?

Winding a Coil Spring

5. What steps are required to make a compression spring using #12 Brown and Sharpe wire size? Include machine settings.

6. How can close-coils be formed at the beginning and the end of a coil spring?

7. What precaution should be taken before cutting the spring wire at the right end of the coil spring?

Unit 44 TRACER ATTACHMENTS

Tracer control of machine tools has become one of the leading metal removal developments in the machine tool industry. It is especially valuable in the machining of intricate forms and contours which are difficult, if not impossible, to produce economically by conventional machining practices. Tracer attachments have found wide acceptance in the aerospace industries where a complexity of forms must be reproduced accurately. The hydraulic universal taper attachment, figure 44-1, can be quickly and easily mounted on a lathe cross-slide. This attachment greatly increases the versatility of the machine.

TYPE OF CONTROL MEDIA

There are three common types of control media used to regulate the path of the tracer slide and cutting tool. They are:

- The *guard rail system* which uses cams and stops.

- The *sequential system* involving switchgear which can be hydraulic, electrical, or pneumatic.

- The *feedback system* which includes both tracer and numerical control equipment.

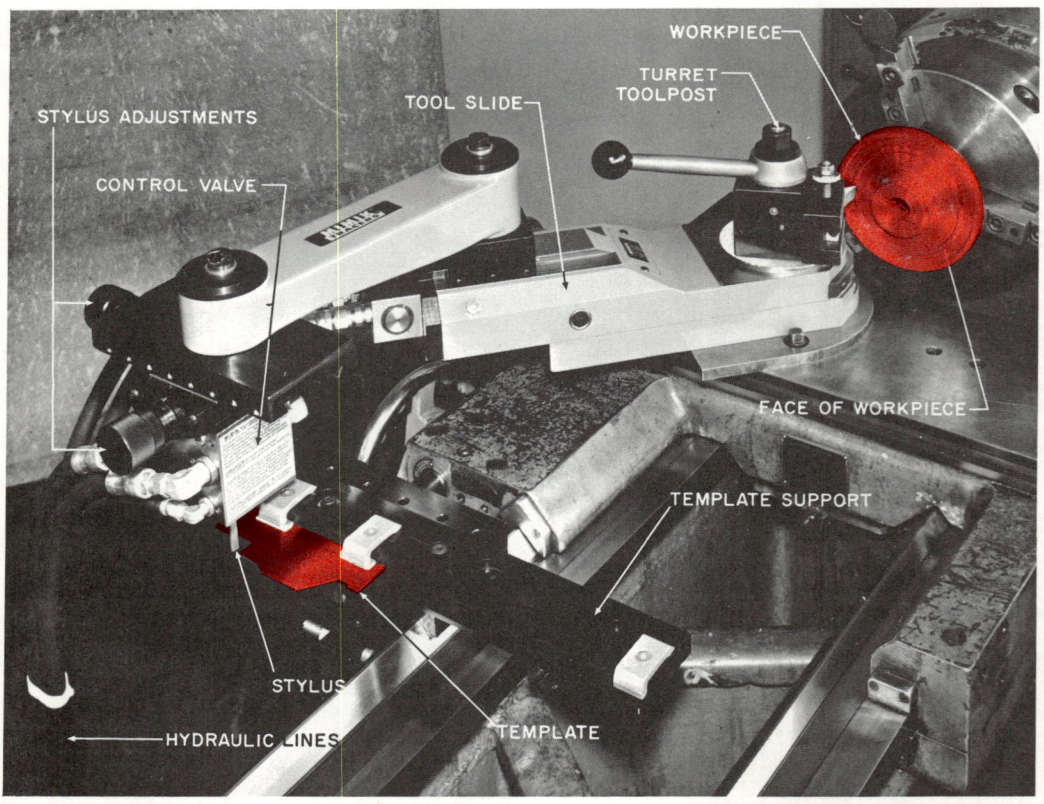

FIG. 44-1 A HYDRAULICALLY OPERATED UNIVERSAL TRACER ATTACHMENT MOUNTED ON THE CROSS-SLIDE OF A LATHE MACHINING AN INTRICATE FORM (Courtesy Mimik Tracers, Inc.)

HYDRAULIC TRACERS

Hydraulic Tracer Principle

Hydraulic tracers provide a means of controlling the movement of a toolslide, upon which the cutting tool is fastened, by controlled oil pressure. A flat or round template having the exact contour of the form desired is mounted on the template support, figure 44-1. The *stylus*, which directs the flow of oil through the control valve, bears against the form of the template and causes the toolslide movement to correspond to the template form. The *control valve*, actuated by the stylus, regulates the direction and flow of oil into a *cylinder* incorporated in the base of the toolslide. A piston, connected to the toolslide, is moved in or out depending upon the direction of the oil flow through the cylinder. The movement of the piston causes the toolslide (and the attached cutting tool) to duplicate automatically on a workpiece the contour of the template being followed by the stylus.

Advantages of Tracer Control Systems

Tracer control systems are continually finding wider use in the machine trade because of the ease of setup and the versatility of the system. Some of the advantages of these systems are:

1. The movement of the toolslide is independent of the movement or accuracy of the machine tool.
2. The accuracy of the machined part does not depend upon the skill of the operator.
3. It provides a means of quickly and accurately reproducing intricate forms which are difficult or impossible to produce by other means.
4. Contours, tapers, diameters and shoulders can all be produced in one setup and cut.
5. Any number of duplicate parts can be produced accurately and with a considerable saving of machining time.

Tracer Attachment Setup

Since there is a wide variety of tracer attachments available, it is impractical to attempt to explain how each attachment should be set up. The instruction manual for each specific tracer should be consulted for setup procedures. Therefore, only the main points which should apply to most tracer attachments are listed below.

1. Be sure that the toolbit has cutting edges at all points where it contacts the work.
2. The point radius on the toolbit should be slightly smaller than the smallest radius on the workpiece.
3. Set the toolslide of the tracer attachment so that it roughly bisects the most acute right- and left-hand angles on the template.
4. Mount the workpiece in the lathe.
5. Mount a stylus having the same form and radius as the point of the cutting tool in the holder.
6. Mount the template support on the lathe bed and align its edge either parallel or at right angles to the cross-slide as required.
7. Locate the template loosely on the rail of the template support in approximate relationship to the workpiece.
8. Bring the point of the cutting tool to the starting point on the workpiece.
9. Position the tracing arm and template so that the stylus is at the same point on the template as the cutting tool is on the workpiece.
10. Clamp the tracing arm and template securely in position.
11. Accurately position the stylus with the template by using the stylus adjustments on the valve assembly.
12. Take roughing and finishing cuts on the workpiece as required.

Section 4 Special Operations

Tracer Attachment Hints

If the following hints are observed, many of the problems associated with the use of tracer attachments should be eliminated.

1. The centerline of the template should be parallel or at right angles to the ways of the lathe, depending upon the workpiece.
2. Any angle or equivalent radius larger than 30° should not be incorporated in the template form, if at all possible.
3. The form of the template must be accurate, smooth, and free of burrs.
4. The included angle of the toolbit should be less than a combination of the smallest right- and left-hand angles on the template.
5. To machine the entire form, the stylus should be set to the point on the template which would produce the smallest diameter on the workpiece.
6. When duplicate parts held in a chuck are required, it is important that each part extends the *same* distance beyond the chuck jaws.
7. When duplicate parts held between centers are required, it is important that each part is the same length and that the center holes are drilled to the same depth.

REVIEW QUESTIONS

1. Why have tracer control systems been widely accepted by the machine industry?

Types of Control Media

2. Name and describe three types of control media.

Hydraulic Tracer Principle

3. Describe the operating principle of a hydraulic tracer.
4. What is the function of the following tracer components:
   a. Stylus?
   b. Control valve?
   c. Toolslide?
   d. Piston?

Advantages of Tracer Control Systems

5. What are three key advantages of tracer control systems?

Tracer Attachment Setup

6. When using a tracer attachment, what shape of cutting tool is required?
7. How should the shape of the stylus compare with that of the cutting tool?
8. How should the cutting tool, stylus, and template be located prior to machining?

Tracer Attachment Hints

9. What are the precautions to be observed when machining duplicate parts held in a chuck; between lathe centers?

Unit 45 DIGITAL READOUT SYSTEM

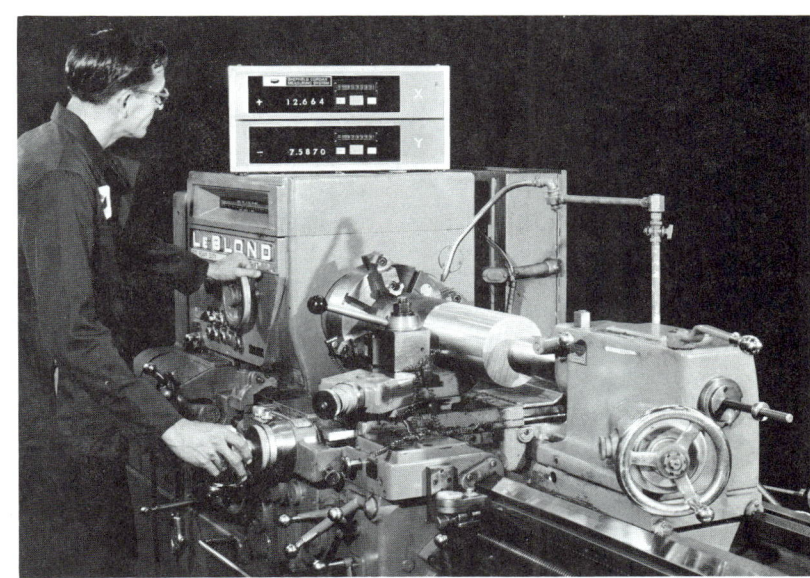

FIG. 45-1 A LATHE EQUIPPED WITH A DIGITAL READOUT SYSTEM (Courtesy Bendix Corp., Automation and Measurement Division)

In the past several years, many industries have equipped various machine tools with digital readout systems. This system can be adapted to most types of machines, including lathes, to improve accuracy and increase production.

The digital readout system, as applied to a lathe, figure 45-1, is a coordinate measuring device which indicates accurate measurements in two directions from a starting point. The length of travel of the toolbit, as well as the depth of cut, can be accurately indicated. As these distances are measured, they are registered on a digital readout box mounted on, or near the lathe, to indicate the exact location of the cutting tool at any time. This continuous numerical readout is achieved by several methods. One of the more popular systems uses the *Moire Fringe Principle*.

MOIRE FRINGE PRINCIPLE

This principle can be easily illustrated by drawing a series of equally spaced lines on two pieces of plastic sheeting. When one sheet is placed over the other at a slight angle, figure 45-2, a series of dark bands appear.

If the top sheet is moved to the right or left, the position of these bands shifts either up or down, depending on the direction in which the sheet is moved. Thus, any longitudinal movement of the set of lines produces a vertical movement of the dark bands. This principle is applied to the digital readout system.

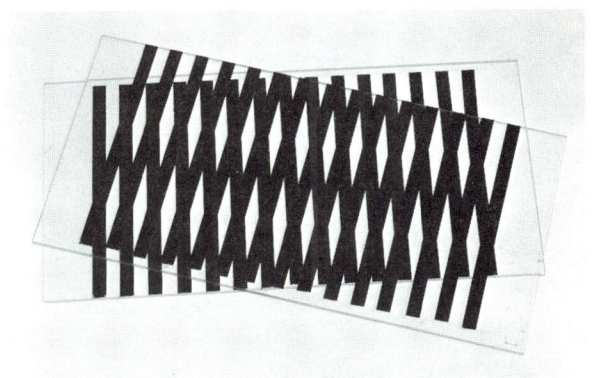

FIG. 45-2 THE MOIRE FRINGE PRINCIPLE (Courtesy Bendix Corp., Automation and Measurement Division)

Section 4 Special Operations

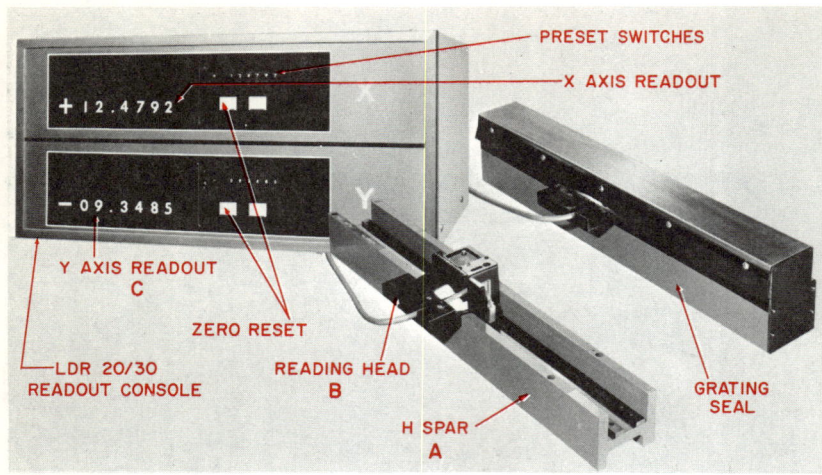

FIG. 45-3 **COMPONENTS OF A DIGITAL READOUT SYSTEM** (Courtesy Bendix Corp., Automation and Measurement Division)

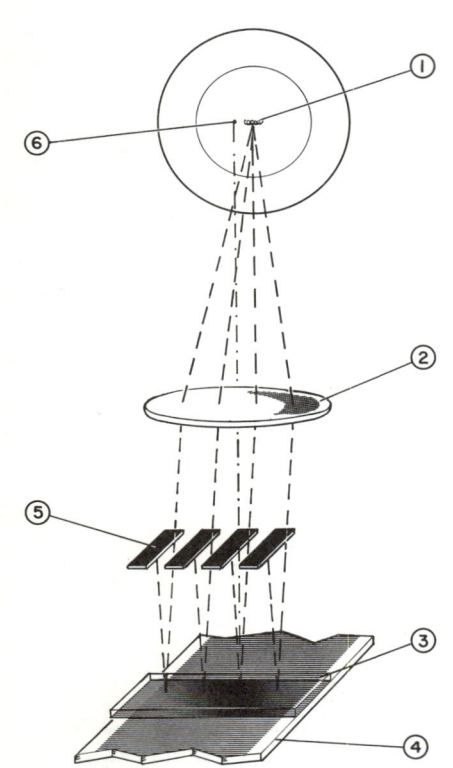

KEY

1. LAMP FILAMENT
2. COLLIMATING LENS
3. INDEX GRATING
4. SCALE GRATING
5. PHOTOCELL STRIPS
6. PRINCIPAL FOCUS OF LENS

FIG. 45-4 **APPLICATION OF THE MOIRE FRINGE PATTERN TO THE COORDINATE MEASURING SYSTEM** (Courtesy Bendix Corp., Automation and Measurement Div.)

THE DIGITAL READOUT SYSTEM

This system has three basic components:

- A *machined stainless steel spar with a calibrated grating* having lines accurately etched .001 apart, figure 45-3A.

- A *reading head,* figure 45-3B, with a similarly etched transparent index grating, is positioned at a slight angle to the lines on the grating spar and a few thousandths of an inch above it. A small light, a collimating lens and four photoelectric cells are mounted in the head, figure 45-4.

- A *counter* with a digital readout display, figure 45-3C.

Operation of the System

Light from the lamp passes through a *collimating lens* and is converted into a parallel beam of light, figure 45-4. This beam strikes the *gratings.* Since the etched lines on the reading head transparent grating are at a slight angle to the lines on the spar, a series of light and dark bands appear, if viewed from directly above. Let us now consider only one band to illustrate the principle. The light shining on the gratings creates a light and dark band which is reflected back up to one of four photoelectric cells which converts the fringe pattern into an electrical signal. As the reading head is moved longitudinally, the pattern

238

Unit 45 Digital Readout System

shifts laterally and is picked up by the next photoelectric cell. As a result, another signal is created, figure 45-5. Output signals from the photoelectric cells are sent to the digital readout box where they indicate accurately the amount and direction of travel (in .0001) of the head at any point.

To indicate accurately the position of the cutting tool, two of the above-mentioned units must be mounted on a lathe. One unit is mounted on the back of the lathe bed to indicate the length of travel of the carriage. This is known as the *X-axis*. Another unit, mounted on the back of the carriage, indicates the travel of the cross-slide. This is known as the *Y-axis*. Both the *X* and *Y* readings are shown on the digital readout box in ten thousandths of an inch.

PRESET TOOLING

To obtain the optimum use from a digital readout system, a preset tooling system is generally used. Lathe tools, figure 45-6, mounted in dovetailed holders, are accurately ground and set in the toolroom on a pre-setting fixture, figure 45-7. When a different tool is required, the dovetailed holder is merely placed on the dovetailed toolpost and accurately locked in place.

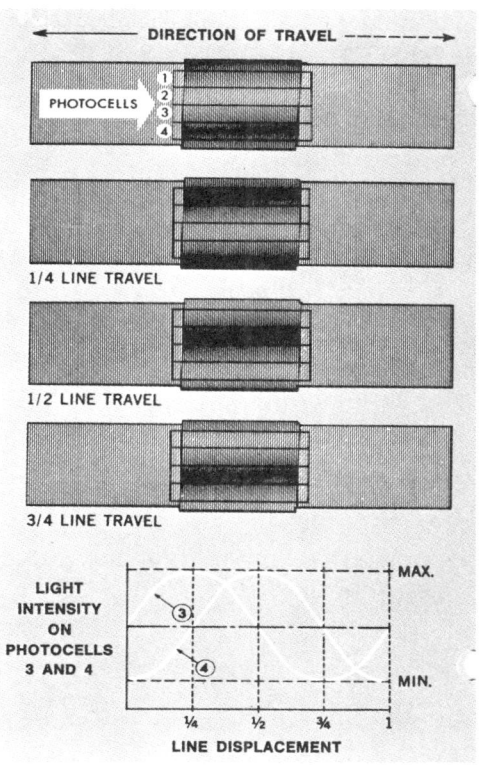

FIG. 45-5 THE FRINGE PATTERN SHIFTS LATERALLY AND CONTINUOUSLY ACROSS THE GRATING PATH. (Courtesy Bendix Corp., Automation and Measurement Div.)

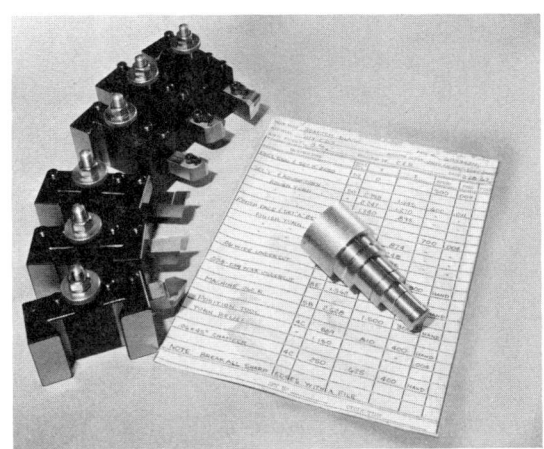

FIG. 45-6 PRESET TOOLS, WHEN MOUNTED ON THE TOOLPOST, ARE AUTOMATICALLY SET FOR USE. (Courtesy Bendix Corp., Automation and Measurement Div.)

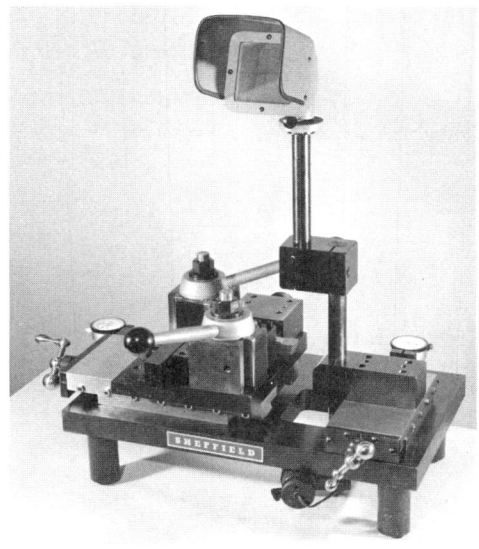

FIG. 45-7 A PRESET TOOLING FIXTURE. (Courtesy Bendix Corp., Automation and Measurement Div.)

Section 4   Special Operations

MACHINE SETUP FOR A LATHE EQUIPPED WITH DIGITAL READOUT SYSTEM

1. Study the print of the part to be machined.
2. Mount the rough stock in the lathe.
3. Set lathe to required speed and feed.
4. Mount a preset facing tool in the toolpost.
5. Face the workpiece and set the digital readout box for the X-axis to zero.

   This is now the starting point for the X-axis and all subsequent measurements on the X–axis are taken from this point.

6. Remove the facing tool and mount a preset turning tool.
7. Start the lathe and take a cut for about 1/4" at the end of the work.
8. Stop the machine and measure this turned diameter with a micrometer.
9. Set the digital reading for the Y-axis to the size indicated on the micrometer.

   The exact diameter of the machined end of the workpiece is now indicated on the Y-axis. Any further depths of cut alter this reading which always indicates the size of the *diameter being turned.*

   The Y-axis system is constructed so that any movement of the toolbit on the Y-axis is multiplied by two, and the reading shown indicates the diameter and not the radius.

10. Continue to turn the workpiece to the sizes shown on the print using the Y-axis readings to indicate the diameters, and the X-axis readings to indicate lengths.

ADVANTAGES OF DIGITAL READOUT SYSTEMS

1. Readout boxes provide clear visible numbers which are more easily read than other measuring devices such as micrometers or dial gages.
2. They provide a constant readout of the tool position.
3. The reading indicates the exact position of the tool and it is not affected by machine wear.
4. The need for other measuring equipment, such as measuring rods, is eliminated when accurate lengths are required.
5. The need for operator calculations is eliminated.
6. Production is increased since the workpiece need only be checked for one diameter.
7. Considerably less work is scrapped than by standard lathe practices.
8. Less operator skill is required.
9. Tooling setup time is greatly reduced by preset tooling.

REVIEW QUESTIONS

1. What is the purpose of the digital readout system on a lathe?

Moire Fringe Principle

2. Describe the Moire Fringe Principle.

The Digital Readout System

3. What are the three basic components in a digital readout system?

Operation of the System

4. How does the digital readout system operate?

5. How is the digital readout system applied to the lathe?

Preset Tooling

6. What is the purpose of preset tooling?

Machine Setup for a Lathe Equipped with Digital Readout

7. Describe the setup procedure required to machine a workpiece on a lathe equipped with a digital readout.

Advantages of Digital Readout Systems

8. What are six advantages of a digital readout system on a lathe?

Unit 46 SURFACE FINISHES

Modern technology requires improved surface finishes to insure long life and the proper functioning of moving components and machines. Since parts with a good surface finish last longer and function more efficiently than those with a poor surface finish, the quality of the surface finish on parts is very important.

Single-point cutting tools are most commonly used in lathe work for a wide variety of external and internal turning operations. The condition and setup of the cutting tool, cutting speed, feed rate, rigidity of the lathe, toolbit radius, and the material of the workpiece, are all factors which affect the quality of the surface finish produced.

1. An *efficient cutting tool,* to give the best cutting action and produce the best surface finish, must:
   a. Contain proper alloying elements to be wear-resistant and maintain a keen cutting edge
   b. Have sufficient toughness to prevent the cutting edge from chipping
   c. Be hard enough to prevent the chip from forming a built-up edge on the cutting tool
   d. Have the proper rake and clearance angles to remove metal quickly and efficiently.
   e. Be set on center and be rigid enough to prevent vibration and chatter during machining

2. The *cutting action* of a single-point toolbit, which affects the surface finish produced, is affected by the following:
   a. *Cutting speed* - too high a cutting speed causes the toolbit to break down rapidly and results in poor surface finish.
   b. *Feed rate* - a coarse feed tends to produce a rough surface finish, while a fine feed rate should produce a good surface finish.
   c. *Rigidity of the machine* - a lathe with closely fitting bearings and machine slides produces fine surface finishes. On lathes with loose bearings or machine slides, chatter occurs during the machining operation and results in a poor surface finish.
   d. *Machinability of the metal* - the ductility and shear strength of a metal affect the surface finish produced. High surface finishes can be produced on free-machining steel, while rougher surface finishes result when cutting hard, stringy metals.
   e. *Cutting fluid* - If cutting fluid is applied to the work and cutting tool during machining, better surface finishes are produced on the workpiece. Cutting fluid not only cools the workpiece, but allows the chip to slide over the cutting tool face freely.

CHECKING SURFACE FINISHES

The surface finish produced on a workpiece by a cutting tool can be visually analyzed and designated fairly accurately as to quality by comparing it to a known standard. If a fingernail is drawn over the standard having a certain microfinish and then the standard is compared with the finish on the work surface, a fairly accurate comparison can be made. If the surface finish of a workpiece must be exact, a

Unit 46 Surface Finishes

SURFACE FINISHES OF VARIOUS LATHE OPERATIONS							
Operation	Material	Speed (r.p.m.)	Feed	Tool	Analyzer Setting Cutoff	Range	Surface Finish (R.M.S.)
FACING	2" diameter aluminum	800	.005	1/32" nose radius	.030	100	30-40
FACING	2" diameter aluminum	600	.010	1/32" nose radius	.030	300	200-225
TURNING	2½" diameter aluminum	500	.007	5/64" nose radius	.030	100	50-60
TURNING	2½" diameter aluminum	500	.010	3/64" nose radius	.030	300	100-120
REAMING	aluminum	500		3/4" diameter reamer	.030	100	25-32
FILING	3/4" diameter machine steel	1200		10" lathe file (2nd cut)	.010	100	50-60
POLISHING	3/4" diameter machine steel	1200		No. 120 abrasive cloth	.010	30	13-15

TABLE 46-1

profilometer or *surface indicator* should be used to measure the surface finish accurately.

The roughness of surface finish is generally measured in microinches (.000, 001). When using a surface indicator, the readings are either in Arithmetical Average (AA) or Root Mean Square (R.M.S.).

Arithmetical Average is determined when total sum of heights and depths of ridges and valleys above and below the mean reference line is divided by the number of measurements.

Root mean square is the most commonly used value for measuring surface finishes. It represents the distances in microinches above and below the mean reference line to the corresponding division points on the irregular contour of the surface; then the square root of the mean of the sum of the squares is taken.

Table 46-1 gives the approximate surface finishes which might be expected for various lathe operations when certain conditions exist.

REVIEW QUESTIONS

1. What are four key factors which affect the quality of surface finish produced on lathe work?
2. What three important characteristics should a cutting tool possess to produce the best surface finish?
3. How do the following factors affect the quality of the surface finish produced?
   a. Rigidity of machine b. Machinability of metal c. Use of cutting fluid

Checking Surface Finishes

4. What is a quick and inexpensive method of comparing surface finishes?
5. Define the following surface finish terms:
   a. Microinch   b. Arithmetical average   c. Root mean square

Unit 47 NUMERICAL CONTROL LATHES

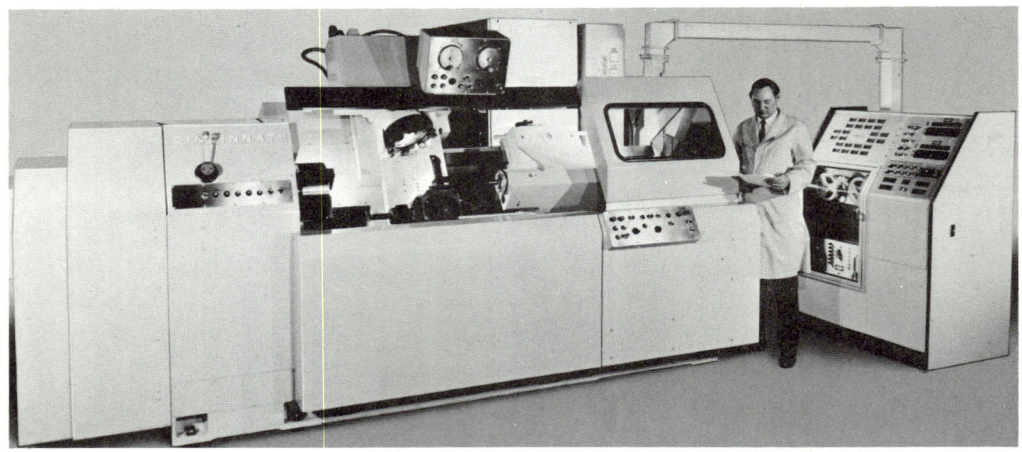

A. A UNIVERSAL TURNING CENTER (Courtesy Cincinnati Milacron Co.)

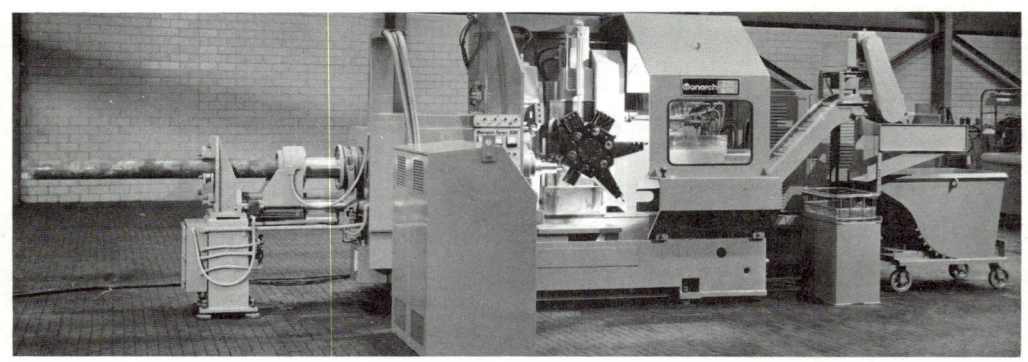

B. A CHUCKING LATHE (Courtesy Monarch Machine Tool Co.)
FIG. 47-1 NUMERICALLY CONTROLLED TURNING MACHINES

Numerically-controlled *turning centers* and *chucking lathes* were developed after an extensive study showed that approximately 40% of all metal machining time represented turning operations. Recognizing this fact, and in order to take advantage of the capability of numerical control, manufacturers developed automatic turning centers and chucking lathes, figure 47-1. These machines can produce workpieces of almost any configuration automatically, in mass-production quantities, or in economic lot sizes as low as one.

CONSTRUCTION

The two common numerically-controlled turning machines are the chucking lathe and the turning center.

Chucking Lathe

The chucking lathe illustrated in figure 47-2A has a single position turret whose indexing and movement is numerical-tape controlled. It contains six tool stations and is used for both internal and external turning operations on work held in a chuck.

Unit 47   Numerical Control Lathes

FIG. 47-2A A CHUCKING LATHE WITH A SIX-POSITION TURRET USED FOR BOTH INTERNAL AND EXTERNAL TURNING (Courtesy Monarch Machine Tool Co.)

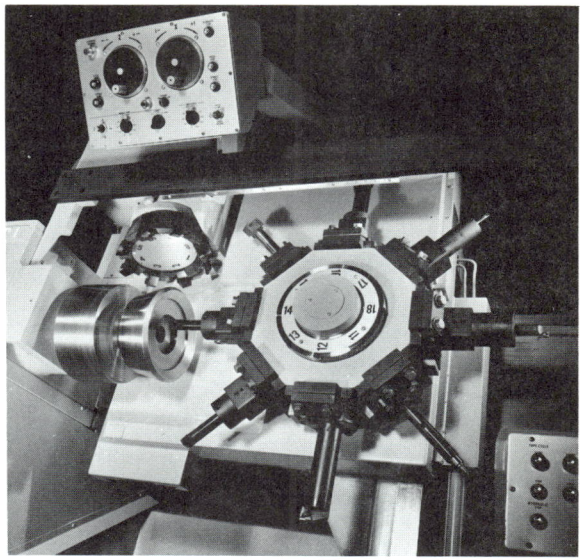

FIG. 47-2B A CHUCKING CENTER WITH TWO TURRETS; ONE FOR EXTERNAL OPERATIONS, ONE FOR INTERNAL OPERATIONS (Courtesy Cincinnati Milacron Co.)

The turret is clamped and unclamped by hydraulic pressure and *exact* peripheral and radial tool positioning is accomplished by a special curvic coupling. The turret is designed so that tools can be located permanently for mass production of duplicate parts, or can be rapidly interchanged. A shoulder stop on toolholders assures that each tool has accurate, positive depth location.

The chucking center illustrated in figure 47-2B has two turrets, each containing eight tool stations. One turret is used for external turning, while the other is used for internal operations. A greater variety of turning operations are possible on this type of machine without having to change cutting tools.

Turning Center

A turning center differs from a chucking lathe in that it is provided with a tailstock which allows turning of work between centers in addition to chuck work. It contains two turrets (one for internal and one for external turning, figure 47-3), mounted on the same cross-slide mechanism. There is a fixed distance between the two turrets and, therefore, they do not interfere with each other. The turning center with the two turrets is very versatile, and allows a wide variety of operations to be performed on a workpiece without the necessity of changing the tools or setup.

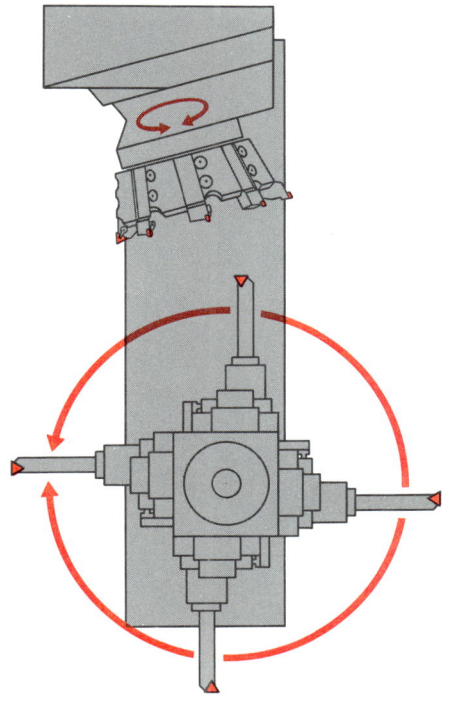

FIG. 47-3 A TURNING CENTER CONTAINS TWO TURRETS (INTERNAL AND EXTERNAL) ON ONE CROSS-SLIDE). (Courtesy Cincinnati Milacron Co.)

245

Section 4 Special Operations

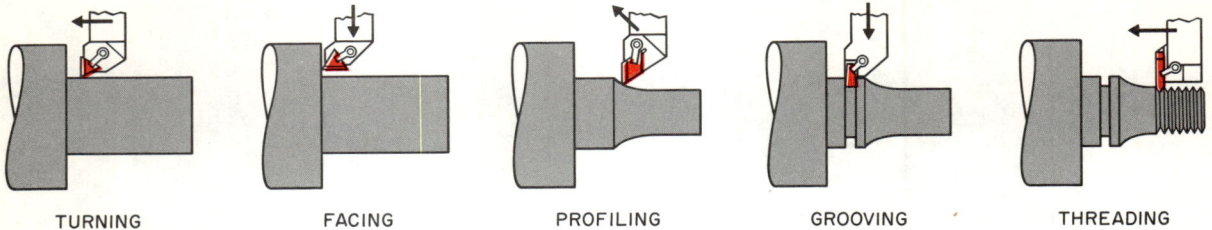

TURNING   FACING   PROFILING   GROOVING   THREADING

FIG. 47-4 THE FIVE BASIC TURNING OPERATIONS INVOLVED TO PRODUCE ROUND WORKPIECES (Courtesy Cincinnati Milacron Co.)

TOOL TURRETS

Almost any external or internal shape can be produced on a workpiece with the five basic turning operations illustrated in figure 47-4.

Replaceable carbide tool inserts, similar to the shapes illustrated in figure 47-4, are mounted in rugged square-shank toolholders. All tools, whether for the internal or external turrets, must be preset so that when they are fastened in the mounting slot of the turret, the cutting-tool point is a specific distance

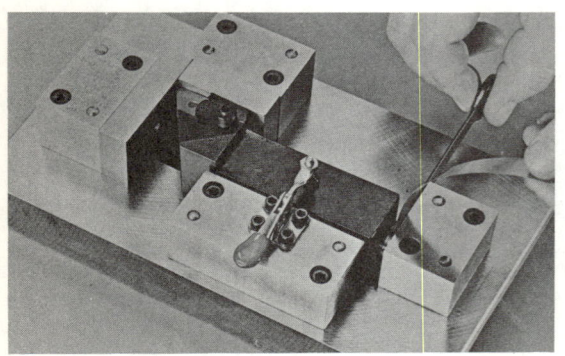

FOG. 47-5 A TOOL-SETTING GAGE USED TO PRESET TOOLS BEFORE THEY ARE MOUNTED IN A TURRET. (Courtesy Cincinnati Milacron Co.)

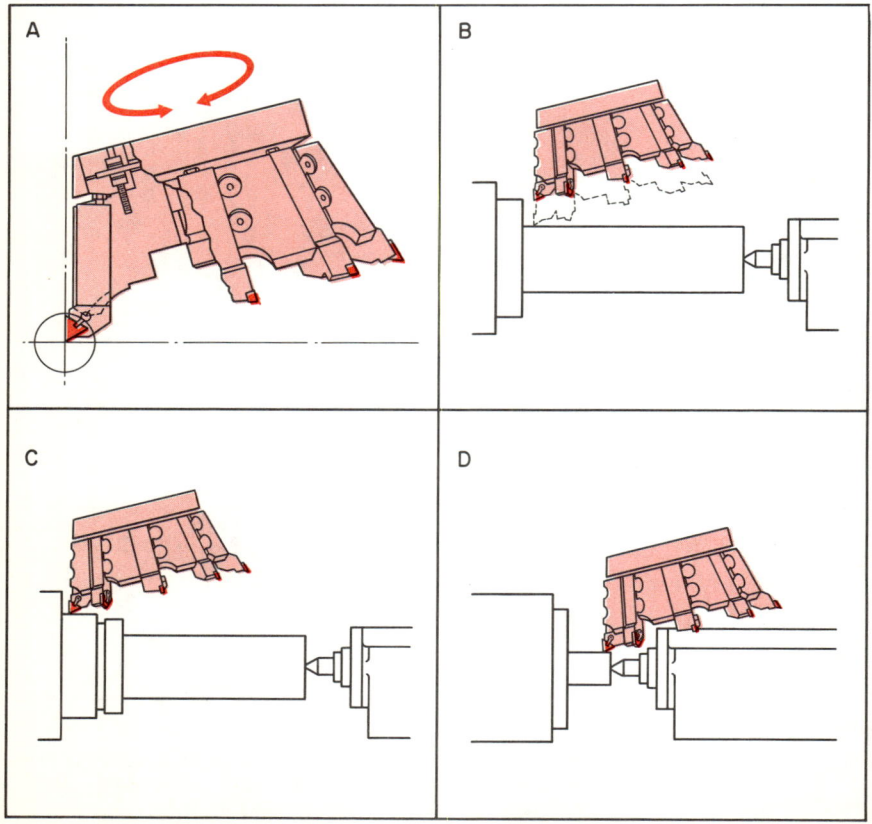

A. EACH CUTTING TOOL IS INDEXED PERPENDICULAR TO WORK CENTERLINE.

B. TOOLS CAN BE INDEXED WITH A MINIMUM RETRACTION FROM THE WORK.

C. CUTTING TOOLS CAN OPERATE CLOSE TO THE CHUCK JAWS.

D. CUTTING TOOLS CAN OPERATE CLOSE TO THE TAILSTOCK CENTER.

FIG. 47-6 ADVANTAGES OF OF THE CROWN TURRET (Courtesy Cincinnati Milacron Company)

from the center of the turret. The tool-setting gage, figure 47-5, enables external turning tools to be accurately preset to within .002 accuracy in the matter of a few minutes. A similar gage is available for presetting internal turning tools.

O.D. Turret

The crown turret supplied with Cincinnati Milacron machines holds up to eight preset external cutting tools ready for immediate indexing to working position. A special curvic coupling, operated by hydraulic pressure, assures precision and positive indexing of the turret for all positions. The axis of the conical turret is tilted to make one edge of the cone perpendicular to the centerline of the workpiece, figure 47-6A. The design of this crown turret allows various tools to be indexed into position quickly with a minimum of retraction from the workpiece, figure 47-6B. It also allows the tools to operate close to the chuck jaws, figure 47-6C, or the tailstock center, figure 47-6D.

I.D. Turret

The internal turret, figure 47-7A, holds eight preset cutting tools, and is indexed to a common point hydraulically in either direction. The eight-position turret allows boring, grooving, and other internal operations to be performed readily. Since all tools are preset to a common point, tool change is rapid and accurate, and the setup time for a variety of workpieces is kept to a minimum.

A. THE RUGGED TURRET ALLOWS MAXIMUM CUTS TO BE TAKEN.

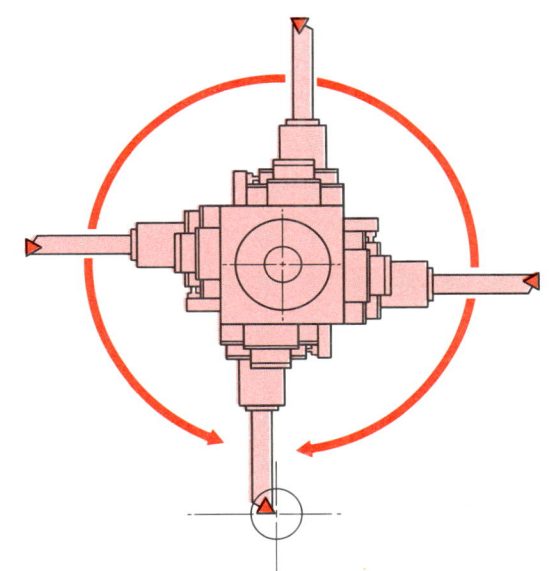

B. THE TURRET IS INDEXED HYDRAULICALLY TO A COMMON POINT.

FIG. 47-7 AN INTERNAL TURRET ON A TURNING CENTER (Courtesy Cincinnati Milacron Company)

REVIEW QUESTIONS

1. Why were numerical control lathes developed?

Construction

2. Name and describe two types of numerical control lathes.

3. How are turrets designed to assure positive and accurate indexing?

Section 4   Special Operations

Tool Turrets

4. Why is it important that tools used on numerical control lathes be preset?

5. Describe the construction of a crown turret.

6. What are the advantages of a crown turret?

7. Describe the construction of an internal turret.

Unit 48 FITS AND ALLOWANCES

One of the phases of lathe work requiring a great deal of skill is that of producing a proper fit on a part or parts. There are generally six classifications of fits. They are as follows:

1. *Running fit* — where a shaft must rotate freely and often at high speeds in a bearing.
2. *Sliding fit* — where a part or parts must be free to slide. The allowances for this type of fit are much the same as for running fits.
3. *Push fit* — where the parts may be assembled by a hand push fit. Parts are prevented from turning by means of a key and held in place by collars or setscrews.
4. *Driving fit* — where parts are assembled by driving or by use of an arbor press. Parts are often prevented from turning by means of a key.
5. *Forced fit* — a tighter fit than a driving fit, where parts are *permanently* assembled with a hydraulic press.
6. *Shrink fit* — a tighter fit than a forced fit where the outer member has to be heated and expanded to fit onto the inner member.

The allowances listed in table 48-1 have been found useful in the production of machined parts and may have to be altered slightly to suit certain conditions.

Shaft Dia.	Running Fits		Sliding Fits	Push Fits	Driving Fits	Forced Fits	Shrink Fits
	Under 600 r.p.m.	Over 600 r.p.m.					
Up to 1/2"	−.0005 to −.001	−.0005 to −.001	−.0005 to −.001	−.00025 to −.00075	+.0005 to +.001	+.00075 to +.001	+.0008 to +.0015
1/2" to 1"	−.00075 to −.0015	−.001 to −.002	−.00075 to −.0015	−.00025 to −.00075	+.0005 to +.001	+.001 to +.002	+.0015 to +.002
1" to 2"	−.0015 to −.0025	−.002 to −.003	−.0015 to −.0025	−.00025 to −.00075	+.0005 to +.001	+.002 to +.003	+.002 to +.0035
2" to 3½"	−.002 to −.003	−.003 to −.004	−.002 to −.003	−.0005 to −.001	+.00075 to +.0015	+.003 to +.004	+.0035 to +.0045
3½" to 6"	−.0025 to −.004	−.004 to −.0055	−.0025 to −.004	−.0005 to −.001	+.001 to +.002	+.004 to +.0055	+.0045 to +.006

Note: The above fits are based on the basic hole size and the allowances listed are made on the shaft diameter.

TABLE 48-1 ALLOWANCES FOR FITS

SECTION 5

Turret Lathe Operations

This entire section, including text and artwork, has been taken from **How to Machine Parts on Turret Lathes,** copyright (c) by the Warner & Swasey Company. Used with permission of the publisher.

Unit 49 FUNDAMENTALS OF THE TURRET LATHE

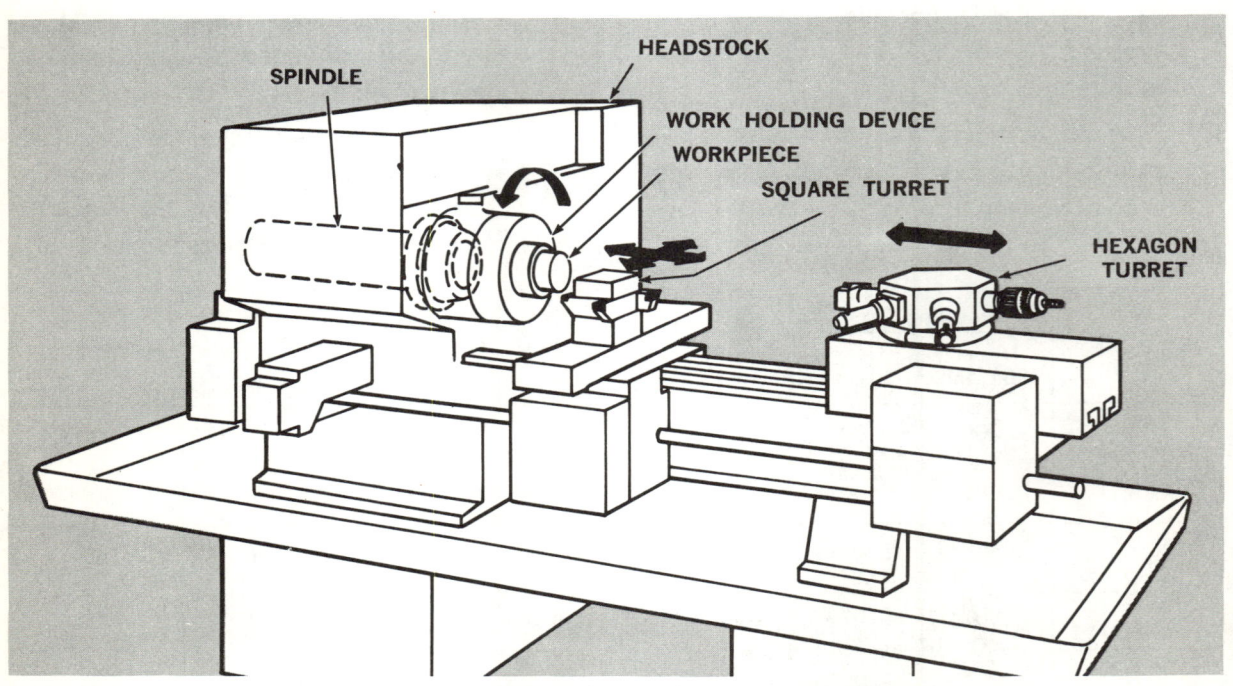

FIG. 49-1 PRINCIPLE PARTS OF A TURRET LATHE

THE TURRET LATHE

The turret lathe is a metal turning machine used to produce a number of identical parts in succession. The heart of this machine tool is a multiple-sided toolholder called a hexagon turret. In addition, turret lathes have a second tool turret called a square turret. Various cutting tools to perform different operations are mounted in these turrets. Then, by rotating or indexing the turrets and moving them back and forth, cutting tools can be brought against a rotating workpiece in a predetermined sequence. By repeating this sequence over and over, any number of pieces can be produced — all exactly alike.

Because workpieces vary so widely in size and shape, turret lathes, too, vary in type and size. There are hand-operated turret lathes, automatic turret lathes, vertical and horizontal types and many others. This section deals specifically with how to machine parts on hand-operated horizontal turret lathes.

The sketch, figure 49-1, shows the basic machine elements common to all turret lathes of this type.

The workpiece or part to be machined is held in a workholding device attached to the end of the spindle of the machine. Power to turn the spindle comes from the main motor through the gears and speed-selecting clutches in the headstock.

The cutters required to perform the necessary operations on the workpiece are set and clamped in the turret lathe's two indexing tool turrets — the hexagon end-working turret and the square turret. After each cut, the turret is indexed to bring the next required cutting tool into working position.

TYPES OF PARTS MACHINED ON TURRET LATHES

Basically, all parts produced on turret lathes can be classified as one of the following two types:

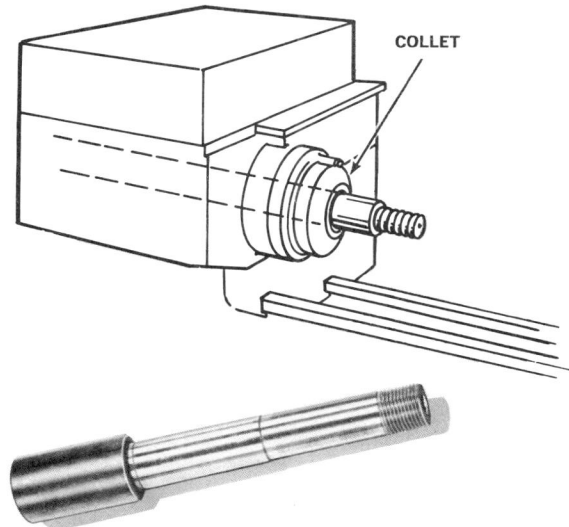

FIG. 49-2 BAR PARTS HELD IN A COLLET FOR MACHINING

Bar Parts

Bar parts are those workpieces produced from a long length of solid bar or tubing stock. The stock passes through the spindle of the machine and is gripped in a collet chuck mounted on the end of the spindle. A portion of the bar — usually just enough for one workpiece — extends through the collet. As this part is completed and cut off, the collet is opened and enough bar for another workpiece is fed forward, figure 49-2.

Although most bar parts are machined from round bar stock, some may be made from square, hexagonal, or even specially shaped extruded stock.

Chucking Parts

Chucking parts are workpieces turned from individual pieces of metal. Chucking parts may start out as rough castings or forgings, or as lengths of precut bar stock called "slugs," figure 49-3.

Chucking parts are gripped in any one of several types of standard or specially designed workholding devices. The three-jaw chuck shown in figure 49-3 is the most commonly used holding device for chucking parts.

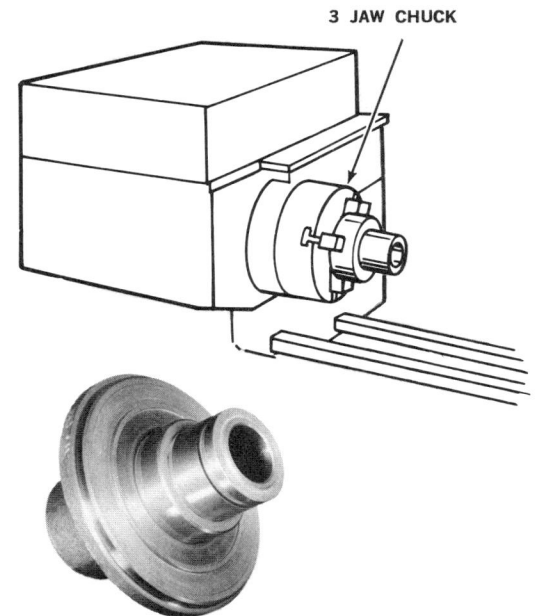

FIG. 49-3 METHOD OF HOLDING SINGLE WORKPIECES

Section 5 The Turret Lathe

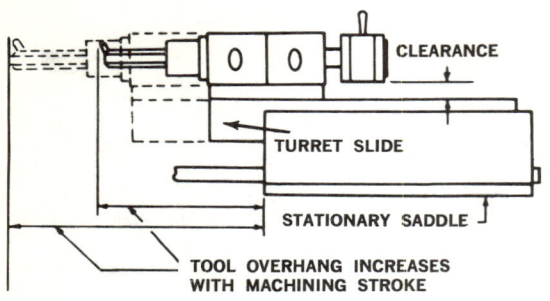

FIG. 49-4 MOVEMENT OF RAM SLIDE

Bar and chucking parts vary in general size and shape. For this reason, many turret lathes are equipped with tooling and work-holding devices for only one of these two types of work. In many shops, a turret lathe may be referred to as either a bar machine or a chucking machine depending on how it is equipped.

TYPES OF TURRET LATHES

Two distinctly different types of horizontal turret lathes are required to produce efficiently today's wide variety of workpieces . . . *ram-type turret lathes* and *saddle-type turret lathes.*

This section describes both types of machines and discusses the important differences between them. The nomenclature charts will help in learning the correct names of all parts of the turret lathe.

Ram-Type Turret Lathe

Ram-type turret lathes are fast, easy-handling machines designed for smaller, compact workpieces. Generally, ram-type turret lathes produce bar parts up to 3" in diameter or chucking work up to approximately 20" in diameter.

The hexagon end-working turret on ram-type machines mounts on a ram slide. The ram slide moves back and forth in the saddle. When a job is set up, the saddle is clamped in a convenient position along the bedways. Then, when producing parts, the saddle remains fixed and only the ram slide is moved, figure 49-4.

Remember, however, the length of the ram slide stroke governs the lengths of cuts from the hexagon turret. Ram slide stroke lengths vary with the size of the machine from 4" to 13".

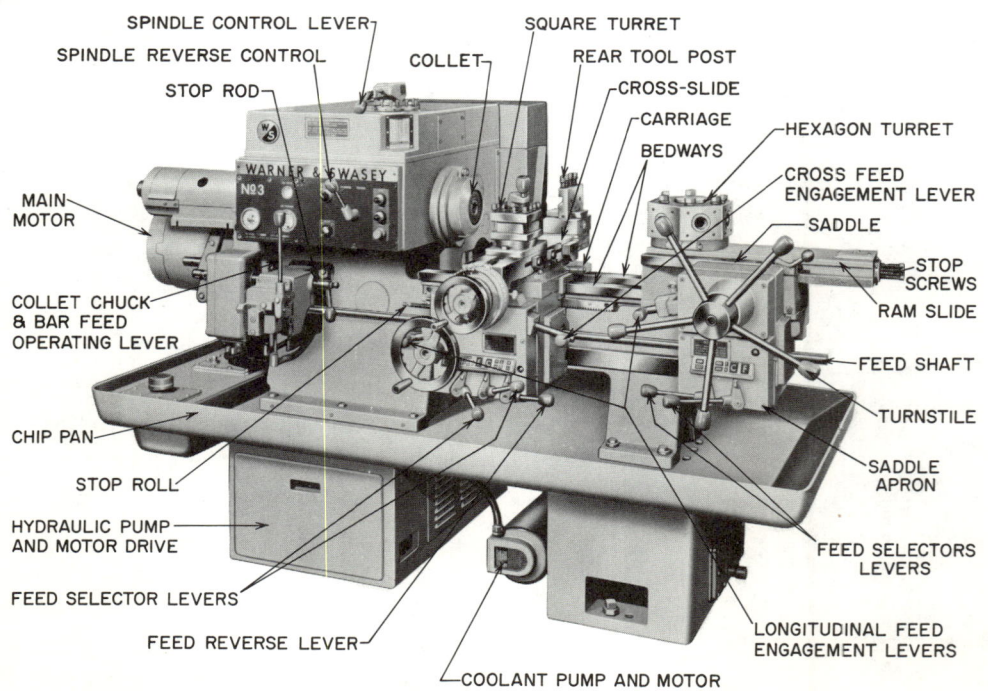

FIG. 49-5 RAM-TYPE TURRET LATHE (Courtesy Warner and Swasey)

The hexagon turret, too, is designed for fast handling. It indexes automatically as the ram slide is moved away from the work at the end of a cutting stroke.

Note that the cross-slide on ram-type turret lathes reaches across both bedways, figure 49-5. Thus, a cutter-holding toolpost can be mounted on the rear of the cross-slide.

Saddle-Type Turret Lathes

Saddle-type turret lathes are basically larger machines ideally suited for heavy-duty jobs — particularly those requiring long, accurate cuts. Standard saddle-type turret lathes can be equipped for bar work up to 12" in diameter and for chucking jobs up to a diameter of approximately 36".

Aside from size, the major difference between ram- and saddle-type machines is in the construction of the saddle which carries the end-working turret, figure 49-6.

In the case of saddle-type lathes, the end-working or hexagon turret mounts directly on the saddle. And when taking cuts, the entire

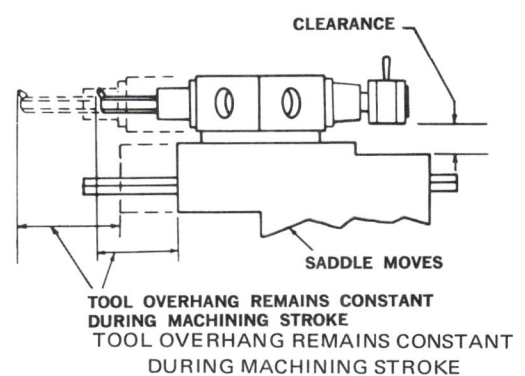

FIG. 49-6 MOVEMENT OF SADDLE UNIT

saddle unit moves along the bedways. Mounting the turret directly on the saddle provides maximum rigidity regardless of the length of cut. Lengths of cuts taken from the hexagon turret on saddle-type machines are, therefore, limited only by the length of the machine's bedways.

There are two different types of saddle-type turret lathes. These are classified as fixed center machines, figure 49-7, and cross-sliding turret machines.

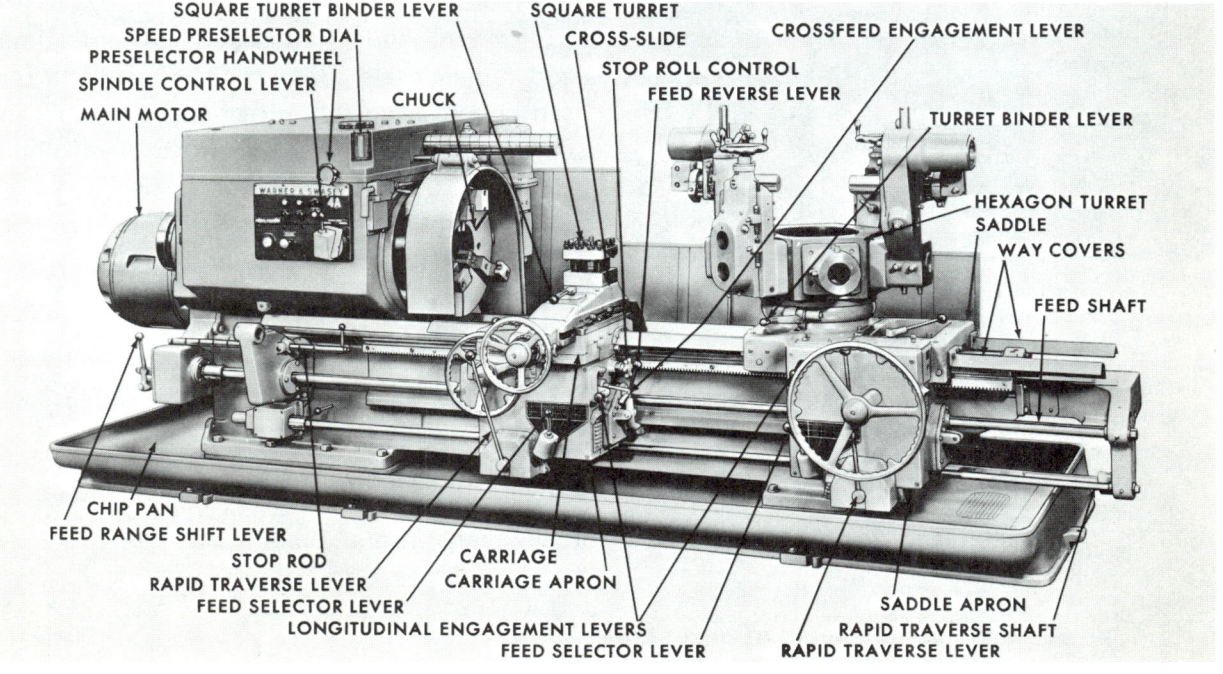

FIG. 49-7 SADDLE-TYPE TURRET LATHE (Courtesy Warner and Swasey)

Section 5 The Turret Lathe

FIG. 49-8 FIXED-CENTER TURRET
(Courtesy Warner and Swasey)

FIG. 49-9 CROSS-SLIDING TURRET
(Courtesy Warner and Swasey)

Fixed-Center Hexagon Turret. The fixed-center machine is so termed because the hexagon turret, figure 49-8, is fixed in line with the center of the spindle. The turret, like turrets on ram-type turret lathes, moves in a line parallel to the axis of the spindle.

Because the hexagon turret travels only longitudinally, adjustments needed to obtain various workpiece diameters are built into the tooling. For example, slide tools and adjustable cutter holders in multiple turning heads are widely used on fixed-center machines.

Cross-Sliding Hexagon Turret. The hexagon turret on cross-sliding turret machines is not mounted directly to the saddle, figure 49-9. Instead, it mounts on a slide which rides on top of the saddle and moves at right angles to the spindle centerline. Thus, the turret on a cross-sliding machine can move at right angles to the spindle axis as well as parallel to the spindle centerline. Should the need arise, a cross-sliding turret machine can be used like a fixed-center machine by placing the turret in line with the center of the spindle and locking the slide.

A cross-sliding turret reduces setup time and tooling costs. Because the hexagon turret can be moved in four directions, one tool can be used to cut several similar surfaces. Slide tools are virtually eliminated.

Several sample setups shown later in this section illustrate the uses and advantages of cross-sliding turret saddle-type turret lathes.

REVIEW QUESTIONS

Fundamentals of a Turret Lathe

1. Define a turret lathe. What is the purpose of this lathe?

2. What features on a turret lathe enable the rapid production of successive parts?

3. What are four types of turret lathes?

4. Explain the basic operation of a turret lathe.

5. What are bar parts and how are they held for machining on a turret lathe?

6. What are chucking parts and how are they held for machining?

Types of Turret Lathes

7. For what purpose are ram-type turret lathes used?

8. Explain the construction of the hexagon end-working turret. How does this turret operate?

9. What governs the length of cut which may be taken with the hexagon end-working turret?

10. What are the major differences between the ram-type and saddle-type turret lathes?

11. Compare the operation of the turret on a saddle-type with a ram-type turret lathe.

12. Compare the construction and operation features of a fixed-center hexagon turret with those of the cross-sliding hexagon turret.

13. What are the advantages of a cross-sliding hexagon turret?

Unit 50 BASIC CUTS

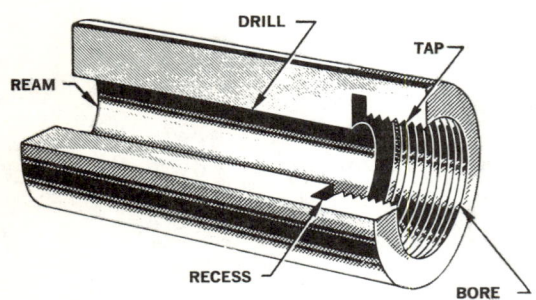

FIG. 50-1 INTERNAL CUTS

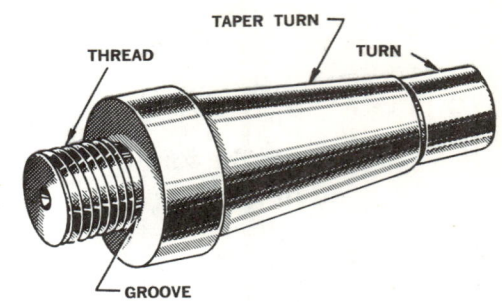

FIG. 50-2 EXTERNAL CUTS

Even the most complex workpieces are produced by combinations of basic internal cuts — cuts inside the workpiece, figure 50-1, and basic external cuts — cuts on the outside of the work, figure 50-2.

Nearly all turret lathe jobs require both internal and external machining operations. Before a job is set up, the types of cuts required and the sequence in which they will be applied to the workpiece must be determined. A sample part setup described on pages 263 and 264, shows how internal cuts are arranged in proper order. Pages 269 and 270 show the proper sequencing of external cuts on a typical job.

Drilling

Drills produce holes in solid stock and remove excess metal from existing holes.

Start Drilling. The accuracy of a drilled hole depends upon its start. Rough and uneven surfaces on bar stock, castings or forgings may cause longer drills to weave at the beginning of a drilling operation. Use a short, rigid start drill, figure 50-3, to spot a true cone in the workpiece.

When turret stations are at a premium the combination stock stop and starting drill (not shown) can be used to position the bar stock to length and start drill.

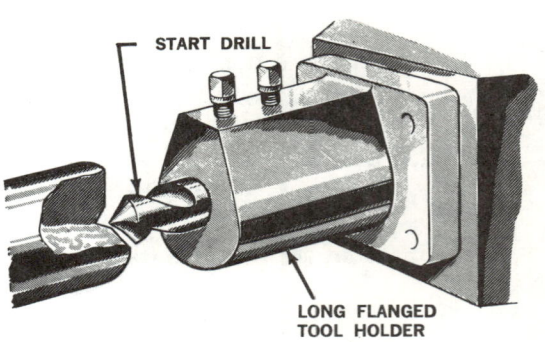

FIG. 50-3 STARTING DRILL

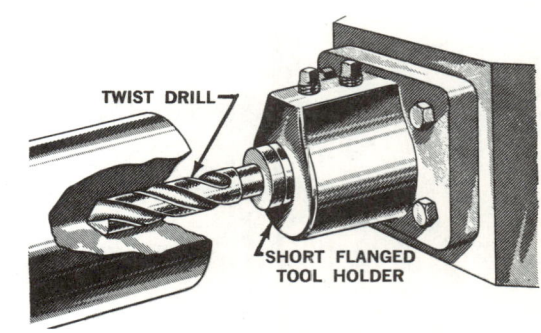

FIG. 50-4 TWIST DRILL

Twist Drilling. Twist drills are essentially two-bladed cutters, figure 50-4. For efficient cutting, keep the drills sharp and correctly ground for the material being machined.

It is good practice to use two drills on extremely large diameter drilling work — a smaller drill to pierce the hole, followed by

Unit 50 Basic Cuts

TOOLHOLDER BUSHING

DRILL CHUCK

PLAIN TOOLHOLDER

HEAVY-DUTY DRILLHOLDER

TAPER-SHANK DRILL SOCKETS

FIG. 50-5 DRILLHOLDERS

a larger diameter drill to enlarge the hole to the desired size.

Twist drills are available with either straight or tapered shanks. Some types of commonly used drill holders are shown in figure 50-5.

Core Drilling. Core drills, figure 50-6 are three- or four-fluted cutters used for enlarging previously drilled holes, or cored or pierced holes. Since cored or pierced holes seldom run true, provide an accurate guide for the core drill by start chamfering the work first with a starting drill or with a chamfering cutter in a boring bar.

When core drilling long holes, use a short boring cut to provide a true start for the core drill. Bore the hole 3/8" to 1/2" deep and to the same diameter as the core drill.

Spade Drilling. A spade drill is made up of a flat cutter firmly clamped in a bar-type holder. Like twist drills, spade drill cutters have two cutting edges.

Spade drills are commonly used to produce large holes over 2" in diameter. Twist drills of this size are generally too long and overhang the turret excessively whereas the spade drill with its straight shank can be snugged-up for minimum overhang, figure 50-7.

Deep Hole Drilling. Any hole longer than four times its diameter is considered a deep hole. Chip removal is a most important factor when

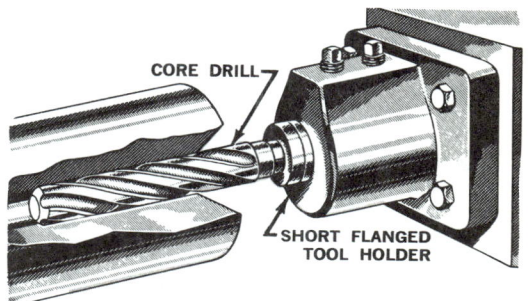
FIG. 50-6 CORE DRILL

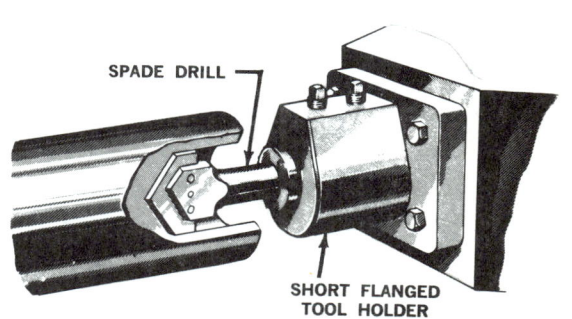
FIG. 50-7 SPADE DRILL

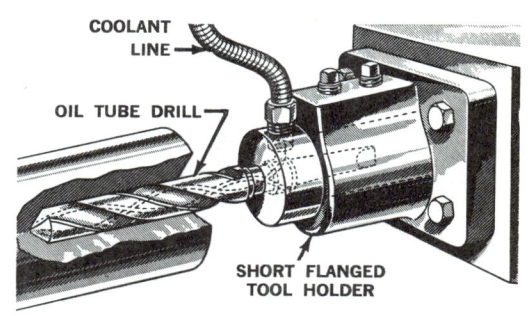
FIG. 50-8 DEEP-HOLE DRILL

Section 5 The Turret Lathe

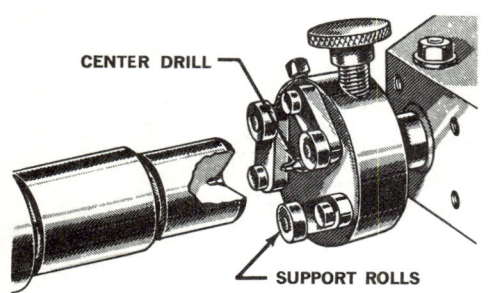

FIG. 50-9 CENTER-DRILLING ATTACHMENT

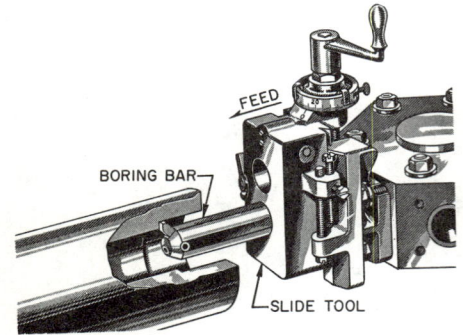

FIG. 50-10 BORING BAR MOUNTED IN SLIDE TOOL

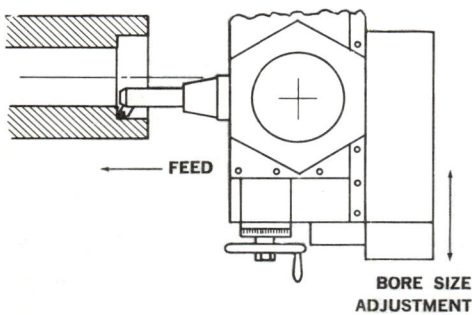

FIG. 50-11 BORING BAR MOUNTED IN TURRET

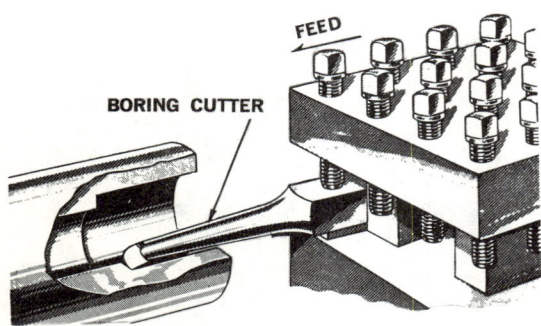

FIG. 50-12 BORING BAR MOUNTED IN SQUARE TURRET

drilling deep holes. To avoid breakage, withdraw smaller drills frequently to allow chips to escape.

Many shops use oil-hole drills, figure 50-8 for deep drilling applications. These drills supply coolant, under pressure, to the point of the drill, washing out chips and keeping the cutting edges cool.

Center Drilling. Center drilling provides a means of supporting the workpiece during later operations on the turret lathe or on other machines. For accurate centers and longer center drill life use a center drilling tool with self-centering rollers which support the end of the part, figure 50-9.

BASIC INTERNAL CUTS

Boring

Boring cuts true up holes and produce accurate size.

Hexagon Turret Boring (Ram- or Fixed-Saddle Type Turret Lathes.) A boring bar held in a hexagon turret mounted slide tool provides the rigidity required for heavy, accurate boring cuts, figure 50-10. The proper bore size can be adjusted quickly by using the slide tool's micrometer dial. Once the cutter is set, lock the slide with the binder clamp.

Hexagon Turret Boring (Cross-Sliding Hexagon Turret Saddle-Type Machines.) Slide tools are not required for boring on saddle-type machines with cross-sliding hexagon turrets. Boring bars are mounted directly in the hexagon turret or in flanged toolholders, figure 50-11. Bore size is set by positioning the turret with respect to the center of the spindle.

Faster setups — requiring fewer tools — are possible with a cross-sliding hexagon turret. Repositioning the turret between cuts permits the boring of several different size holes with the same boring bar.

Square Turret Boring. Though less common, boring can also be done using the square

Unit 50 Basic Cuts

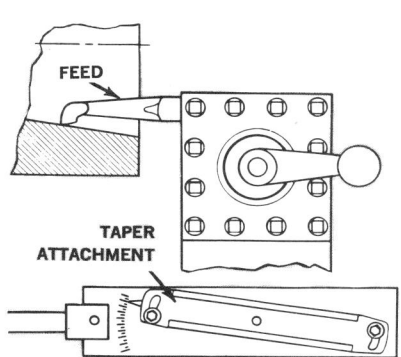

FIG 50-13 BORING WITH TAPER ATTACHMENT — TOOL IN SQUARE TURRET

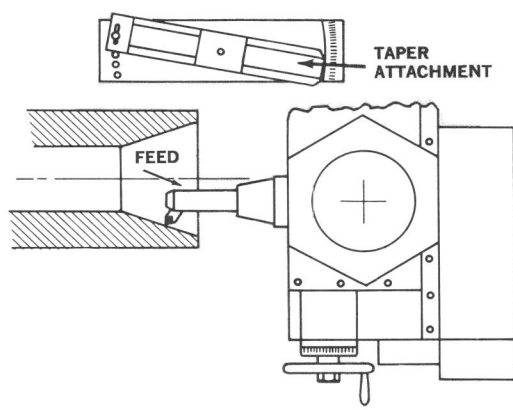

FIG. 50-14 BORING WITH TAPER ATTACHMENT — TOOL IN HEXAGON TURRET

turret, figure 50-12. However, square turret boring is usually limited by the rigidity of the tool. In most cases, this method is confined to boring short holes using light cuts and lower feeds.

Square Turret Taper Boring. Internal tapers can be cut from the square turret using the taper attachment, figure 50-13. Set the attachment for the desired angle and establish bore size with cross-slide cross-motion handwheel.

For accurate taper boring, feed the cutter *away* from the spindle, or in the direction of the greater bore.

Cross-Sliding Hexagon Turret Taper Boring. Because the cross-sliding hexagon turret provides both cross and longitudinal motion it, too, can be arranged with a taper attachment, figure 50-14.

The cross-sliding hexagon turret taper attachment is preferred for big work requiring heavy cuts since larger, more rigid boring tools can be used.

Boring Bars. Boring bars like the one shown, figure 50-15, are versatile tools. On one end an enclosed slot holds cutters for rough or semifinish boring, or for inside facing. An open slot on the opposite end of the bar holds cutters for counterboring, chamfering, and similar operations.

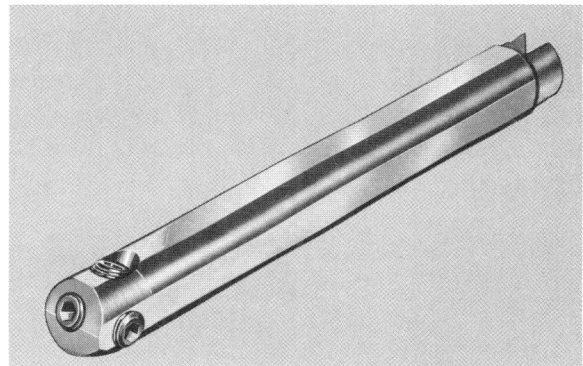

FIG. 50-15 REVERSIBLE BORING BAR

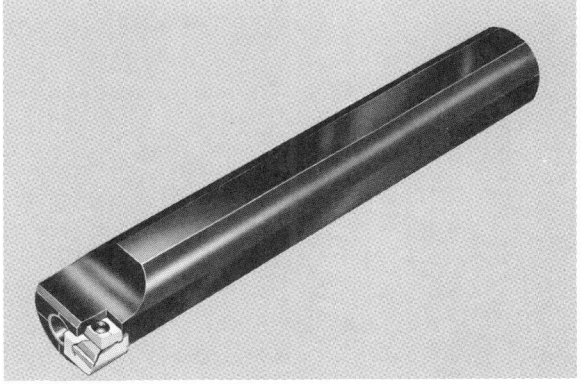

FIG. 50-16 FINE-ADJUST BORING BAR

Fine-adjust boring bars, figure 50-16, are designed for accurate finish boring. Position the bar for approximate size, using a slide tool or the cross-sliding hexagon turret. Then, set final bore size by using the graduated adjustment built into these precision bars.

Section 5 The Turret Lathe

Reaming

Holes are reamed when an accurate size and smooth finish are required.

Fluted Reamers. Reamers are finishing cutters designed to remove .003 to .015 inches of metal from a previously drilled or bored hole, figure 50-17.

Fluted reamers, found in most shops, can be either solid or expandable. Solid reamers are ground to the proper hole size and cannot be adjusted. Expanding reamers are also ground to size. However, they can be expanded a few thousandths of an inch to compensate for wear or regrinding.

To produce straight, properly sized holes, mount the reamer in a floating reamer holder, figure 50-17. The float in the holder permits the reamer to align itself with the existing hole. Guide the reamer by hand into the drilled or bored hole. Use coarse feeds and slow speeds.

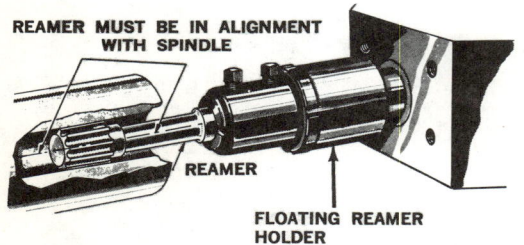

FIG. 50-17 REAMING WITH FLUTED REAMER

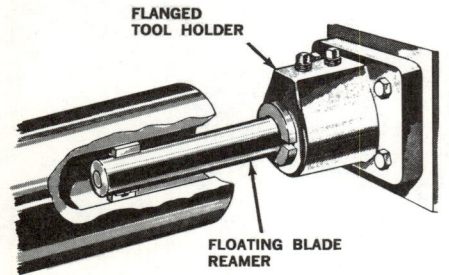

FIG. 50-18 REAMING WITH FLOATING-BLADE REAMER

Floating-Blade Reamers. Adjustable floating-blade reamers, figure 50-18, are practical for large holes. Mount these reamers solidly in an adjustable toolholder or directly in a flanged toolholder. The two cutting blades provide the floating action necessary for proper alignment.

For smooth finish and accurate size, keep the reamer blades sharp. Use coolants on steel workpieces.

Taper Reamers. Taper reamers are often used to finish small-diameter taper bores, figure 50-19. Steep tapers may require rough and finish taper reaming operations to obtain the desired size and finish. Hold taper reamers in floating holders or adjustable toolholders to assure accurate alignment.

Recessing, Facing, and Chamfering

Certain internal cuts, such as recessing and facing, require a combination of toolslide motions.

Hexagon Turret. Internal grinding reliefs, thread clearances, and chamfers all require a longitudinal toolslide motion to position the cutter at the proper point inside the workpiece and, then, either a cross or vertical feed stroke to make the actual cut.

In the case of ram-type or fixed-center saddle-type machines, a quick-acting slide tool mounted on the hexagon turret provides the vertical motion for these internal cuts, figure 50-20. A recessing cutter ground to shape and held in a boring bar gives maximum rigidity for heavier cuts, figure 50-21A, B, C. Depth of cut is accurately determined by stop screws on the slide tool.

On cross-sliding hexagon turret saddle-type machines, mount the cutter-holding bar directly in the turret or in a flanged toolholder. Use the cross-feeding motion of the turret to take the cut.

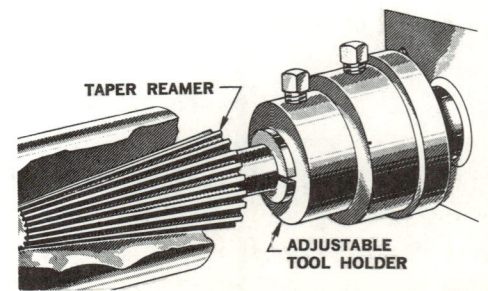

FIG. 50-19 REAMING WITH TAPER REAMERS

Unit 50   Basic Cuts

Square Turret. It is also possible to recess, face, or chamfer internally using a cutter held in the square turret, figure 50-22. Since rigidity is limited by the cross section of the cutter and its overhang from the turret, take lighter cuts using hand feed.

Internal Threading

Internal threads may be cut by tapping or by single-point threading.

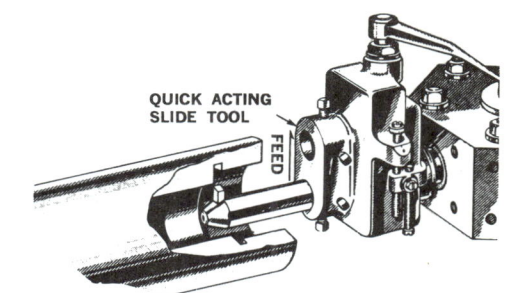

FIG. 50-20 RECESSING WITH TOOL MOUNTED IN HEXAGON TURRET

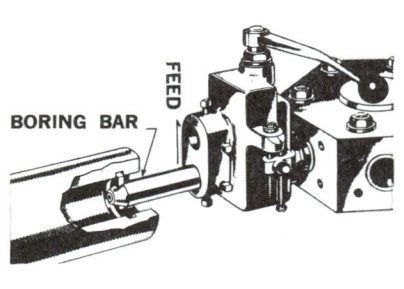 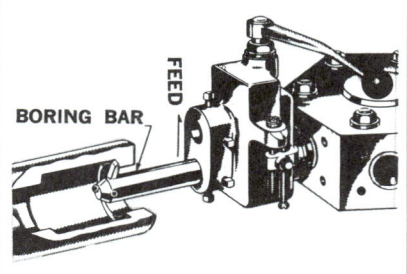 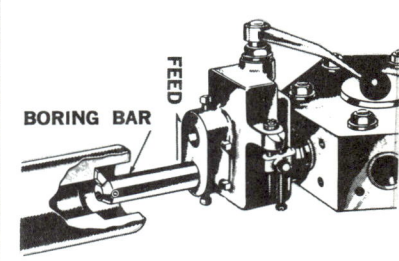

A. INTERNAL RECESSING   B. BACK FACING   C. INTERNAL FACING

FIG 50-21   SOME COMMON INTERNAL CUTS

Tapping. Small-diameter internal threads are most commonly produced by using a solid tap held in a releasing tap holder, figure 50-23. Figure 50-24 shows the nomenclature of a typical solid tap.

To tap a hole you must advance the hexagon turret and provide sufficient pressure to start the tap in the hole. Once the tap starts cutting, its own helix angle and the rotation of the workpiece combine to pull it into the work at the proper rate.

As the tap feeds into the workpiece, carefully apply pressure to the turnstile or handwheel to keep the turret and tap holder moving at the same rate as the tap. Never force the tap faster than it normally moves into the work. Never allow the tap to drag and pull the turret along as it feeds in. Both forcing and dragging produce defective threads.

Set a turret stop to end the forward motion of the turret approximately 1/8" before the tap reaches the required length of

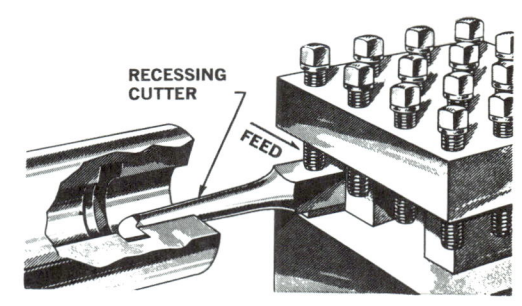

FIG. 50-22 RECESSING WITH TOOL MOUNTED IN THE SQUARE TURRET

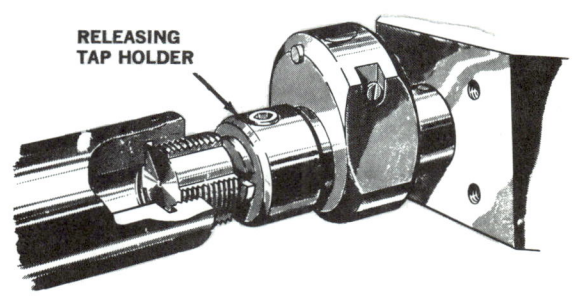

FIG. 50-23 TAP HOLDER

261

Section 5 The Turret Lathe

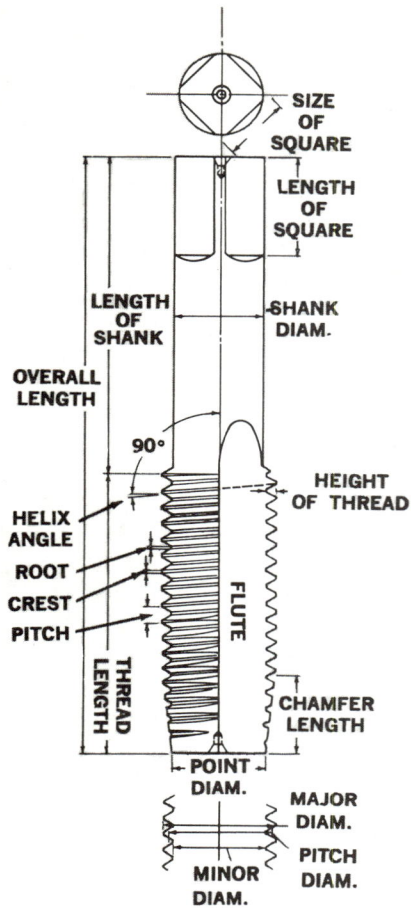

FIG. 50-24 SOLID TAP

FIG. 50-25 LEADING-ON ATTACHMENT

FIG. 50-26 THREAD-CHASING ATTACHMENT

thread. When the forward progress of the turret and tap holder is stopped, the tap holder's automatic release allows the tap to revolve with the workpiece.

To withdraw the tap, reverse the direction of spindle rotation. As the tap follows the thread back out of the hole, keep the turret and tap holder moving with the tap.

Larger diameter holes — approximately 1 1/2" and greater — are often threaded with collapsing taps. These taps are designed so that the thread-cutting edges pull away from the workpiece at the end of the forward tapping stroke. It is not necessary to reverse the spindle to withdraw a collapsing tap from the threaded hole. Collapsing taps can be held directly in the turret or in a flanged toolholder.

Accurate Threading Lead. Accurate internal threads require that the tap advance into the work at precisely the required rate or lead. The leading-on attachment, shown in figure 50-25, feeds the hexagon turret on ram-type machines by means of leaders and half nuts to the correct pitch for the thread being cut.

An automatic knock-off mechanism, operated by the turret stop screws, permits the maintaining of accurate depths on tapping cuts to shoulders or in blind holes.

Saddle-type turret lathes can be equipped with thread-chasing attachments, figure 50-26, to advance the taps accurately into the work.

Single-Point Threading. Single-point threading produces accurate threads which are concentric with other machined diameters.

Unit 50 Basic Cuts

FIG. 50-27 THREADING TOOL AND WORKPIECE

Single-point threading requires making several thread-cutting passes. The exact number depends on the work material and the pitch of the thread. For each pass, the cutting tool is set to cut a little deeper until the desired thread size is reached, figure 50-27.

The threading cutter is commonly held in a bar mounted in a slide tool. In the case of cross-sliding hexagon turret machines, the bar is held in a flanged toolholder. When single-point threading from the cross-slide use a square turret thread-chasing cutter holder or a special holder to which a bar can be adapted.

To feed the threading tool at the desired pitch, use either the thread-chasing attachment or the lead screw threading attachment.

BASIC INTERNAL CUTS — REVIEW

Sequence of Internal Cuts. Internal cuts must be set up in the proper sequence because the performance of each basic machining operation depends upon the form and accuracy produced by a previous cut.

For example, producing the threaded adaptor, figure 50-28, requires start drilling, drilling, boring, reaming, internal recessing, and tapping. The setup drawing, figure 50-29,

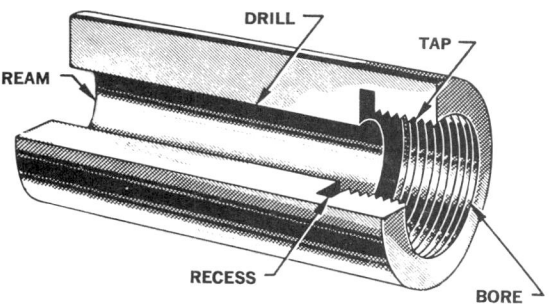

FIG. 50-28 THREADED ADAPTOR

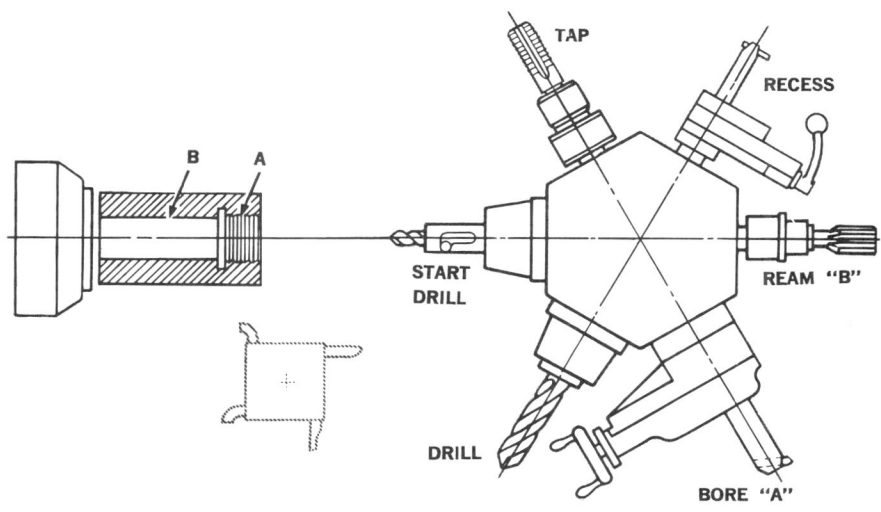

FIG. 50-29 SEQUENCE SETUP

263

Section 5 The Turret Lathe

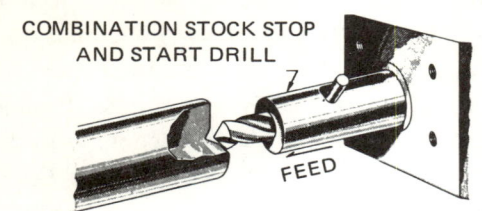

A. Start drill bar to provide true, clean start for drill in next station. Use either a start drill or, if part length is critical, a combination stock stop and start drill.

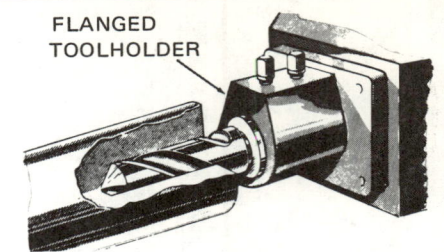

B. Drill the hole through the part. Use a drill socket to adapt the drill to a flanged toolholder.

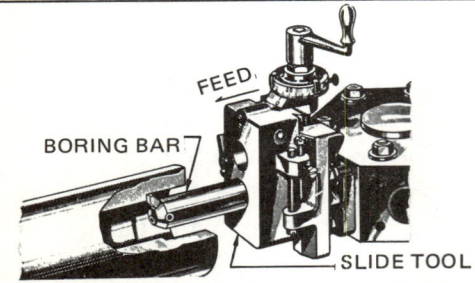

C. Single-point bore thread diameter to proper size. For quick adjustment to size, hold boring bar in a slide tool.

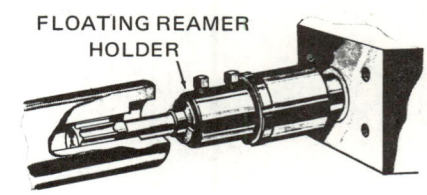

D. Ream previously drilled hole. To assure proper alignment with drilled hole, mount fluted reamer in floating reamer holder.

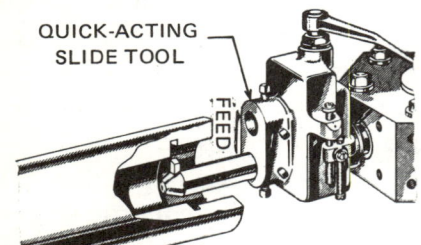

E. Recess thread clearance groove. The boring bar with a recessing cutter is mounted in a quick-acting slide tool.

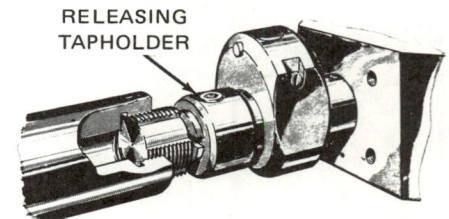

F. Cut thread with tap held in a releasing tapholder. If this were an odd-size thread it would be necessary to single-point it from a square turret or hexagon turret.

FIG. 50-30 BASIC CUTS REQUIRED TO MACHINE THE THREADED ADAPTOR

and the step-by-step illustrations, figure 50-30 A, B, C, D, E, F, show the proper sequence of these cuts.

In most cases, standard sizes are specified for internal work, permitting the use of standard drills, reamers, and taps. However, for odd-size bores or threads, special tools must be used or these surfaces should be machined by single-point boring and single-point threading.

BASIC EXTERNAL CUTS

Turning

The turning process produces round, cylindrical shapes.

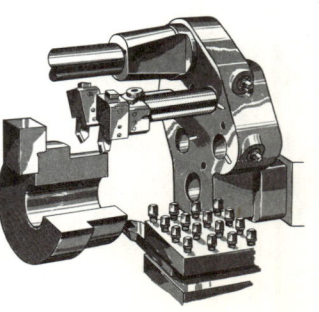

FIG. 50-31 MULTIPLE TURNING HEAD

Unit 50 Basic Cuts

Overhead Turning. Turning cuts from the hexagon turret assure rigid tool support for heavy metal removal.

Turning cutter holders are mounted in multiple turning heads or in single turning heads. The multiple turning head, figure 50-31, permits the mounting of several cutter holders in the same turret station.

Use reversible plain cutter holders for rough turning cuts. Reversible adjustable cutter holders are preferred for finishing cuts because they have a built-in micrometer adjustment which makes it easy to set cutters to size.

Side Turning. Turning cuts, also, are commonly taken from the square turret, figure 50-32. When combining machining operations, turning cuts from the square turret can be performed at the same time as drilling or boring from the hexagon turret.

FIG. 50-32 TURNING FROM THE SQUARE TURRET

To duplicate sizes accurately, always make the final adjustments by advancing the square turret towards the workpiece to remove all backlash between the cross-slide screw and nut. Always index the turret in the same direction.

Bar Turning. Turn long bar work — which overhangs the collet — with a bar turner. The bar turner's rolls support the workpiece permitting heavier cuts and higher feed rates, figure 50-33.

For smooth surface finish, set the turner rolls behind the tip of the turning cutter to burnish the turned surface.

To produce concentric diameters, mount the rolls ahead of the turning cutter on a previously turned diameter, figure 50-34.

Bar turners have a built-in cutter relieving feature. At the end of a turning cut, use the cutter-relieving lever to move the cutter away from the workpiece. With the cutter retracted, the bar turner can be moved back off the workpiece without leaving cutter withdrawal marks on the finished surface. Push the lever forward to reset the cutter for the next piece.

Taper Turning. A taper attachment makes it possible to machine internal or external tapers from the square turret on ram-type of fixed-center-saddle-type turret lathes, figure 50-35.

FIG. 50-33 ROLLS SET FOR FINISH CUT

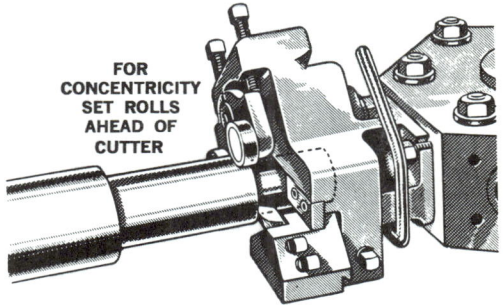

FIG. 50-34 ROLLS SET FOR CONCENTRIC CUT

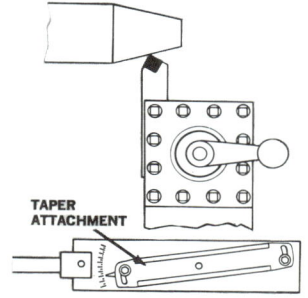

Section 5 The Turret Lathe

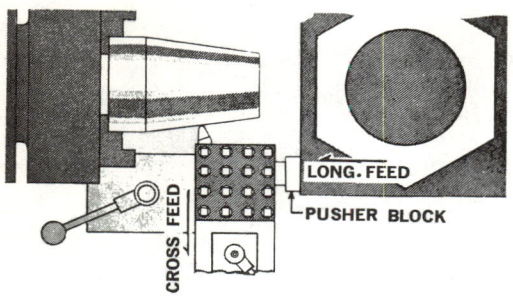

FIG. 50-36 TAPER TURNING WITH COMBINED FEED

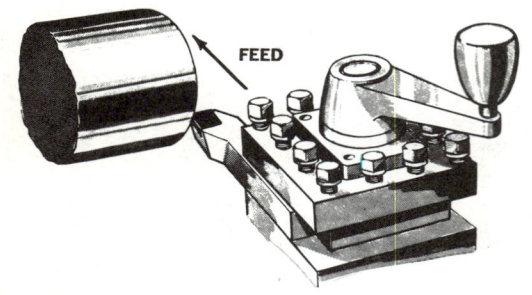

FIG. 50-37 FACING WITH SQUARE TURRET

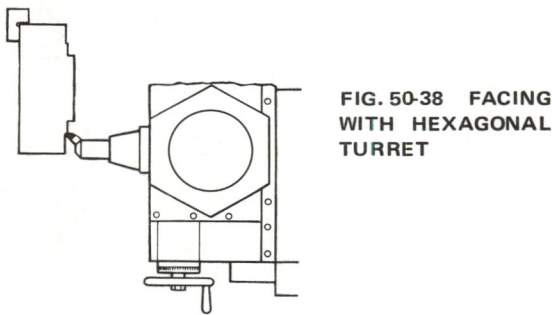

FIG. 50-38 FACING WITH HEXAGONAL TURRET

Taper attachments are available for cross-sliding hexagon turret machines for either the cross-slide or the cross-sliding turret.

The guide plate of the taper attachment when set to the desired taper angle, controls the travel of the cutter to generate the taper on the workpiece.

Combined-Feed Taper Turning. Tapered surfaces can be roughed out by combining square turret crossfeeds and hexagon turret longitudinal feeds. The hex turret moves the square turret forward — longitudinally — as the cross-slide feeds under power — at right angles to the spindle. As shown in figure 50-36, a pusher block is used between the two slides.

Taper angles may be varied by increasing or decreasing the crossfeed or longitudinal feed.

Facing

Facing produces flat surfaces at right angles to turned diameters.

Square Turret. Most facing cuts are taken with single-point cutters mounted in the square turret, figure 50-37. Workpiece length is established by positioning the cross-slide carriage along the bedways. Make sure the cross-slide carriage is firmly clamped to the bedways during facing cuts.

Facing cuts are often combined with drilling or boring cuts from the hexagon turret.

Hexagon Turret. When square turret stations are at a premium, short facing cuts can be taken from the hexagon turret by using the quick-acting slide tool, figure 50-38. The cutter is fed at right angles to the spindle with the hand operating lever of the slide tool.

In the case of cross-sliding turret machines, facing from either turret is common. For cuts from the hexagon turret, hold the cutter in a three-slot cutter holder, boring bar, or in a tool-base cutter block.

End Facing. Face the end of longer bar parts with an end former, figure 50-39, or with a combination turner and end former. Both of

FIG. 50-39 FACING END OF BAR STOCK

Unit 50 Basic Cuts

these tools have rolls which support the workpiece and keep it from springing away from the cutter. Be sure to set the rolls ahead of the cutter and feed the tool on by hand. Also, to get a smoother finish, use slower cutting speeds.

Forming and Cutoff

Forming is a fast method of producing finished diameters and shapes.

Square Turret. Most single chamfering, necking and grooving cuts are taken from the square turret, figure 50-40. Some forming cutters that are commonly used are shown in figure 50-41. If several necking and grooving cuts are needed, mount the cutters in a necking cutter block at the rear of the cross-slide.

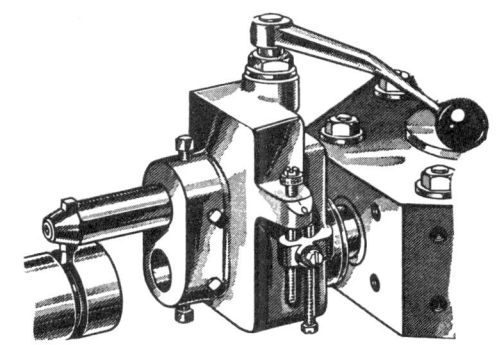

FIG. 50-42 GROOVING FROM A HEXAGON TURRET

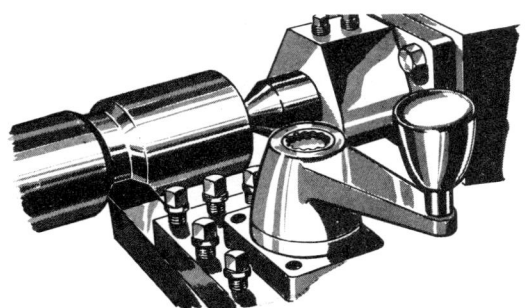

FIG. 50-40 CHAMFERING, NECKING, AND GROOVING CUTS WITH SQUARE TURRET

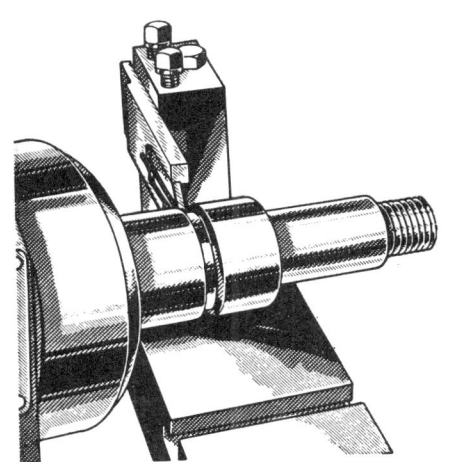

FIG. 50-43 CUTTING A FINISHED PIECE FROM BAR

Hexagon Turret. Although the practice is not too common, grooves or chamfers may be cut from the hexagon turret, figure 50-42. Use a bar held in a quick-acting slide tool. On cross-sliding turret machines, mount the forming cutter in a boring bar or use the tool base cutter block.

Cutoff. After a bar part is machined, it must be cut off the end of the bar.

A carbide cutoff holder with interchangeable carbide cutoff blades gives maximum rigidity for fast, smooth cutoff, figure 50-43. Mount the holder in the square turret or in a rear cutter block.

Cutoff feed rates are very important when carbide tools are being used.

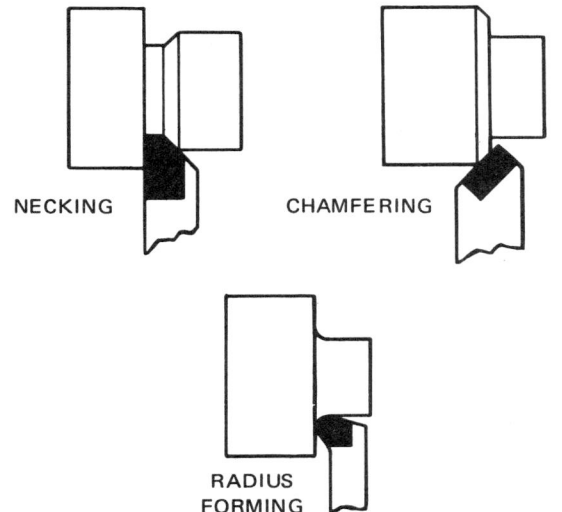

FIG. 50-41 COMMON TYPES OF FORMING CUTS

Section 5 The Turret Lathe

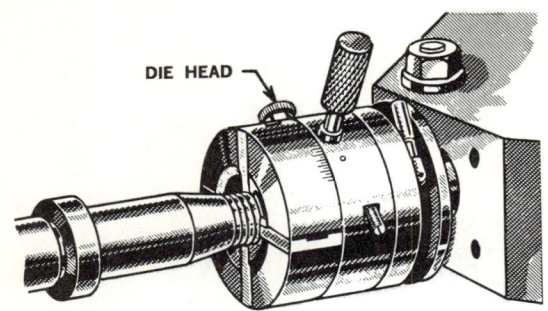

FIG. 50-44   SELF-RELEASING DIE HEAD

FIG. 50-45   METHODS USED TO SECURE ACCURATE THREADS

FIG. 50-46A   THREAD-CHASING ATTACHMENT FOR CROSS-SLIDES ON RAM- AND SADDLE-TYPE MACHINES

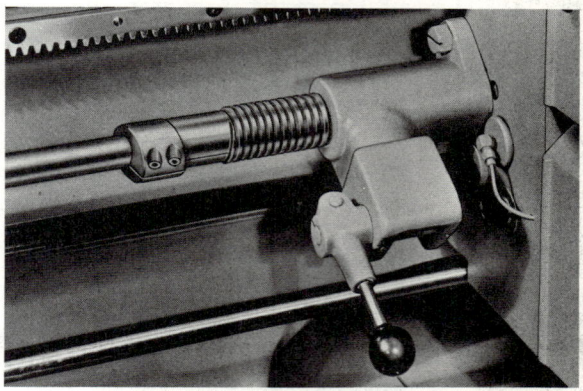

FIG. 50-46B   THREAD-CHASING ATTACHMENT FOR HEX TURRET ON SADDLE-TYPE TURRET LATHES

Threading

The final choice of threading methods depends on the size, type, and accuracy of the thread.

Die Heads. Small-diameter, standard size external threads generally are cut with self-opening die heads, figure 50-44. These die heads spring open and release automatically when the desired thread length has been machined. Start the die head on the work and provide an even pressure on the turnstile to keep the hexagon turret moving with the die head.

Accurate Threading With Die Heads. Leading on a die head by hand is satisfactory for most threads. But, long or accurate threads require a more positive means of feeding the die head over the work, figure 50-45. On ram-type machines, use a hexagon turret leading-on attachment. Or, if the machine is equipped with a cross-slide thread-chasing attachment, positive lead is provided by linking the cross-slide to the hexagon turret.

To provide positive die head lead on saddle-type machines, use leader- and follower-type hexagon turret thread-chasing attachment.

Single-Point Threading. Especially accurate threads, threads with unusual specifications, or threads which must be concentric to other machined diameters are cut by the single-point method, figures 50-46A and B.

A thread-chasing attachment, set up on the cross-slide carriage, leads on the square turret at the desired rate for each revolution of the spindle. The threading cutter is held in the square turret. As with internal single-

Unit 50 Basic Cuts

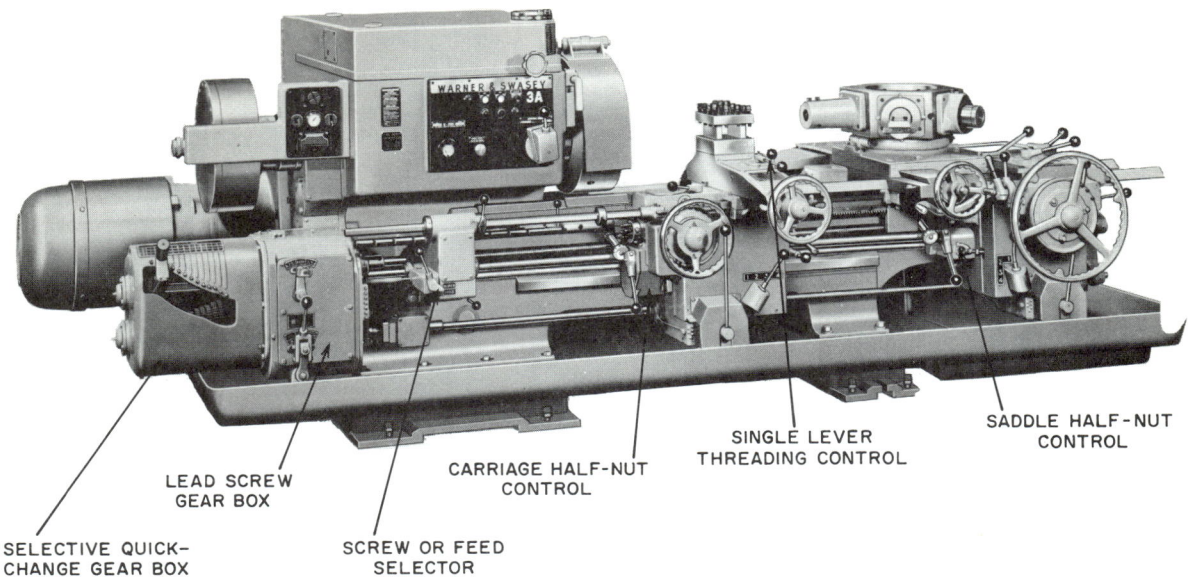

FIG. 50-47 MACHINE EQUIPPED WITH LEAD SCREW THREADING ATTACHMENTS (Courtesy Warner and Swasey)

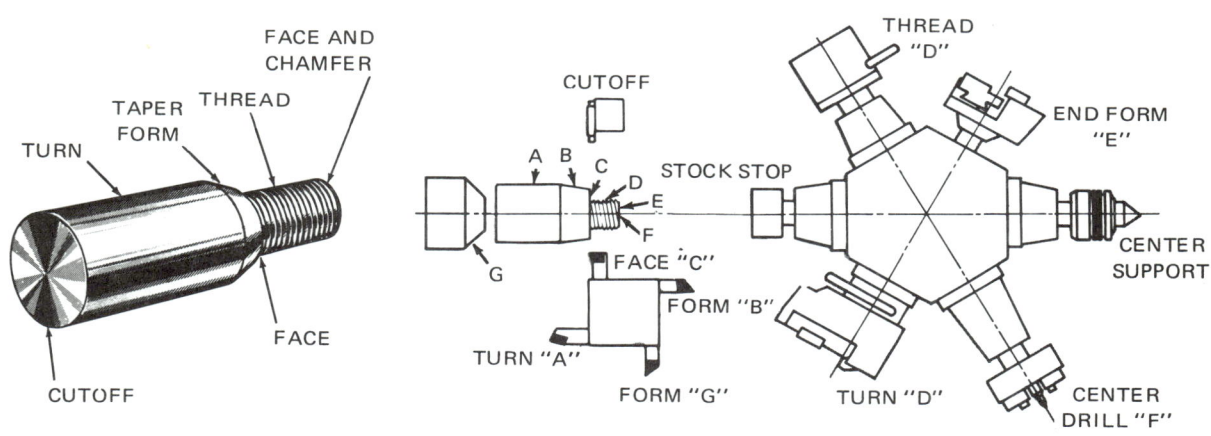

FIG. 50-48 THREADED SHAFT   FIG. 50-49 SEQUENCE OF OPERATIONS FOR MAKING A THREADED SHAFT

point threading, a series of cutting passes is required to complete the thread.

Threads can also be single-pointed on turret lathes equipped with lead screw threading attachments, like the one on the machine in figure 50-47. The lead screw is an independent shaft used exclusively for threading. Threads can be cut from either turret.

Sequence of External Cuts. Although not always as critical as the sequence of internal cuts, the order in which external cuts are made is important. Certain cuts must follow a set sequence, figure 50-49. For example, before a thread can be cut, the thread diameter must be turned to size. Other cuts, however, can be taken at almost any point in the cycle. Remember, a small amount of time spent planning the best sequence of cuts can save effort and time on each piece produced.

The threaded shaft, figure 50-48, requires a number of typical external cuts. The sequence of these operations is indicated in figure 50-49. They are illustrated in detail in figure 50-50A through I, page 270.

Section 5   The Turret Lathe

A. After feeding out the stock to proper length against a revolving bar stop, turn the thread diameter. Use a bar turner mounted on the hexagon turret. The chamfered bar end helps give the turning cutter a gradual start.

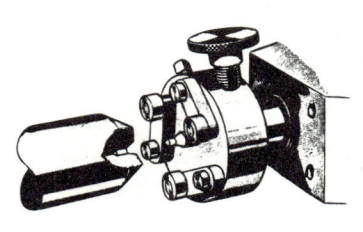

B. Center drill the end of the bar with a center drilling tool. The rolls of this tool support the work and assure an accurate center.

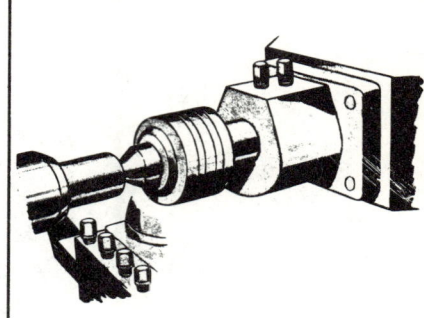

C. Support the shaft with a revolving center and form the tapered section from the square turret.

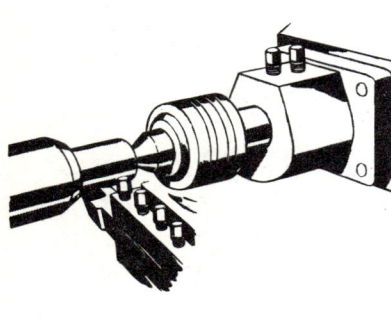

D. Without withdrawing the revolving center, face the shoulder at the end of the thread diameter.

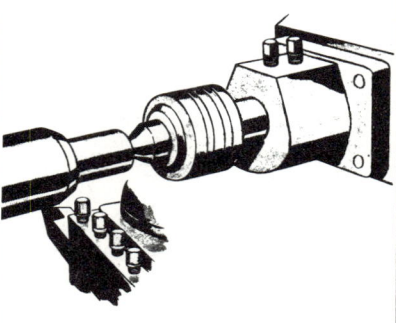

E. Still supporting the part, turn the large diameter from the square turret.

F. Face and chamfer the end of the shaft with an end former or a combination turner and end former. Both of these tools have rolls which hold the work firmly against the cutter.

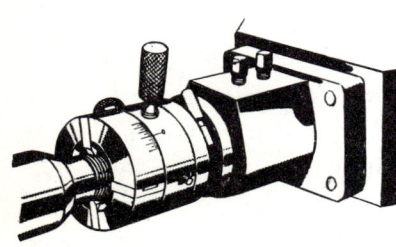

G. Cut the thread using a self-opening die head arranged with the proper size thread-chasing cutters.

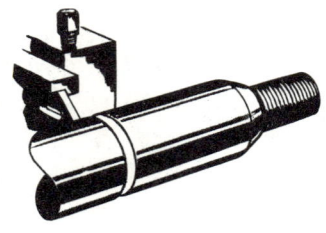

H. Cut off the workpiece with a carbide cutoff mounted in the rear toolpost.

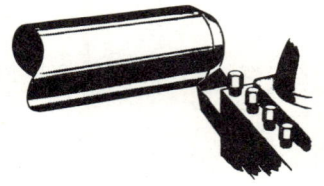

I. After completed part has dropped off, chamfer the end of the bar with a cutter in the square turret.

FIG. 50-50   THE SEQUENCE OF CUTS REQUIRED TO MACHINE THE THREADED SHAFT

REVIEW QUESTIONS

1. What eight machining operations can be performed on a turret lathe?

Basic Internal Cuts

A. Drilling

2. Explain the purpose of start drilling. What type of drill is used for this operation?

3. What procedure should be used on large diameter drilling work?

4. Describe a core drill. What is its purpose?

5. What procedure should be used when core drilling long holes? Explain why.

6. Describe a spade drill. What is the purpose of this drill?

7. What are two methods of deep-hole drilling?

8. What is the purpose of center drilling?

9. What are four devices used to hold drills in a turret lathe?

B. Boring

10. How does the adjustment of a cutting tool for bore size on a ram-type turret compare with that on a cross-sliding hexagon turret?

11. What are two methods of boring tapers?

12. Describe the construction of a fine-adjust boring bar. What is its purpose?

C. Reaming

13. For what purpose are reamers used?

14. What are three types of reamers used on turret lathes?

15. What is the purpose of a floating reamer holder?

16. Describe an adjustable floating-blade reamer.

17. What procedure is followed when reaming steep taper holes?

D. Recessing, Facing, and Chamfering

18. What are three common internal cuts?

19. What two methods can be used to cut internal recesses?

20. What precautions should be taken when cutting internal recesses using a square turret?

E. **Internal Threading**

21. How are small-diameter threads generally produced?
22. What is the procedure for tapping a hole?
23. How is a tap withdrawn from a hole?
24. Describe the operation of a collapsing tap.
25. What is the purpose of the leading-on attachment?
26. List all the operations in proper sequence required to produce the threaded adaptor.

Basic External Cuts

A. **Turning**

27. What is the purpose of a multiple turning head?
28. What is the advantage of machining a diameter by side turning?
29. What is the purpose and function of the bar turner?
30. What two methods can be used to cut external tapers?

B. **Facing, Forming, and Cutoff**

31. What are three methods of facing the end of a workpiece?
32. What two methods may be used to groove or chamfer?
33. Name three types of forming cuts. What is the purpose of each cut?

C. **Threading**

34. How does a die head function?
35. How are long, accurate threads produced with a die head?
36. What are two methods of single-point threading on turret lathes?

Basic External Cuts — Review

37. List, in the proper sequence, all the operations required to produce the threaded shaft.

APPENDIX

TABLE 1: DECIMAL EQUIVALENTS OF PARTS OF AN INCH

Fraction	Decimal	Fraction	Decimal
1/64	.015625	33/64	.515625
1/32	.03125	17/32	.53125
3/64	.046875	35/64	.546875
1/16	.0625	9/16	.5625
5/64	.078125	37/64	.578125
3/32	.09375	19/32	.59375
7/64	.109375	39/64	.609375
1/8	.125	5/8	.625
9/64	.140625	41/64	.640625
5/32	.15625	21/32	.65625
11/64	.171875	43/64	.671875
3/16	.1875	11/16	.6875
13/64	.203125	45/64	.703125
7/32	.21875	23/32	.71875
15/64	.234375	47/64	.734375
1/4	.25	3/4	.75
17/64	.265625	49/64	.765625
9/32	.28125	25/32	.78125
19/64	.296875	51/64	.796875
5/16	.3125	13/16	.8125
21/64	.328125	53/64	.828125
11/32	.34375	27/32	.84375
23/64	.359375	55/64	.859375
3/8	.375	7/8	.875
25/64	.390625	57/64	.890625
13/32	.40625	29/32	.90625
27/64	.421875	59/64	.921875
7/16	.4375	15/16	.9375
29/64	.453125	61/64	.953125
15/32	.46875	31/32	.96875
31/64	.484375	63/64	.984375
1/2	.5	1	1.

TABLE 2: LETTER DRILL SIZES

Letter	Size in Decimals	Letter	Size in Decimals
A 15/64	.234	N	.302
B	.238	O	.316
C	.242	P	.323
D	.246	Q	.332
E 1/4	.250	R	.339
F	.257	S	.348
G	.261	T 23/64	.358
H 17/64	.266	U	.368
I	.272	V 3/8	.377
J	.277	W	.386
K 9/32	.281	X	.397
L	.290	Y 13/32	.404
M	.295	Z	.413

Appendix

TABLE 3: NUMBER DRILL SIZES

No.	Size in Decimals	No.	Size in Decimals	No.	Size in Decimals	No.	Size in Decimals	No.	Size in Decimals
1	.2280	17	.1730	33	.1130	49	.0730	65	.0350
2	.2210	18	.1695	34	.1110	50	.0700	66	.0330
3	.2130	19	.1660	35	.1100	51	.0670	67	.0320
4	.2090	20	.1610	36	.1065	52	.0635	68	.0310
5	.2055	21	.1590	37	.1040	53	.0595	69	.02925
6	.2040	22	.1570	38	.1015	54	.0550	70	.0280
7	.2010	23	.1540	39	.0995	55	.0520	71	.0260
8	.1990	24	.1520	40	.0980	56	.0465	72	.0250
9	.1960	25	.1495	41	.0960	57	.0430	73	.0240
10	.1935	26	.1470	42	.0935	58	.0420	74	.0225
11	.1910	27	.1440	43	.0890	59	.0410	75	.0210
12	.1890	28	.1405	44	.0860	60	.0400	76	.0200
13	.1850	29	.1360	45	.0820	61	.0390	77	.0180
14	.1820	30	.1285	46	.0810	62	.0380	78	.0160
15	.1800	31	.1200	47	.0785	63	.0370	79	.0145
16	.1770	32	.1160	48	.0760	64	.0360	80	.0135

TABLE 4: TAP DRILL SIZES
AMERICAN NATIONAL FORM THREAD

NC National Coarse			NF National Fine		
Tap Size	Threads per Inch	Tap Drill Size	Tap Size	Threads per Inch	Tap Drill Size
# 5	40	#38	# 5	44	#37
# 6	32	#36	# 6	40	#33
# 8	32	#29	# 8	36	#29
#10	24	#25	#10	32	#21
#12	24	#16	#12	28	#14
1/4	20	# 7	1/4	28	# 3
5/16	18	F	5/16	24	I
3/8	16	5/16	3/8	24	Q
7/16	14	U	7/16	20	25/64
1/2	13	27/64	1/2	20	29/64
9/16	12	31/64	9/16	18	33/64
5/8	11	17/32	5/8	18	37/64
3/4	10	21/32	3/4	16	11/16
7/8	9	49/64	7/8	14	13/16
1	8	7/8	1	14	15/16
1 1/8	7	63/64	1 1/8	12	1 3/64
1 1/4	7	1 7/64	1 1/4	12	1 11/64
1 3/8	6	1 7/32	1 3/8	12	1 19/64
1 1/2	6	1 11/32	1 1/2	12	1 27/64
1 3/4	5	1 9/16			
2	4½	1 25/32			

The major diameter of an NC or NF number size tap or screw = (N x .013) + .060
EXAMPLE: The major diameter of a #5 tap equals
(5 x .013) + .060 = .125 diameter

Appendix

MATERIAL	RAKE AND RELIEF ANGLES IN DEGREES FOR SINGLE-POINT LATHE TOOLBITS							
	Side Relief		End Relief		Side Rake		Back Rake	
	H.S.S.	Carbide	H.S.S.	Carbide	H.S.S.	Carbide	H.S.S.	Carbide
Aluminum	12	6 to 10	8	6 to 10	15	10 to 20	35	0 to 10
Brass	10	6 to 8	8	6 to 8	5 to -4	8 to -5	0	0 to -5
Bronze	10	6 to 8	8	6 to 8	5 to -4	8 to -5	0	0 to -5
Cast Iron	10	5 to 8	8	5 to 8	12	6 to -7	5	0 to -7
Machine Steel	10 to 12	5 to 10	8	5 to 10	12 to 18	6 to -7	8 to 15	0 to -7
Tool Steel	10	5 to 8	8	5 to 8	12	6 to -7	8	0 to -7
Stainless Steel	10	5 to 8	8	5 to 8	15 to 20	6 to -7	8	0 to -7

MATERIAL	CUTTING SPEEDS IN F.P.M. USING HIGH-SPEED STEEL TOOLS					
	TURNING, BORING, FACING		THREADING	REAMING	PARTING	KNURLING
	Rough Cut	Finish Cut				
Aluminum	200	300	50	50	200	40
Brass	180	220	45	20	150	40
Bronze	180	220	45	20	150	40
Cast Iron	70	80	20	35	80	30
Machine Steel	100	110	40	40	80	30
Tool Steel	70	80	25	35	70	30
Stainless Steel	75-90	100-130	35	35	70	20

MATERIAL	CUTTING SPEEDS IN F.P.M. FOR CARBIDE AND CERAMIC TOOLS			
	ROUGH CUT		FINISH CUT	
	Carbide	Ceramics	Carbide	Ceramics
Aluminum	300-450	400-2000	700-1000	600-3000
Brass	500-600	400-800	700-800	600-1200
Bronze	500-600	150-800	700-800	200-1000
Cast Iron	200-250	200-800	350-450	200-2000
Machine Steel	400-500	250-1200	700-1000	400-1800
Tool Steel	300-400	300-1500	500-750	600-2000
Stainless Steel	250-300	300-1000	375-500	400-1200

INDEX

A

Accessories, 11-20
Accidents, safety precautions, 29, 30
Accuracy, checking for, 55
Acme Threads, cutting, 133, 134
    measurement, 147
Adjustable strip mandrel, 156
Alignment, centers, 60-63
    faceplate work, 215
Allowances for fits, table, 249
Aluminum file, 88
American National Acme Thread, 121
American National Thread, 120
    measurement, 146
American Standard
    tapered spindle-nose chuck, 165
Angle of keenness, 40
Angle plate, 15, 213
Angular shoulder, 83
Apron, 8
    handwheel, 8
Arithmetical average surface finishes, 243
Automatic feed lever, 8

B

Back rake, 40
Bar parts,
    produced on turret lathes, 251
Bar turning, turret lathes, 265
Basic cuts, 256-270
Bed, 6
Bell center punch method, 55
Bench lathe, 1
Bent-tail lathe dog, 66
Beveled shoulder, 83
Bored hole, squaring shoulder of, 193
Boring, 191-193
    cuts on turret lathe, 258
Boring toolholders, 34, 35, 191
Brass file, 88
British Standard Whitworth Thread, 121
Brown and Sharpe tapers, 102

Brown and Sharpe Worm Thread, 122

C

Cam-lock spindle-nose chuck, 166
    mounting, 169
    removing, 168
Carbide toolbits, 31
    toolholders for, 33
Carbide tools, cutting speeds for, 49
Carriage, 8
Carriage accessories, 17, 18
Cast nonferrous alloy toolbits, 31
Cemented carbide tools, 31
    sharpening, 43
Center drills, 56
Center head method, 53
Center holes, drilling and truing, 178, 179
    locating, 53-58
Center punch mark, truing, by
    faceplate work, 215
Center rest chuck, mounting rotors, 223
Centering work, 53-56
Centers, dead, 15-17
    alignment, 60-63
    facing work between 71-73
    mounting work between, 64-69
Ceramic tools, 32
    cutting speeds, 49
Chalk method, checking trueness, 64, 173
Chamfering, 260
Chucking lathe, numerically-controlled, 244
Chucking parts, produced on turret lathes, 251
Chucks, 11-14
    cutting-off, 185, 186
    drilling, 188-190
    facing, 180-182
    mounting and removing, 165-169
    mounting work in, 171-176
    turning in, 183, 184
Clamp lathe dog, 66
Cleaning lathe, 22
Clearance angles, 39, 40

Coil springs, winding, 231, 232
Collet chucks, 13, 14
    mounting work in, 174-176
Combination chucks, 13
Combined-feed taper turning, 266
Commutator, turning 222-224
Compound rest, 8
    cutting a taper, 108-110
    depth setting for grinding, 229
    special form turning, 218
    taper turning, 207
Compound rest graduated collar, 74-76
Concave face radius, 219
Concentric cuts, 265
Control media, tracer attachments, 234
Convex face radius, 219
Core drilling, 257
Counterboring, 194
Countersink, truing center hole, 178
Crankshafts, 161-164
Cross-sliding hexagon turret lathe, 254
    boring cuts, 259
    taper boring, 259
Crossfeed screw handle, 8
Cross-slide, 8
Crown turret, 247
Cutting, centering work, 53-56
    depth, 74-76
    eccentrics, 161-164
    facing, 71-73
    form turning, 115-117
    grooving, 97-99
    knurling, 93-95
    mandrel, 155-159
    parallel turning, 78-80
    shoulder turning, 82-85
    taper turning, 100-112
    thread cutting, 126-140
    threads, 119-124
    turret lathes, basic cuts, 256-270
Cutting-off, 185, 186, 267
Cutting-off toolholders, 34
Cutting speeds, 48-51
Cutting tools, 31, 32
    single-point, nomenclature, 37, 38
    types and applications, 39

D

Dead centers, 15-17
    See also Centers
Deep hole drilling, 257
Depth of cut, 50, 74-76
    National form thread, settings for, 131
Dial indicator, faceplate work, 215, 216
    trueness checking, 64, 174
Diamond-tipped cutting tools, 32
Die heads, 268
Digital readout system, 237-240
Draw-in collar attachment, 14
Draw-in collet chucks, mounting work in, 175
Dressing grinding wheel, 43
Drill press, drilling center holes on, 57
Drilling, drill chuck, 188
    turret lathe, 256
Driveplate, 14
Driving fit, 249

E

Eccentric shafts, 161-164
End-cutting edge angle, 39
End facing, 180
    turret lathes, 266
End-relief angle, 39
Engine lathes, 1, 2
Expansion mandrels, 156
Expansion reamers, 196
Expansion-stud mandrel, 157
External cuts, sequence, 269, 270
    turret lathes, basic, 264
External grinding, 226

F

Face concave or convex turning, 218
Facing, 71-73, 180-182
    compound rest for, 75
    turret lathes, 260, 266
Faceplate, 14
Faceplate work, 213-216
Feed directional plunger, 8
Feed lever, 8
Feed rods, 7
Feed reverse lever, 7
Feeds and threads, 7
Feeds, cutting speeds, 50
Filing, 87-90
Filleted shoulder, 84

Index

Fine-adjust boring bar, 259
Finish turning, 80
Finishing reamers, 196
Fits and allowances, 249
Fixed-center hexagon turret lathe, 254
Fixed-saddle type turret lathe, boring cuts, 258
Flanged mandrel, 156
Floating-blade reamers, 260
Fluted chucking reamer, 196
Fluted reamers, 260
Follower, use in special form turning, 220
Follower rest, 17, 152, 153
Forced fit, 249
Form turning, 115-117
Forming, 267
Four-jaw independent chuck, 13
    mounting rotors, 222
    mounting work in, 172-174
    turning eccentrics in, 163
Freehand form turning, 115
Function, 1

G

Gages, thread measurement, 143, 144
Gang mandrel, 156
Gap-bed lathe, 2
Gearbox, 7
General-purpose toolbit, grinding, 44
    sharpening, 46
Gibs, 25
Graduated micrometer collar, 74
Grinders, preparing, 42
Grinding, 225-229
Grinding cutting tools, 42-46
Grinding wheel, dressing, 43
Griptru chuck, 12
Grooving, 97-99

H

Half-center, 16
Hand reaming, 200
Hand taps, 202
Headstock, 6
    accessories, 11-15
    aligning centers, 60-63
    center trueness, 64, 65
Headstock chuck and mounting rotors, 223

Headstock spindle, 7
Heavy-duty boring bar holder, 35, 192
Heavy-duty toolpost, 36
Hermaphrodite caliper method, 54
Hexagon turret, 260
    boring cuts, 258
    facing, 266
    forming, 267
High-speed steel toolbits, 31
    toolholders for, 32
High-speed steel tools, cutting speeds, 49
    feeds, 50
High-speed steel centers, 16
Holding devices,
    toolholders and toolposts, 32-37
Holes, deep hole drilling, 257
    tapping by hand, 203
    tapping by power, 204
        See also Center holes
Hydraulic tracers, 235

I

I.D. turret, 247
Internal cuts, basic, turret lathe, 258, 263
    recessing, facing, and chamfering, 260
Internal grinding, 225, 227
Internal threading, 210, 211
    turret lathe, 261
Internal threads tapping, 202
Internal turret, 247
International Metric Thread, 122

J

Jacob's chuck commutator turning kit, 222
Jacob's spindle-nose collet chuck, 14
    mounting work in, 175
Jarno taper, 102
Jobber's reamer, 196

K

Knurling, 93-95

L

Lathe bed, 6
    accessories, 18
Lathe center, grinding, 227
Lathe center method, checking accuracy, 55
Lathe dogs, 66

Index

Lathes, 1-5
    accessories, 11-20
    drilling center holes, 58
    maintenance, 22-27
    parts, 5-9
    safe operation, 29, 30
    size, 5
Lead screw, 7
Left-hand offset toolholders, 32
Live center, truing, 65
Long-angle lathe file, 87
Lubrication, 23-25

M

Machine reamers, 196-200
Machining time, calculating, 50
Machining work on mandrel, 158
Magnetic lathe chucks, 14
Maintenance, 22-27
Mandrels, 155-159
    spring winding, 231
Manufacturing lathes, 3
Metric threads, 139
Micro-Set adjustable center, 17, 63
Micrometer collars, 9
    depth of cut, 74-76
Micrometer carriage stop, 20
Micrometers, checking taper, 110
    thread measurement, 144
Mill file, 87
Moire fringe principle, 237
Morse taper, 102
Mounting chucks, 166-168
Mounting faceplate, 214
Mounting work in chuck, 171-176
Multiple-start threads, cutting 136-139

N

National form threads, depth settings for, 131
    tap drill size, 203
Numerical control lathes, 5, 244-247

O

O.D. turret, 247
Offset tailstock method, taper turning, 106-108
Optical comparator, 146
Overhead turning, turret lathes, 265

P

Parallel turning, 78-80
Parting, 185, 186
Parts, 5-9
Plain taper attachment, 19
Polishing, 90, 91
Preset tooling system, for digital readout system, 239
Production lathes, 4, 5
Push fit, 249

Q

Quick-change gearbox, 7, 126
Quick-change lathe, cutting metric thread on, 139
Quick-change toolholder, 36

R

R.P.M. calculations, cutting speeds, 48-51
Radius turning attachment, 218
Rake angles, grinding, 42
Ram-type turret lathe, 252
    hexagon turret boring, 258
Reaming, 196-200
    turret lathes, 260
Recessing, 260
Relief angle of toolbit, checking, 45
Removing chucks, 167
Revolving dead center, 16
Right-hand offset toolholders, 32
Ring gage, checking taper, 110
Root mean square, surface finishes, 243
Rose reamer, 196
Rotors, mounting, 222, 223
Rough turning, 78
Roughing reamers, 196
Round grooves, 97
Round toolpost, 17
Rubber-flex chuck, 14
Running fit, 249

S

Saddle, 8
Saddle-type turret lathes, 253
    boring cuts, 258
Safety clamp lathe dog, 66

Index

Safety devices, shear pins and slip clutches, 26
Safety precautions, 29-30
    grinder operation, 43
Screw thread micrometer, 144
Screw threads, 119-124
Self-holding tapers, 100, 101
Shear pins, 25
Shoulder turning, 82-85
    compound rest for, 75
Shrink fit, 249
Side-cutting edge angle, 39
Side rake angle, 40
Side-relief angle, 39
Side-turning, turret lathes, 265
Sine bar, checking taper, 111
Single-point cutting tools nomenclature, 37, 39
    types and applications 39, 40
Single-point lathe toolbits, rake and relief angle for, 42
Single-point threading, 262, 268
Single-spindle automatic lathe, 4
60° right-hand thread, 130
Size, 5
Sliding fit, 249
Slip clutches, 27
Solid forged boring tool, 191
Solid-type toolholders, 34
Spade drilling, 257
Special form turning, 218-220
Special-purpose taps, 202
Speed lathe, 2
Spindle-nose tapers, 100, 101
Spindle-type chucks, 165, 166
    mounting work in, 176
Split-nut lever, 8
Spring collet chucks, 14
    mounting work in, 175
Spring-type toolholders, 34
Spring winding, 231, 232
Square grooves, 97
Square shoulder, 82, 193
Square taper turret boring, 259
Square thread, 122, 134-136
Square turret, 261
    boring cuts, 258
    facing, 266
    forming, 267
Standard solid mandrel, 155

Standard taper pins, 102
Standard toolpost, 35
Steady rest, 20, 150-152
    mounting rotors, 222
Steep tapers, 100
Straight gibs, 26
Straight hole reaming, 198
Straight-tail lathe dog, 66
Straight toolholder, 32
Super-shear file, 88
Surface-finishes, 242, 243
Surface-gage method, 54
    truing work, 173

T

Tailstock, 7
    accessories, 15-17
    aligning centers, 60-63
    calculating taper turning, 103
Tailstock spindle, 7
    drilling with drill mounted in, 189
Taper attachments, 18, 104-106
    setting to accurate angle, 208
Taper micrometer, checking taper, 112
Taper reamers, 197, 260
Taper-shank mandrel, 157
Taper spindle-nose chuck, 165
    mounting, 169
    removing, 168
Taper turning, 100-112, 206-209, 266
    turret lathes, 265
Tapered gibs, 26
Tapered hole reaming, 199
Tapers, measuring, 110-112
Tapping, 202-204
    turret lathes, 261
Telescopic taper attachment, 19, 104
Template, use in special form turning, 220
Thread-chasing dial method,
    cutting multiple threads, 138
Thread comparator micrometer, 145
Thread cutting, 126-140
    compound rest for, 75
Thread measurement, 143-148
Thread plug gages, 143
Thread ring gages, 143
Thread snap gages, 144

Index

Threaded mandrel, 156
Threaded shaft, sequence of
    operations, 269, 270
Threaded spindle-nose chuck, 165
    mounting, 169
    removing, 167
Threaded-stud mandrel, 157
Threading, internal threading, 210, 211
    turret lathes, 261-263, 268
Threading tool, resetting, 131-133
Threading toolholders, 34
Threads, 119-124
Three-jaw chuck, 11
    mounting work in, 171, 172
Three-wire thread measurement, 145
Throwaway carbide inserts, 31
    toolholders for, 33
Tool turrets, numerical control lathes, 246
Toolbits, grinding, 42-45
    sharpening, 46
Toolholders, 32-35
    machining between centers, 66-69
    truing work, 174
Toolpost grinder, 225-228
Toolposts, 17, 35-37
Toolroom lathe, 2
Tracer attachments, 234-236
Tracer lathes, 3

Trueness of centering, 64, 65
    centerhole, 178-179
    faceplate work, 216
    four-jaw independent chucks, 172-174
Turning, parallel, 78-80
    in chuck, 183, 184
    shoulder, 82-85
    special form, 218-220
    turret lathes, 265
Turning centers, numerically-controlled, 244, 245
Turning cuts, turret lathes, 264-266
Turret lathe, 4
    basic cuts, 256-270
    fundamentals, 250-254
Turret-type toolpost, 36
Twist drilling, 256
Two-jaw universal chuck, 11
Types of lathes, 1-5

U

Unified thread, 121

V

V-block, 213
V-shaped grooves, 97

W

Ways
Wiggler, faceplate work, 213
Work-holding devices, chucks, 11-14

677 (6C1700)